Middle School
Life Science

Third Edition

JEFFCO
PUBLIC SCHOOLS
Building Bright Futures

Kendall Hunt
publishing company

Front cover image: Copyright © Geoffrey Kuchera, 2009. Used under license from Shutterstock, Inc.
Back cover image: Robert Lindholm/Visuals Unlimited, Inc.

Kendall Hunt
publishing company

www.kendallhunt.com
Send all inquiries to:
4050 Westmark Drive
Dubuque, IA 52004–1840
1-800-542-6657

JeffCo Public Schools
1829 Denver West Drive #27
Golden, Colorado 80401
303-982-6500
www.jeffcopublicschools.org

The material from the first edition was based upon works supported by the National Science Foundation under grant number MDR-8550202. Any opinions, findings, conclusions, or recommendations expressed in this publication are those of the author(s) and do not necessarily reflect the views of the Foundation.

Contents

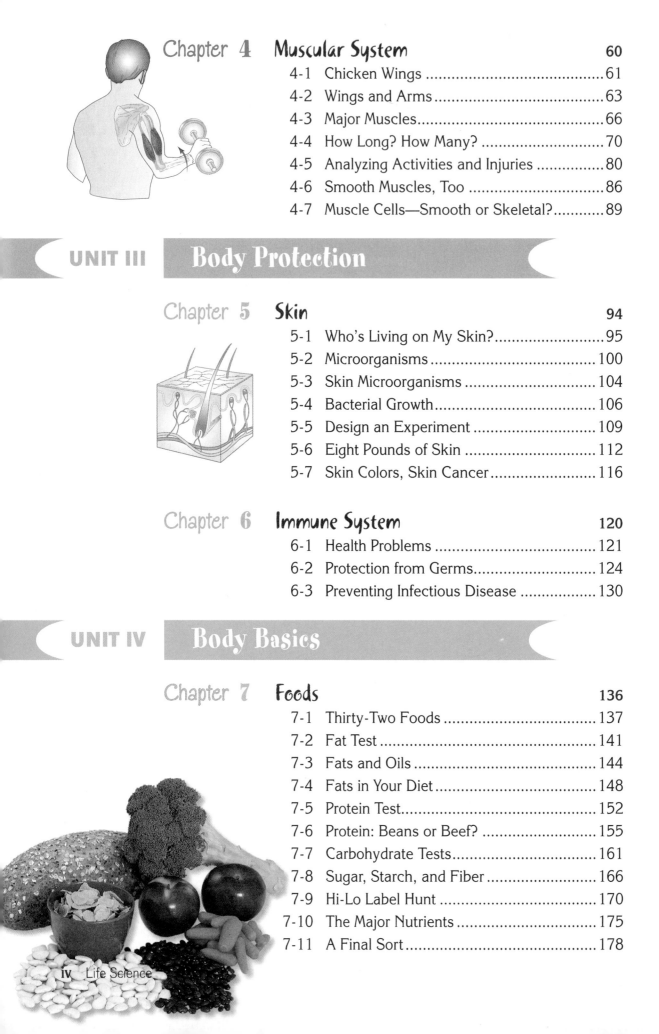

UNIT III — Body Protection

UNIT IV — Body Basics

Unit VII Cells and Genetics

Unit VIII Ecosystems and Ecology

Acknowledgments to the Third Edition

AUTHORS

Judy Capra *Retired, Jeffco Public Schools*
Michelle Garland *Secondary Science Specialist, Jeffco Public Schools*

PROJECT COORDINATOR

Michelle Garland *Secondary Science Specialist, Jeffco Public Schools*

PROJECT ADVISOR

Tammy Weatherly *Director of Science, Jeffco Public Schools*

CONTRIBUTORS

Chris Cornejo *Drake Middle School Colorado*
Nicole Knapp *Private Consultant Colorado*
Cyndi Long *Douglas County High School Colorado*
Carolyn Kirk *Carmody Middle Schoo Colorado*
Elizabeth Hudd *Secondary Science Specialist, Jeffco Public Schools*

ADVISORS – MIDDLE SCHOOL SCIENCE TEACHERS

Sheila Askham *Oberon Middle School*
Veronica Barbosa *North Arvada Middle School*
Erin Brassil *O'Connell Middle School*
Lisa Kring *Falcon Bluffs Middle School*
Susan Lamb *Summit Ridge Middle School*
Miu Lee *Oberon Middle School*
Debbie Marino *Deer Creek Middle School*
Erin Mayer *Oberon Middle School*
Susan McMahon *Bell Middle School*
Mika Melvin *Dunstan Middle School*
Debbie Miller *Evergreen Middle School*
Andrea Pless *North Arvada Middle School*
Katy Pope *Retired Middle School Teacher*
Marilyn Stephens *Oberon Middle School*
Tonia Tamburini *Summit Ridge Middle School*

and the many other teachers in Jeffco Public Schools who volunteered suggestions and assistance.

CONTENT ADVISORS

Carolyn Brewer *Secondary Science Specialist, Jeffco Public Schools*
Judy Hindman *Coordinator, Tobacco/Wellness, Jeffco Public Schools*
Vivian "VJ" Johnston *Coordinator, Safe & Drug-Free Schools, Jeffco Public Schools*
Linda Morris *Elementary Science Coordinator, Denver Public Schools*
Betsy Thompson *Director, Student Services, Jeffco Public Schools*

CONTENT REVIEWERS

James Platt, PhD *Department of Biological Sciences, University of Denver*
David Hanych, PhD *Ecologist/Science Educator, Division of Research on Learning, National Science Foundation*

PRODUCTION ASSISTANT

Jennifer Shirley *Administrative Assistant, Jeffco Public Schools*

Acknowledgments to the Second Edition

AUTHOR

Judy Capra

ADVISORS-MIDDLE SCHOOL SCIENCE TEACHERS

Sharon Coons, *Summit Ridge Middle School,* Colorado
Bruno DeSimone *Mandalay Middle School,* Colorado
Doug Fishman *Burgundy Farm Country Day School,* Virginia
Merikay Haggerty, *Ken Caryl Middle School,* Colorado
Kathy Hawkins, *Wayland Middle School,* Massachusetts
Kate Hayne, *Soroco Middle School,* Colorado
Debbie Loker *Dodge City Middle School,* Kansas
Sue Loomis, *North Arvada Middle School,* Colorado
Connie McLaughlin, *Ken Caryl Middle School,* Colorado
Veronica McLaughlin, *Mandalay Middle School* Colorado
Vicki Mitchell, *Wheat Ridge Middle School,* Colorado
Katie Pope, *Dunstan Middle School,* Colorado
Dana Sakowski, *Moore Middle School,* Colorado
Bette Schwartz, *Deer Creek Middle School,* Colorado
Wynne Simpson, *Moore Middle School,* Colorado
and the many other teachers in the Jefferson County (CO) Schools as well as in other schools across the United States who volunteered suggestions and assistance

CONSULTANTS

Marsha Barber, *Senior High Earth Science Specialist*
Diane Bergeron, *Sixth Grade Teacher*
Vicki DeHoff *Parent, Registered Nurse*
Jeff Ginsberg, *Assistant Principal, Moore Middle School*
Alison Graber, *Science Education Specialist*
Terry Kwan, *Science Education Specialist,* Massachusetts
Francie Marbury, *Reading Specialist*
Jan Myers, *Eighth Grade Physical Science Teacher*
Melinda Reed, *Senior High Biology Teacher*
Bill Rowley, *Director, Division of Instruction,* Jefferson County
Jennifer Sestrich, *Student, Golden Senior High School*
Jonathan Sestrich, *Student, Bell Middle School*
Jason Shin, *Student, Summit Ridge Middle School*
Janet Shin, *Parent, Secondary Science Teacher*
Betsy Thompson, *Safe and Drug Free Schools,* Jefferson County

CONTENT ADVISORS

Lisa Bassow, *Family Practitioner, private practice*
James Hanken, *Anatomy, University of Colorado, Boulder*
Joanne Newton, *Gastroenterology, University of Colorado Health Sciences Center*
Tom Ranker, *Botany, University of Colorado, Boulder*
Gregory Snyder, *Biology, University of Colorado, Boulder*
Diane Sweeney, *University of Colorado Sports Medicine Clinic*
John Van Ryan, *Denver Metro Wastewater Management*

CONTENT REVIEWERS

David Armstrong, *Ecologist, University of Colorado, Boulder*
Jeff Calder, *Monterey Bay Aquarium*
John Cohen, *Immunologist, University of Colorado Health Sciences Center*
Leticia Delarosa, *Pharmacist, Rocky Mountain Drug Center*
Kathleen Fugerstone, *National Wildlife Research Center, USDA*
David Godfrey, *Sea Turtle Survival League*
Therese Johnson, *Wildlife Management Biologist, Rocky Mountain National Park*
Malcolm Macmillan, *School of Psychology, Deakin University, Australia*
Nancy Riordan, *Nurse Practitioner, Children's Hospital, University of Colorado Health Sciences Center*
Christopher Shaw, *Collections Manager, George C. Page Museum,* Los Angeles, California
Greta Wilkening, *Neuropsychologist, Children's Hospital, University of Colorado Health Sciences Center*

PRODUCTION ASSISTANT:

Nancy Kembel

STAFF EDITOR:

Fern Wilson

Acknowledgments to the First Edition

AUTHOR
Judy Capra

PROJECT ADVISOR
Harold Pratt

STAFF WRITER
Kathleen Ranwez, *Moore Junior High, Jefferson County, Colorado*

CONTRIBUTORS
From Jefferson County, Colorado Schools

Carol Bollig	Jeff Ginsberg	Connie McLaughlin
Mary Braukman	Merikay Haggerty	Bette Schwartz
Sharon Close	Bonnie Hayes	
Bruno DeSimone	Mary Ann Larsen	

FIELD TEST TEACHERS
From Jefferson County, Colorado

Ray Atkinson	Ralph Hancock	Vicki Mitchell
Gary Ballengee	Rod Hayes	Jim Mundell
Linda Behm	Bruce Hogue	Barb Nelson
Gary Blubaugh	Diderick Iversen	John Nelson
Gary Buker	Dennis Karsten	Bruce Nicholls
Betsy Cantrall	David Kim	Sabina Raab
Bernard Ciarvella	Frank Krein	Jim Rothrock
Peg Coats	Bob LeCour	Wynne Simpson
Joe Creber	Walt Loeblein	Erle Swanson
Kelly Curran	Gayle LoPiccolo	Margaret Suzukida
Jim Ellis	Julie McClellan	Jerry Trebilcock
Jan Ensminger	Sue McCool	Marian True
Ken Evridge	Mike McGraw	Vicki Weaver
Larry Franca	Cathy McIntosh	Jim Whitmore
Joan Fretz	Tim Middle	Chris Williams
Dennis Giullian	Ruth Mitchell	Lindon Wood

From other school districts in Colorado

Dale Brinker, *Douglas County Schools, Parker Junior High*

Lori Bryner, *Adams County School District #12, Westlake Village Junior High*

Donna Dale, *Adams County School District #12, Eastlake Campus*

Margo Kramer, *Boulder Valley Schools, Broomfield Heights Middle School*

Mike Mancini, *Adams County School District #12, Westlake Village Junior High*

Scott Miller, *Boulder Valley Schools, Broomfield Heights Middle School*

Linda Morris, *Sheridan School District #2, Sheridan Middle School*

ADVISORY COMMITTEE
Ronald Anderson, *Professor of Science Education, University of Colorado*

Mary Braukman, *Junior High Science Teacher, Jefferson County Schools*

Faye Hudson, *Middle School Science Teacher, Denver Public Schools*

Pat Patrick, *Senior High Science Teacher, Jefferson County Schools*

Gary Ranck, *Biologist, Regis University, Denver*

John Selner, *Pediatric Allergist, private practice*

Eva Sujansky, *Pediatric Geneticist, University of Colorado Health Sciences Center*

CONTENT ADVISORS
David Armstrong, *Ecologist, University of Colorado, Boulder*

Sue Brandon, *Elementary Science Resource Specialist, Jefferson County*

Daniel Chiras, *Ecologist, University of Colorado, Denver*

Kurt Cunningham, *Environmental Educator, Colorado State University Extension Services*

Pat DeRosia, *Language Arts Resource Specialist, Jefferson County*

Cindy Fite, *Senior High Science Teacher, Jefferson County*

John Gapter, *Botanist, University of Northern Colorado*

Barbara Gentry, *Junior High Science Teacher, Jefferson County*

Karen Hollweg, *Environmental Educator, Denver Audubon Society*

Susan Loucks-Horsley, *Education Researcher, The NETWORK, Andover, Massachusetts*

Roger Pool, *Secondary Science Resource Specialist, Jefferson County*

Carse Pustmueller, *Ecologist, Colorado Natural Areas Program*

Rick Spitzer, *Naturalist, Rocky Mountain National Park*

Al Swanson, *Secondary Science Teacher, Jefferson County*

Robert Tully, *Wildlife Manager, Colorado Division of Wildlife*

CONTENT REVIEWERS
David Armstrong, *Ecologist, University of Colorado, Boulder*

Roberta Beech, *Pediatrician-Adolescent Medicine, Westside Clinic, Denver, Colorado*

Daniel Chiras, *Ecologist, University of Colorado, Denver*

Amy Cronister, *Genetic Associate, The Children's Hospital, Colorado*

Gordon Guist, *Bioengineer, FMC Corporation, Maine*

Ann Halbower, *Pediatrician, University of Colorado Health Sciences Center*

Joanne Johnson, *Dietician, The Allergy and Respiratory Institute of Colorado*

David Kessel, *Pediatrician-Sports Medicine, private practice*

Ted Ning, *Urologist, private practice*

Marylou Rottman, *Botanist, University of Colorado, Denver*

Sharon Schilling, *Health Educator, Denver Public Schools*

Reginald Washington, *Pediatric Cardiologist, The Children's Hospital, Colorado*

Jan Watson, *Obstetrician, University of Colorado Health Sciences Center*

Message to

Welcome to Life Science! You are about to begin a year-long study of life. In this course, you will focus on living systems, particularly on ecosystems and body systems. You will learn more than facts about the systems. You will do science! Flip through the pages of this book. Notice that you will do a variety of activities, record the data, and then analyze your data. The readings will help you understand the scientific concepts that form the basis of the activities. Throughout the course, you will practice using your new skills and applying your new knowledge to other situations. You will see how science is used in your everyday life and you will be better able to understand news stories about science. The information and skills you learn can help you make wiser decisions.

More than anything else, the purpose of this course is to help you learn to think. Keep in mind that if, after thinking, you still don't understand something, it is all right to ask questions. Asking questions is the basis of all science.

When we visit classrooms, one of the questions students always ask is, "How did you decide to write this book?" Actually, no one person sat down and thought, "I'm going to write a Life Science book." Instead, this program was a group effort involving many people—scientists, teachers, parents, other educators, and, of course, students. Students suggested topics, gave us their papers to read, and turned in their questions. They also tested all the activities and experiments. Based on their experiences, the activities were reworked and rewritten until we got them right.

Students

You can be part of this process. Anytime during the coming year, if you are struck by an idea about how to improve this program, share it with us. Write down your thoughts and send them to the publisher. (The address is at the bottom of the title page.) The publisher will forward your letter to us. We would like to hear from you, and we promise to answer your letters. Your input could make the next edition of this book even better. Some of the students who provided written feedback for this edition are named on the acknowledgments page.

Have a good year. We hope you enjoy using this book as much as we enjoyed writing it for you.

Introduction to Life Science

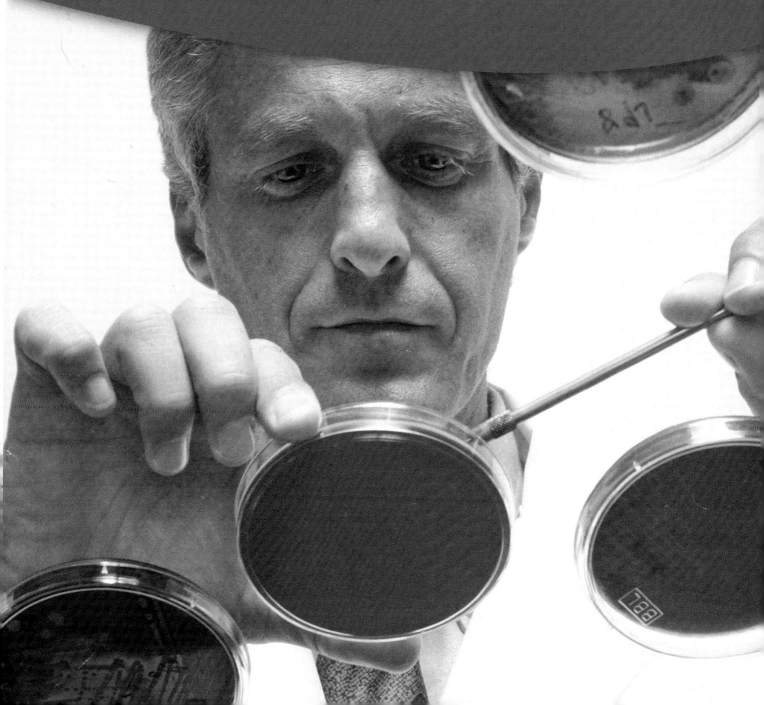

Chapter 1
Introduction to
Microscopes and Systems

In this course, you will study living systems and you will use a variety of tools to help you learn about the systems. How many systems do you see pictured here? What are they? What tools can you identify in this picture?

Introduction to Microscopes

Do you recognize what these word pairs have in common?

pen: writer
camera: photographer
telescope: astronomer

scalpel: surgeon
wrench: mechanic
microscope: biologist

The first word in each pair names a tool; the second word is the worker who uses the tool. Scientists often use tools to collect data or to extend their senses. By using a microscope, a biologist can see structures and organisms that are too small to see otherwise. What other tools do people use to extend their sense of sight? What are examples of tools scientists use to collect data?

In this course, you too, will use a variety of tools to help you understand science, and you will use microscopes for observing extremely tiny structures. It is not easy to use this tool correctly, and it can be challenging to decide what you are seeing. However, with time, practice, and patience, you will learn this important skill.

What is the correct way to carry a microscope?

What is the correct way to focus a microscope?

There are two parts to this activity. First, learn the parts of a microscope. Then, while working with a partner, practice the basic rules for using a microscope. Keep the focusing questions in mind while you work.

Includes *SciLINKS*®
NSTA

Topic: Science and Technology
Go to: www.scilinks.org
Code: MSLS3e3

Materials

- microfilm #1
- microscope
- Handout, *Microscope Parts*
- Handout, *Using a Microscope*

Procedure

Part A: Microscope Parts

Label the diagram of the microscope on the handout, *Microscope Parts,* while your teacher discusses each of its parts.

Part B: Using a Microscope   *(Work with a partner.)*

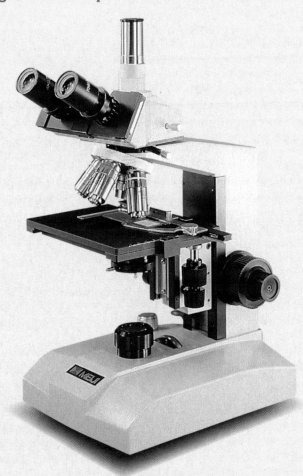

1. One of you should bring a microscope to your table. When you do so, carry the microscope with both hands. Grasp the arm with one hand while the other hand supports the base. Place the microscope near the center of the table, not by the edge. You do not want to bump it accidentally.

2. Each of you will have a role as a checker or a learner. Decide who will do what.

a. *Checker*: reads directions, makes sure learner understands procedure

b. *Learner*: listens to directions and does as instructed, asks questions if something is unclear

c. You will switch roles later so you will get to work in both roles.

3. Take turns completing the first part of the handout, *Using a Microscope*.

   a. The *checker* should read from the *learner's* handout.

   b. While the *checker* reads, the *learner* follows directions.

   c. Each time the *learner* successfully completes a step, the *checker* initials that step and reads the next one.

   d. When the *learner* completes the first section, the *checker* signs the *learner's* paper.

   e. Trade roles and repeat this procedure.

4. When you both know how to set up a microscope, switch roles again and complete the focusing section of the handout.

5. When both you and your partner have completed the handout, use the following questions to quiz each other:

   a. Why is it important to look at the microscope from the side, while lowering the body tube with the coarse adjustment knob?

   b. When can you see more of an object—under low power or high power?

   c. Why are there two adjustment knobs on the microscope?

   d. Why are you supposed to start with low power?

Memorize the rules for using a microscope. (Your teacher will test you on these!)

Microscope Rules

Always

1. Carry a microscope with both hands, one on the base and one on the arm.

2. Place the microscope near the center of the table, not near the edge.

3. Always begin with low power.

4. Watch from the side when you lower the body tube with the coarse adjustment knob.

5. Never use the coarse adjustment knob with high power.

6. Return the microscope to the low power lens when you are finished.

7. Use lens paper to clean microscope lenses.

Never

Use direct sunlight for a light source.

1-2 Microscope Skills: Magnification and Drawings

Now that you know the basic rules for using a microscope, you are ready to learn how to figure magnification and how to make a useful sketch of what you see.

A microscope contains two lenses that magnify objects. One lens is in the eyepiece and another one is on the nosepiece. If you want to find out how many times the object you are viewing is enlarged, you need to figure the **total magnification**. To calculate total magnification, multiply the magnification of the two lenses.

eyepiece magnification $\times$ low power magnification $=$ total magnification at low power

The number of times the lens magnifies is printed on the lens itself. Suppose the lens in the eyepiece magnifies 5 times and the low power lens also magnifies 5 times. The total magnification would be 25 times. When you look through a microscope, what you see is called your **field of view**. A circle such as the one on the right is used to represent the field. To make a drawing of what you see through a microscope, draw a circle, and then sketch what you see in the field of view. When you do so, do not move the microscope slide.

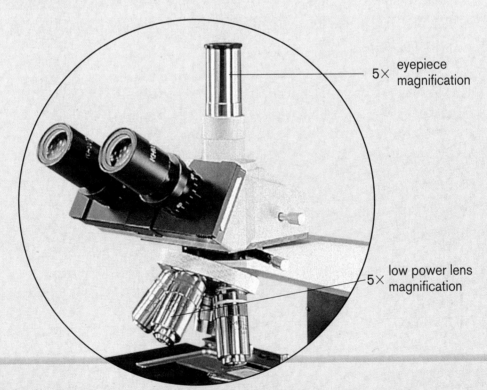

5× eyepiece magnification

5× low power lens magnification

What is the magnification of your microscope?

What should you include in every microscope drawing?

There are two parts to this activity. You will practice calculating magnification, making drawings, and measuring the objects you view with a microscope. You will also **solve** some problems that are designed to help you learn these skills. Keep the focusing questions in mind while you work.

- microfilms #2 and #3
- microscope
- clear plastic metric ruler
- Handout, *Magnification and Drawings*

Procedure

When you look at the handout, you will see that there are two parts to this activity. You need a microscope for Part A, but not for Part B. It does not matter which part you do first.

As you work, you will need to make sketches while looking through the microscope. Memorize these five "rules" and follow them each time you make a microscope drawing.

A Microscope Drawing: Five "Rules"

1. Neatly sketch the object so it takes the same amount of space in the circle as the actual object takes in your field of view.

2. Include enough detail so someone else can recognize the object. Try to be accurate.

3. Title the drawing so the viewer will know what you were looking at.

4. Label everything you recognize on your drawing.

5. Record the total magnification next to the drawing.

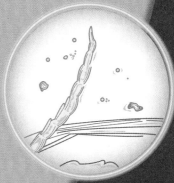

1-3

Systems in Life Science

In this class, you will study systems. There are all kinds of systems. There are legal systems, political systems, and educational systems. There are also sprinkler systems, sewer systems, and exhaust systems in cars. In the kitchen, you may have a coffee-making system or a system for making popcorn. In this class, you will be studying life science and examples of living systems. Think about it ...

What is a system?

Why are petunias, ponds, and people examples of systems?

Work in a team with several other students to illustrate a system and begin to think about living systems.

Materials

- chart paper
- colored markers or crayons

Procedure

Part A: A System   (Work in a small group of 3 or 4 students.)

1. Look over the steps in this procedure and plan how you will spend your time. You will have 12 minutes to complete Part A of this activity. The chart below shows one way you might want to divide your time.

Time	Task
2 min	plan work
5 min	illustrate a system
3 min	write definition
2 min	hang chart, clear desk
12 min	**total time**

Includes SCLINKS NSTA

Topic: Biological Systems
Go to: www.scilinks.org
Code: MSLS3e8

2. Choose a system to illustrate. Choose any system—it does not have to be a living system. Use words, pictures, or both to sketch the system on the chart paper. Be sure to include the following things:
   a. the name of the system
   b. the main parts of the system
   c. what the system does (or how the parts interact)
3. Think of a definition for the word "system." When you all agree, write it across the bottom of the chart paper.
4. Hang up your chart and put away the markers.

Part B: Systems   *(Work as a class.)*

1. One student from each group is to tell about the system their group illustrated. In less than one minute, the presenter should explain:
   a. the name of the system
   b. the main parts of the system
   c. what the system does (or how the parts interact)
2. Listen carefully to each presenter. Look for a pattern—what is the same for all systems? If the presenter forgets to tell you something about the system, ask a question.
3. Of the systems that were presented, count the number that are living systems.
4. As a class, agree on a definition of a system. Write down the definition so you can refer to it later.

Part C: Systems in Life Science   *(Again, work in small groups.)*

Take five minutes to answer the following three questions. Be prepared to discuss your answers in class.

In what ways are petunias and people alike?

In what ways are people and ponds alike?

Why are petunias, ponds, and people examples of systems?

Chapter 2

Introduction to
Cells

What living things do you see in this photo? How do you know they are living?

2-1

Observing Pond Water

This year you will be studying Life Science, the study of living organisms. What do you think it means to be alive? Most of us know that animals and plants are living, but what about the bacteria that may have caused you to get strep throat last winter? Is it a living organism? What about water or rocks? Are they alive? In this unit, you will be introduced to what scientists consider to be the basic unit of life—the cell. Some organisms are made of only one cell while others, like humans, are made of trillions of cells.

What do single-celled organisms look like?

In this activity, you will have the opportunity to see living, single-celled organisms called protozoa. Protozoa are a very diverse group of organisms that vary widely in size, shape, features, and habit.

- microscope
- samples of pond water with droppers
- microscope slide
- cover slips

Procedure   *(Work with a partner.)*

1. Look at the sample of pond water your teacher gave you. Answer analysis question 1.
2. Prepare a microscope slide with pond water.
   (See **Figure 2.1.**) Place a drop of pond water on the slide and cover it with the cover slip.
3. Examine the slide under low power. If you need to, move your slide around to see all areas of the drop of pond water.

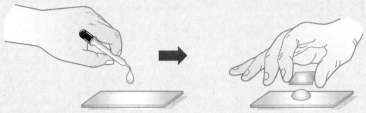

Figure 2.1

Making a slide of pond water

4. Switch to high power and look again.

5. Choose the magnification you think is best for observing protozoa and list as many protozoa as you can find. For each protozoon, draw a picture and label it with a name that describes the organism, for instance, "round with hair-like projections."

6. Complete a microscope drawing, following the microscope drawing rules you learned in *Chapter 1-2 Microscope Skills: Magnification and Drawings.*

7. Answer analysis questions 2 through 4 and write your conclusion.

Analysis Questions

1. Without using a microscope:

   a. What are five things you observe about the pond water?

   b. What can you see that you think might be living? What can you see that you think might *not* be living?

2. How are protozoa like animals? Think of different kinds of animals, such as dogs, birds, or insects. How are protozoa different from such animals?

3. How did different protozoa move? If possible, describe the movement of several different protozoa.

4. What do you think protozoa use for food? What might eat protozoa?

Conclusion

Write a paragraph that answers the focusing question.

Extension Activity

Research: Protozoa are part of a group of organisms called protists. Find out more about this group of organisms and the important role protists play in the environment.

2-2

Plant and Animal Cells

Plants are made up of parts, such as leaves, stems, and roots. Animals are also made up of parts. Different types of organisms have different types of parts. For example, a dog has legs, ears, and a tail but a tomato plant has leaves, roots, and fruit. All of these parts, whether they make up a plant, a dog, or even you, are all made of cells.

What structures can you see inside a cell?

How is a plant cell different from an animal cell?

In this lab, you will have the opportunity to observe cells from both plants and animals. Cells are made of different structures. Some of these structures you can see when you look at the cell with your school microscope, but other cell structures are too small to see unless you have a more powerful microscope.

- *Elodea*
- onion
- flat toothpicks
- microscope slides
- bucket containing 10% bleach solution to rinse microscope slides
- pop bottle containing 10% bleach solution to dispose of used toothpicks
- Handout, *Microscope Drawings of Living Things*

- slide covers
- stain
- droppers for water
- soap and paper towels

Procedure

(Work with a partner.)

1. Since you will be making your own microscope slides, you need to know how to prepare a wet mount. Read the procedure in the box to find out what you need to do.

Wet Mounts
1. Clean the microscope slide and cover slip.
2. Select a specimen to study, such as a thin piece of a plant, or cells scraped from the inside of your cheek.
3. Place the specimen on the microscope slide. **Light must be able to pass through the specimen in order for you to see anything.**
4. Put 2 or 3 drops of water onto the specimen.
5. Set the cover slip on its edge on the slide and carefully lower it to cover the specimen.
6. If the cover slip does not rest flat on the slide, tap it gently with your eraser to smash the specimen. Be careful not to crack the cover slip. Your specimen may be too thick if it does not flatten easily. If so, you may need to start over with a smaller or thinner specimen.

2. Following your teacher's instructions, prepare wet mounts of onion, *Elodea,* and cheek cells.

   a. Onion: Peel a small piece of tissue from inside a layer of onion.

   b. *Elodea*: Find an area on the slide that contains only one or two layers.

   c. Cheek Cells: Your sample may appear folded and might be more difficult to draw.

3. For each slide:

   a. Use the low power lens to look for the specimen. Beware! Air bubbles are not living things! Bubbles are round and easy to see. If you find an air bubble, dirt, or scratches, try again.

   b. Switch to high power and look again.

   c. Remove the slide from the microscope to stain it.

      ▪ Place a drop of stain along one edge of the cover slip.

      ▪ Hold a piece of paper towel next to the opposite edge of the cover slip. This will "pull" the stain under the cover slip. (See **Figure 2.2.**)

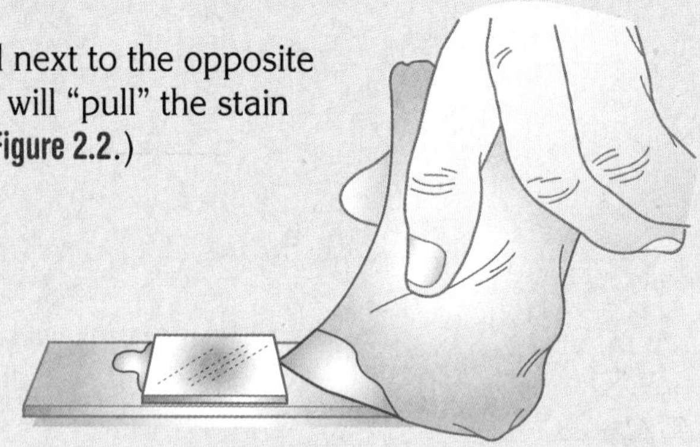

Figure 2.2
Adding stain

d. Draw what you see using low power. *Make sure you use the microscope drawing rules!*

e. Now switch to high power, refocus, and draw what you see.

4. When you are finished, wash your microscope slide and cover slip. Rinse the slide and cover slip used for cheek cells in the 10% bleach solution to kill any remaining bacteria or viruses. Throw away the used toothpick in the discard bottle. (It also contains a bleach solution.)

> ● **Caution**
> - Prevent the spread of germs! Follow directions for discarding items that come into contact with body fluids.
> - Stains can permanently discolor things. Be careful when you use them.

Analysis Questions

1. The three samples you looked at were from different living things: an onion, a plant, and an animal (you).

    a. How were the cells from each sample alike?

    b. How were the cells from each sample different?

    c. Complete a Venn diagram to help you compare and contrast the cells.

2. What characteristic do you think would be most helpful for identifying an onion cell? A plant cell? An animal cell?

3. Why do you think the directions said to put a drop of water on the microscope slide?

4. How did the stain affect the appearance of the cells?

Conclusion

Write complete answers to the focusing questions.

Extension Activity

Going Further: Compare cells from different parts of the same plant. For instance, you could look at cells from an apple peel, pulp, and stem; the upper and lower surface of a leaf; a flower petal; or from the surface of a twig. Sketch the cells. Try to think of reasons for any differences you observe.

2-3

Cells: The Basic Unit of Life

Cells. Without these structures, there wouldn't be life on Earth. The following section summarizes basic information about the structure and function of cells.

Levels of Organization

It is impossible to count all of the cells in a human body, but if you could, you'd find that it's made up of about one hundred trillion cells. That's hard to imagine, isn't it?

A jumble of trillions of cells would not make a human being. The cells must take on specific jobs, "work" with other cells, and be located in the right places. In other words, the cells must be organized. A group of similar cells that have a similar function is called a **tissue**. During this coming year, you will study several kinds of tissues. For example, you will observe muscle cells in muscle tissue and blood cells in blood tissue. When you think about the parts of the body, however, you are more likely to think about organs than about tissues.

An **organ** is a group of tissues that carries out a specific function. For example, the heart is an organ. It is made up of blood tissue, muscle tissue, nerve tissue, and connective tissue. Together, these tissues act as a muscular pump that keeps blood moving. An eye, a kidney, and a stomach are each an organ, as is each bone in the body.

A group of organs that work together is called a **system**. If you continue to think about the heart as an example, you can say that the heart (an organ) is part of the circulatory

20 trillion	red blood cells
75 billion	white blood cells
	skeletal muscle cells
10 million	smell receptors
100 million	nerve cells in brain
?	connective tissue cells
?	bone cells

100 trillion cells = a human body

system. (See **Figure 2.3**.) Examples of organs that make up the digestive system include the stomach, tongue, and esophagus. In all, there are ten systems. The systems cannot work by themselves; they must work together to make an **organism**.

Think "backward" now: An organism is made up of systems; systems are made up of organs; organs are made up of tissues; and tissues are made up of cells. Thus, a **cell** is the smallest living level of organization. For this reason, it is called the basic unit of life.

Includes sci LINKS. NSTA

Topic: Cells, Tissues and Organs
Go to: www.scilinks.org
Code: MSLS3e16

Ten Systems

Each body system has a function. Which systems do you think are responsible for each of these activities?

- breaking down food
- exchanging gases
- providing support
- movement
- fighting infection
- transporting small molecules
- removing waste molecules from blood
- creating new life
- coordinating body functions and actions
- producing hormones

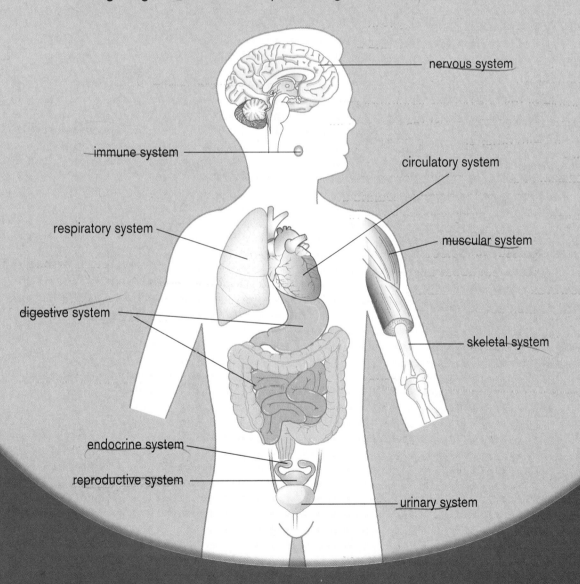

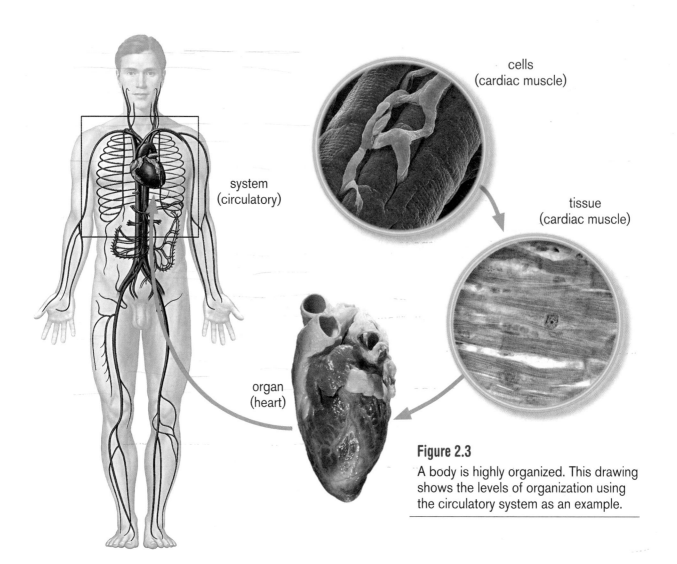

cells
(cardiac muscle)

system
(circulatory)

tissue
(cardiac muscle)

organ
(heart)

Figure 2.3
A body is highly organized. This drawing shows the levels of organization using the circulatory system as an example.

The Discovery of Cells

If you were a student 300 years ago, you wouldn't be studying cells because cells were just being discovered. In the late 1600s, a scientist by the name of **Robert Hooke** was experimenting with a new invention, a microscope that magnified 30 times. Using his penknife, he sliced a very thin section off a piece of cork (from the bark of an oak tree) and observed it under his microscope. In his notes, he wrote about observing extremely small, irregular structures that he called cells. His drawing of cork cells is shown in **Figure 2.4**. Hooke was intrigued by the tiny "boxes" he could observe with his microscope, but he had no idea that all plant and animal tissues are made of cells.

Over the next 150 years as the quality of microscopes improved, people learned more and more about cells. By the end of the 1800s, their findings were summarized into what we now know as the cell theory. The cell theory has two basic ideas:

- all organisms are made of one or more cells, and
- all new cells come from other cells.

Figure 2.4
Robert Hooke made this drawing of cork cells using a microscope that magnified 30x.

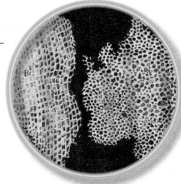

Since this theory was stated, thousands, maybe even millions of experiments have been done on cells, but these basic ideas have never been disproved. That's why these statements are called a theory.

Cell Size and Shape

Most cells are extremely small. Sperm and red blood cells are so tiny that they are hard to see with the light microscopes you have at school. The egg cells produced by birds are the largest type of cell. The largest human cells are also eggs, but even they are hard to see without magnification. Most human cells are neither extremely large nor exceptionally small. Most can be viewed when they are magnified 100 to 400 times. (You can understand why Hooke could never have viewed cells from *all* tissues, since his microscope only magnified 30 times.)

The shapes of cells vary even more than their size. That's not surprising, since the shape of a cell usually reflects what the cell does. Nerve cells are long and skinny, extending from one part of the body to another. This shape is ideal for carrying messages, but it would be very impractical for carrying oxygen to specific organs. The small round shape of a red blood cell is a much better design for flowing through blood vessels. Each cell is specialized for the job it does. (See **Figure 2.5**.)

Cell Structure

At first, scientists thought cells were like "blobs" of jelly. As the quality of microscopes improved, they learned more and more about the structure of cells. They realized that, even though cells differ in their sizes and shapes, every animal cell has a membrane, cytoplasm, and a nucleus. Refer to **Figure 2.6** while you read this section.

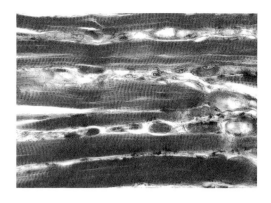
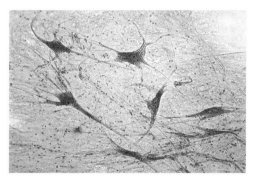

Figure 2.5
Each cell type has a characteristic shape.

—nerve cells

Figure 2.6
Animal cell

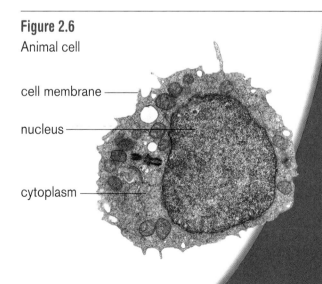

cell membrane

nucleus

cytoplasm

The **cell membrane** holds a cell together. It also controls what enters and leaves the cell. Food molecules must diffuse through the membrane to enter the cell. Once inside, the nutrients are broken down. The waste molecules that are formed diffuse out of the cell by crossing the membrane. If the membrane is damaged in any way, the cell may die. A membrane is extremely thin; it would take 100,000 membranes to make a stack 1 mm thick. This is too thin to see with the light microscopes you use in school. Inside the cell membrane are the nucleus and the cytoplasm.

The **nucleus** is the largest structure inside a cell. You can usually see it with a light microscope, but sometimes it's necessary to stain the slide before the nucleus will be visible. The nucleus is the "control center" that directs the cells' activities. It controls all of the chemical reactions that go on in the cell. If it were possible to visit a nucleus, you'd find molecules moving everywhere. Some would be entering the nucleus, some would be leaving, and some would be under production. There's lots of activity in a nucleus.

Mitochondria are often called the "power plant" of cells since they provide the energy needed by cells to divide, contract, and move. The process of creating energy for a cell is called *cellular respiration*. Most of the chemical reactions in cellular respiration happen in the mitochondria. The number of mitochondria in a cell varies depending on what a cell needs to do and its activity level. Many cells have only a few hundred mitochondria while other types of cells may contain several thousand. For example, each of your liver cells contains 1,000 to 2,000 mitochondria, making up a considerable portion of the entire volume of the liver cell. Mitochondria are very small structures so you will not be able to see them with the light microscopes you have at school.

The **vacuole** is a membrane-lined sac that may store nutrients the cell needs to survive or waste products that eventually get sent out of the cell. An animal cell may have several

small vacuoles. Plant cells, on the other hand, usually have only one large vacuole that can take up more than half the volume of the cell. Vacuoles play an important role in the structure of a plant. When a plant is well-watered, water collects in the vacuole causing the plant to become rigid. However, if a plant does not have enough water, then the pressure in the vacuole is reduced, causing the plant to wilt.

Cytoplasm is a watery gel that contains all the molecules and structures found within a cell, such as mitochondria and vacuoles. Water makes up most of the cytoplasm. About 75% to 85% of a cell by weight is water. Most of the structures in the cytoplasm are too small to see with school microscopes. These tiny structures are called **organelles**, meaning "little organs." They do the work of the cell. Some organelles contain enzymes and are sites for important chemical reactions such as making proteins, or taking apart molecules that are no longer needed. Still others function during cell division. Each organelle has its own function.

Plant Cells

In addition to cytoplasm, mitochondria, vacuoles, a nucleus, and a cell membrane, plant cells have cell walls and chloroplasts. (See Figure 2.7.) A **cell wall** gives a plant cell its shape and provides support for the entire plant. Cell walls are important, since plants do not have skeletons. They are made of long cellulose molecules.

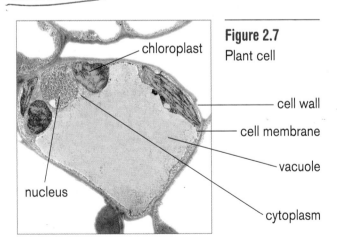

Figure 2.7
Plant cell

chloroplast
cell wall
cell membrane
vacuole
cytoplasm
nucleus

When you eat, your digestive system digests (breaks down) food into much smaller pieces that can be transported to your cells. Humans, however, cannot digest cell walls because we don't produce enzymes that can break down cellulose. Instead, cellulose moves right through our digestive tracts. Nutritionists refer to this indigestible plant material as "fiber." Plant-eating animals such as cattle, deer, and horses cannot break down cellulose either, but they have bacteria in their digestive tracts that digest it for them.

Chloroplasts and Photosynthesis

All living organisms get their energy from food. Unlike animals, plants do not eat food; they produce it in their **chloroplasts**. A single green plant cell may contain as few as 20, or more than 100 of these small green structures. Chloroplasts are the food production center of a plant cell. In order to produce food, they need light, water, and carbon dioxide. Chloroplasts use the energy from light to combine water with carbon dioxide to make sugar. This process is called **photosynthesis**. When chloroplasts make sugar, some oxygen is left over. This oxygen is released into the atmosphere. In fact, all of the oxygen in the air is produced by chloroplasts. Photosynthesis can be summarized like this:

carbon dioxide + water + light ↦ sugar + oxygen
(CO_2) (H_2O) $(C_6H_{12}O_6)$ (O_2)

Photosynthesis

Figure 2.8

The chloroplasts in green plants use energy from the sun to combine carbon dioxide (CO_2) with water (H_2O) to make sugar. In the process, some oxygen is released into the atmosphere.

You can see that plants do not get their food directly from the soil, water, or air. Instead, they use materials from these sources to make their own food (sugar). Plants can't make food, however, unless they have light and, for most plants, this light comes from the sun.

Once a cell has a supply of sugar, it can use it in three general ways. First, it may store the sugar to use later. It does so by producing starch. When you eat a food such as a potato, you are eating food that the potato plant stored. Second, a cell may use the sugar molecules to build other cell structures. For instance, the sugars may be connected in long chains to make cellulose. Third, a cell may break down the sugar during cellular respiration to get the energy it needs to stay alive.

Alive and Active

A single living cell can do many of the same things that a much larger, multicellular organism can do. Like large organisms, a cell can move, respond to its environment, grow, and reproduce. In addition to these basic functions, some cells have specialized functions. For example, muscle cells contract, nerve cells carry messages, and white blood cells destroy disease-causing microorganisms.

These activities all require energy. That's why cells, like large organisms, need food, oxygen, and water. Cells use oxygen when they break down nutrients for energy. At the same time, they produce waste in the form of carbon dioxide and heat.

A cell is certainly not a "blob." As you can see, cells are quite complex. So complex that some scientists spend their entire careers studying these small but important structures. The more you learn about them, the more intrigued you will be to learn more.

Analysis Questions

1. Create a chart comparing the structure and function of the following cell structures. Indicate on chart whether you would find the structure in a plant cell only, an animal cell only, or in both types of cell.

   nucleus   cytoplasm   cell wall
   chloroplast   cell membrane   vacuole
   mitochondria

2. List the following structures in order of size—from smallest to largest. (Try to answer from memory before you look up the answer.)

   cell   organ   organism
   system   tissue   organelle

3. What important discovery is Robert Hooke known for?

4. What are the two parts of the cell theory?

5. Which of these things do you eat? Explain.

   nuclei   cytoplasm   membranes
   chloroplasts

6. Choose one of these questions to answer:
   a. Are nerve cells smarter than other cells? Explain.
   b. Do lung cells breathe? Explain.

7. Why do you think animal cells do not have cell walls? Why do animal cells not have chloroplasts?

Extension Activity

History: Learn more about Robert Hooke. He was an ingenious British physicist who explored many fields of science. What topics did he study?

Dichotomous Key

When you clean your bedroom, do you organize all your stuff so you know where you can find it later? Do you put all your books on a book shelf and all your clothes in the closet or in a dresser? Grouping similar things together helps us to be organized and to keep track of things. Scientists organize living things the same way. They like to know what each plant, animal, and other organisms are and have developed methods to group or classify them. One tool that scientists use is called a **dichotomous key**. A dichotomous key allows you to determine the identity of items in the natural world, such as trees, wildflowers, mammals, reptiles, rocks, and fish. Keys consist of a series of choices that lead you to the correct name of a given item. "Dichotomous" means "divided into two parts." Therefore, dichotomous keys always give two choices in each step.

Why do scientists classify organisms?

How does a dichotomous key work?

Keep these questions in mind while you create your own dichotomous key. Then use a prepared dichotomous key to identify protozoa.

Materials

- Handout, *Dichotomous Key Flowchart*
- prepared *Protozoa* microscope slides
- microscope

Procedure

Part A: Create a Dichotomous Key *(Work in a group of 4 students.)*

1. Have everyone take off their right shoe and put it in the middle of your group.
   a. Look at all the shoes and discuss with your group members what is similar and what is different about the shoes.

b. Divide the shoes into two groups using a characteristic that can be observed. For example, shoes that are white and shoes that are not white. You do not need to have an equal amount of shoes in each group.

c. Record the characteristic in the first level of the *Dichotomous Key Flowchart.*

d. Continue to divide the shoes into groups based on one observable characteristic until each shoe is in a subgroup by itself. Record the characteristics on the flow chart.

2. When your teacher instructs you, pass your shoes to the group next to you.

Try to classify the new shoes with your flowchart. Do you need to make changes to your flowchart so that the new shoes can be classified? If so, go ahead and make the necessary changes.

3. Once again, when your teacher instructs you, pass the shoes you have now to the group next to you.

Again, try to classify these new shoes with your flowchart. Make any changes necessary so that these shoes can also be classified.

4. Answer analysis questions 1 through 3.

Part B: Use a Dichotomous Key to Classify Protozoa

(Work with a partner.)

1. Observe the prepared protozoa slides under the microscope.
   a. Find and focus the protozoa under low power.
   b. Switch to high power and focus.

2. Use the *Protozoa Dichotomous Key* to determine what type of protozoa you are viewing.

3. Answer analysis questions 4 and 5.

Protozoa Dichotomous Key
1. The microorganism is rod shaped .... Go to 6
1. The microorganism is not rod shaped .... Go to 2
2. The microorganism is circula .... Go to 5
2. The microorganism is oval .... Go to 3
3. Has an obvious nucleus .... Go to 4
3. No nucleus is seen .... *Diatoma hiemale*
4. The nucleus is round, and the cell is smaller in size .... *Euglena*
4. The nucleus is long and narrow,   and the cell is larger in size .... *Paramecium*
5. Circular colony of many cells.................................... *Volvox*
5. Single cell .... Go to 4
6. Organism has horizontal stripes................................ *Lyngbya*
6. Organism has a spiral design along its length .... *Spirogyra*

Analysis Questions — — — — — — — —

1. What characteristics did you and your partners decide to use to classify your shoes?

2. When you received a new set of shoes the first time, did you need to change your key? If yes, explain why. What about the second set of shoes?

3. What did you find to be the most difficult as you were writing your key? What was the easiest part of writing the key?

4. List two characteristics that all the protozoa had in common. List two characteristics that were different for all the protozoa.

5. Can more than one protozoon occupy the final place in the dichotomous key? Explain your reasoning.

Conclusion

Write a few sentences to answer each of the focusing questions.

Extension Activity

Going Further: Choose the magnification you think is best for observing the protozoa on the prepared microscope slides, and sketch what you see. Label any cell parts that you recognize.

Cell Model

Bone or blood, lung or leaf, they all come from living organisms so they're all made of cells. Throughout the rest of the year, you'll be looking at cells, all kinds of cells. Some of them will be easier to see than others but they all contain similar structures with similar functions.

How can a model of a cell be used to understand the functions of cell structures?

Think about this question as you build a model of a cell using everyday household objects.

Materials

- common household objects
- butcher paper
- glue or tape

Procedure   *(Work in a group with three or four other students.)*

Part A: Collecting Your Cell Parts

1. With your group, you will be collecting common items from around your houses that represent the *function* of different cell parts and organelles. Your group will use these items to create a three-dimensional cell model in your classroom.

   Read the entire procedure so you understand what to do.

Includes *sci*LINKS®
NSTA

Topic: Cell Structure and Function
Go to: www.scilinks.org
Code: MSLS3e26

2. On a piece of paper, copy the following table:

Cell Scavenger Hunt			
Cell Structure	Function	Household items representing each *function*	Responsible group member
Cytoplasm			
Cell membrane			
Nucleus			
Mitochondria			
Vacuole			
Chloroplast			
Cell wall			

3. Discuss with your group the function of each of the cell structures listed in the table. If you need help remembering the function, refer back to *Chapter 2–3 Cells: The Basic Unit of Life*.

   a. Fill in the function of each structure on your table.

   b. With your group, think of common items around your house that perform a similar function as the cell structure. For example, a cell membrane filters what enters and leaves a cell. Therefore, a coffee filter could represent the function of a cell membrane. In the appropriate column in your table, fill in ideas for items your group comes up.

   c. Decide with your group members, who will bring what items from home to build your cell model. Be sure to bring in the items you are responsible for on the date designated by your teacher.

Part B: Building Your Cell Model

1. Place your group's "cell parts" on the paper provided by your teacher. Next to each item, write the following information on the paper:

   a. Name of the cell part

   b. Function(s) of the cell part

   c. Why your item represents the function of the cell part.

2. When your teacher tells you it is time, move around the room to look at other groups' cell models.

   a. What items did they used in their models that are different from your model?

   b. What items are the same?

   c. Can you recognize plant cells versus animal cells?

Analysis Questions

1. Analyze the model.
   a. List at least three ways this is like a real cell.
   b. List at least three ways this differs from a real cell.
2. Explain why your skin cells don't need chloroplasts.
3. How would you decide whether an unknown cell was an animal cell or a plant cell?
4. What might happen to a plant if the chloroplasts in its cells quit functioning?
5. What might happen if the nucleus of a cell became damaged?
6. Muscle cells tend to have many more mitochondria than other cells. Why do you think there are more mitochondria found in muscle cells than in other cells?

Conclusion

Write a complete answer to the focusing question.

Extension Activity

Technology: As microscopes improved, scientists learned more about cells. Look for pictures taken through an electron microscope. (It magnifies much more than a light microscope.) Find pictures of cell structures other than the ones you can see with your light microscope. What are some functions of these cell structures?

UNIT II

Body Structure

Chapter 3

Skeletal System

Joints allow for movement. How many joints do you think are in a hand?

3-1

Meet HB

This lesson is titled "Meet HB" but actually, you already know HB. HB is a human body, and we suspect you've been living inside a human body for quite a few years. You've learned some things about the human body by experience, some things by accident, and you've probably learned some things by reading, watching television, or studying the body in school. This year, you will learn even more about the body. But, before you start, think about some of the things you already know.

How many parts make up the human body? How many can you name?
Which parts can you do without?

Your challenge: Working as a class, then in a team, show what you already know about the human body by "building a body."

- HB outline
- colored paper
- scissors
- tape

Procedure

Part A: The Human Body   (Work as a class.)

Think about what parts are *inside* the human body. As a class, make as complete a list as you can. With all the students in class, you should be able to name most of the parts of the body. *Try not to look through this book for ideas!*

Part B: The Beginning of HB   *(Work in a small group.)*

1. Join your group and read the directions for steps 2 through 6.
   You are working in a cooperative group, which means you must decide
   who will be in charge of what. Everyone should have a job. Share
   responsibilities!

2. Sketch the body parts that your teacher assigns to your group.
   They should be about the size that they would actually be if HB were real.

3. Cut out the parts.

4. As a group, decide how important each
   part is for a functioning body. Decide if it is
   very important, somewhat important, or not
   important.

5. Label each body part (two stars, one
   star, or X) to show your group's decision.
   Make sure that you can all explain the
   reason for your decision.

   **   very important
   *   somewhat important
   X   not important

6. Attach the body parts to
   HB in the correct locations.

Part C:
Thinking About HB

(Again, work as a class.)

1. Listen as one person
   from each group reports
   the group's decisions to the
   class. Ask questions if you want further
   information, or disagree with what they
   decided.

2. Continue to think about HB. You will
   be studying the human body in the next
   five units. When you finish studying the
   body, you will build HB-2. How different
   do you think HB-2 will look compared
   with this HB?

Figure 3.1

Analysis Questions

1. Do you think one body part is more important than another? If so, which do you think is the most important? Why?
2. Which body parts do you think you know the most about?
3. Considering that this may be the last time you study the human body in school, what do you think you should learn? What will you need to know in order to make wise decisions about taking care of your body?
4. The human body is made up of systems, such as the digestive system and the skeletal system, but do you think it is correct to think of the total body as a system? Explain.

Conclusion

Answer the focusing questions with a sentence or two that explains what you learned by doing this activity.

Body Systems

When studying the human body, it is useful to think about it in "sections." For instance, you could study the arm, the leg, or the head, but scientists often "section" the body differently. They study the body by looking at systems. In this class, you will take a scientific look at each body system.

Figure 3.2

This may be the last time you study the human body in school. Of course, you may learn more about it in high school or college, and, if you become a health professional like the people shown here, you will study the body for years. Even if you never study the human body again in school, you will continue to learn more about it because you will live in a body for the rest of your life.

Whose Bones?

Imagine that you've been hiking for several hours, and you finally reach your destination—a beautiful mountain lake. Immediately you feel as if the lake belongs to you. No one else is around. All you hear is the wind in the trees, the chatter of a squirrel, and the occasional quack of a duck.

Walking slower, you scan the shore of the lake, looking for a place to rest. In the distance, you notice some white objects lying on the sand. As you approach them, you can tell that they aren't flowers, litter, or rocks. You wonder if they could be pieces of driftwood. Then it dawns on you, these are bones! Bones of various sizes and shapes are scattered all around the area. Curious, you pick up one, then another, and another. No two are alike. You wonder, "How did these bones get here? What animals are they from?"

What can you learn about an animal by observing one of its bones?

Test your powers of observation by studying four real bones. Imagine that you picked them up from the lake shore. Keep the focusing question in mind as you inspect the bones.

- four different bones
- Handout, *Whose Bones?*

Procedure
(Work in a group with two or three other students to inspect the four bones.)

1. Decide which member of your group will be the recorder. You also need a time-keeper to make sure you stay on task.
2. Start with the bone that your teacher gives your group. (It does not matter which one you inspect first.)
   a. Fill in the appropriate column on the handout while you inspect the bone. Discuss your answers.
   b. Your teacher will tell you how much time you will have with each bone.

   Caution: Handle the bones with care. They can be fragile!

3. When your teacher calls time, pass the bone to another group and get a new bone to inspect. Complete the handout for this bone. Remember to watch the time!
4. Repeat this procedure with the third and fourth bones.
5. Answer the analysis questions on your own paper. Use complete sentences and be prepared to discuss your answers.

Analysis Questions _ _ _ _ _ _ _ _ _

1. List at least three ways that all the bones were the same.
2. List at least three ways that the bones differed from each other.
3. What do you think are some of the most helpful clues for identifying a bone?
4. What do you think a person can tell about an animal by looking at one of its bones? Copy this list of characteristics, and then circle the ones that you think can be determined by looking at an animal's bones. Be prepared to explain your answers.

size	the sounds it made
eye color	whether it was male or female
age	weight
diet	if it was sick or healthy
intelligence	if it lived in water or on land
favorite food	if it could fly

5. List at least three careers in which you might need to know something about bones.
6. If you really did find an assortment of bones on the shore of a lake, what would you think about how they got there. Write at least two possible explanations.

Conclusion

Write a short paragraph that answers the focusing question. Remember to use details to explain your answer.

Extension Activities

1. **Community Resources:** Visit a museum and study the bones that are on display. Become an expert on one type of bone. For instance, observe only the bone that forms the upper part of the leg. Record your observations, noting how all upper legs bones are alike and how they are different.
2. **Careers:** Learn more about one of the careers you listed in question 5. Find an article that describes this career, or interview someone who works in this field. With your teacher's permission, invite this person to speak to your class.

Bones and Skeletons

Every bone has a story to tell, if you know how to "read" it. The longer people study bones, the better they become at revealing their stories.

Rancho La Brea is famous location in Los Angeles, where scientists have collected and study many fossils. Christopher Shaw, a scientist, uses fossilized bones from Rancho La Brea to understand the stories of animals that lived there tens of thousands of years ago. At that time, crude oil seeped up along deep cracks in the ground until it reached the surface. It mixed with the sand and formed shallow pools on the ground. The heat from the sun caused some of the liquid to evaporate away, leaving behind sticky asphalt. Dust and leaves settled on top of these asphalt pits. When animals unknowingly stepped into the sticky "traps," they could not free themselves, and so they died and were preserved in the asphalt. In these pits, researchers have found the bones of over four hundred kinds of animals ranging from tiny insects to huge mammoths. (There are not any dinosaur bones in this collection. The earliest animals in the pits died about 40 thousand years ago. By then, dinosaurs had already been extinct for 65 million years.)

Mr. Shaw is most interested in the bones of saber-toothed cats. More than 166,000 bones from about 2,500 of these big cats have been unearthed. The cats' sturdy, limber bones indicate that they probably pounced on their prey and held them down with their powerful legs. Every bone tells something about the cats, but Mr. Shaw is particularly intrigued by the misshapen bones. Why? Because they tell him about the big cats' behaviors. When he observes the bones, he looks for signs of injuries, and interprets the clues. For instance, he may find holes in the scapula that tell him that one saber-toothed cat was bitten by another. One pelvis was so badly damaged, Mr. Shaw thinks it was evidence that saber-toothed cats were social animals and got help from other cats. Why would he predict this? Because the pelvis shows signs of healing, but there was no way an animal with such a serious injury could have survived unless other cats brought it food, and protected it from other hungry carnivores.

Figure 3.3

Scientists working with some of the bones from the La Brea pits. The map shows the location of the site.

Los Angeles

Figure 3.4

Saber-toothed cat skeleton and a reconstruction of what it probably looked like.

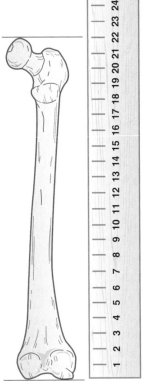

Observing Bones

You may not yet be able to figure out the injuries that saber-toothed cats suffered, but you can use clues to identify a bone and predict something about the animal from which it came. Look at the femur in **Figure 3.5**. Make a mental list of several observations about this bone.

Did you think of observations? Remember, an observation is something you can see, feel, smell, or taste. When you observed the photograph, the only sense you could use was your sight. So one of the first observations you probably made was its color. If you looked closely, you might have noticed smaller details such as fine gray streaks, a tiny crack, or small holes. You should also have observed the overall shape of the bone—it is long with bulges at each end. A better observation would have been that there is a rounded knob sticking up from one end of the bone. Why is this better?

Figure 3.5

This is a femur, the long bone from the upper part of the leg.

Whenever possible, an observation should include actual sizes. It is much better to make a measurement instead of using words like big or small. (Scientists always try to use measurements when making observations.) How long is this femur? What other measurements could you make? What other types of observations could you make if you could actually hold this bone?

Interpreting Observations

Once you have made your observations, you can try to interpret them. When you use your observations as evidence to reach a conclusion, you are making an inference. Some observations are more useful than others when you want to "read" a bone. For example, color is not a very helpful observation since all dry bones are similar in color. Size is a much more useful clue. When you observe the size of the bone, you can infer the size of the animal from which it came. Another observation, the round knob at one end of the femur, is a clue about how the bone fits together with other bones. When you observe a knob, you can infer that it probably fits inside another bone. You can use each of your observations to make an inference.

Bones: Parts of Skeletons

If you put together a whole set of bones, you would have a skeleton. A skeleton is the structural support system of a body. When you study bones, you are studying parts of an internal skeleton. Humans and many other animals have internal skeletons; that is, their support system is inside their bodies. Every bone fits in a specific place in the skeleton and is adapted to do a different job. When attached to muscles, some bones allow for movement, and others protect internal organs. Together, the bones that make up an internal skeleton support the body, give it shape, and protect internal organs.

Includes sci **LINKS** NSTA

Topic:	Observation and Inference
Go to:	www.scilinks.org
Code:	MSLS3e41

Figure 3.6

Like humans, snake eels and polar bears have internal skeletons.

Figure 3.8 Body Structure

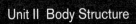

Analysis Questions

1. What was one observation that Mr. Shaw made about a saber-toothed cat bone? How did he interpret his observation?

2. List at least eight observations (not inferences) for the bone shown in **Figure 3.7**. Circle the three that you think would be the most useful if you were trying to identify the bone.

3. List four functions of bones and skeletons.

   Look at the drawings of the four skeletons in **Figure 3.8** to answer questions 4 through 7.

4. For each skeleton, identify the type of animal you think it came from. Use the letters A, B, C, and D to label your answers. List one observation that you used to make each identification.

5. List or sketch three bones or structures that you can see in all four skeletons. (Note: They may not look the same in all the skeletons but they have the same names.)

6. List one characteristic that makes each skeleton unique (different from all of the others).

7. Compare the legs of skeletons A, B, and C (look at one of C's back legs).

   a. List at least three ways they are alike.

   b. List at least three ways they are different from each other.

8. Look at the "mystery skeleton" in **Figure 3.9**.

   a. Describe at least five things that you can tell about this animal by looking at its skeleton. List more if you can.

   b. Look at the things you described above. Mark the observations with an O and the inferences with an I.

Extension Activities

1. **Careers:** Call a local university or museum and ask to speak with a paleontologist. Ask what he or she likes about the career and find out what training was required. With your teacher's permission, invite this person to speak to your class.

Figure 3.7

Figure 3.9

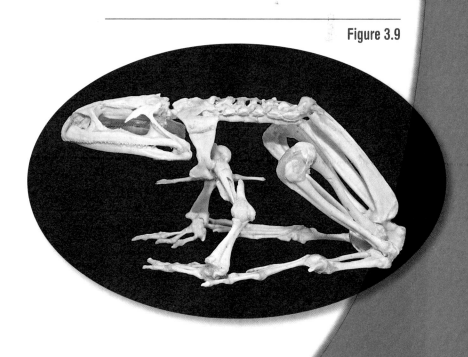

Figure 3.10
Visitor Center at Dinosaur
National Monument

2. **Research:** Find out more about the Rancho La Brea site or learn about a collection of dinosaur bones. Dinosaur National Monument in northeastern Utah has a large excavation area, but there are many others that you could research.

Animal Skeletons

All animals have some kind of support system, but they don't all have bones. Some animals, such as jellyfish, worms, and anemones, are supported by the pressure of the fluid within their bodies.

Lobsters, barnacles, and grasshoppers are examples of animals with external skeletons. Instead of bones, they have hard outer coverings that are tough and water resistant. The shells of sea animals are so strong that they can resist the continuous battering of waves. In order to grow, some of these animals must shed their skeletons and wait for new ones to harden.

3-4

The Human Skeleton

Your internal skeleton is made up of 206 bones. How many of them can you name? It is not important to memorize the names of all the bones, but it is helpful to know the names and functions of some of them. As you read about the human skeleton, find each bone in **Figure 3.12** and look at its structure. You can often tell something about the function of a bone by looking at its size, shape, and other characteristics.

The Skull

At the top of the skeleton is the skull. It is made up of two main parts, the cranium and the mandible. The cranium protects your brain. It is made of more than twenty bones that are fused together. Hollow spaces in the bones on each side of the cranium contain the smallest bones of the body. These are the three bones that are important for hearing. The mandible is the largest and strongest facial bone. Your dentist may talk to you about your mandible when you have your teeth examined.

The Trunk

The backbone, or vertebral column, protects the nerves in the spinal cord and supports the skull and body. It is made up of vertebrae. These bones are separated by discs that act as miniature shock absorbers. The vertebral column is not perfectly straight; it curves naturally at the neck and lower back. If it curves too much or in the wrong way, it may need to be corrected.

The clavicles are important for posture because they brace the shoulders and help hold the scapulas in position. Do you remember what the rat scapula looked like? The shape of yours is very similar. Its broad, flat shape provides a large surface area for the attachment of muscles that move your arm.

Figure 3.11

At birth, the cranium is not one solid bone. It is made up of smaller bones that are separated by tough tissue. As a baby grows, the bones grow and fill in these tissue areas.

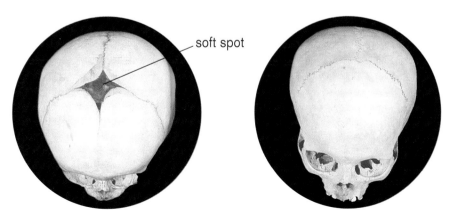

soft spot

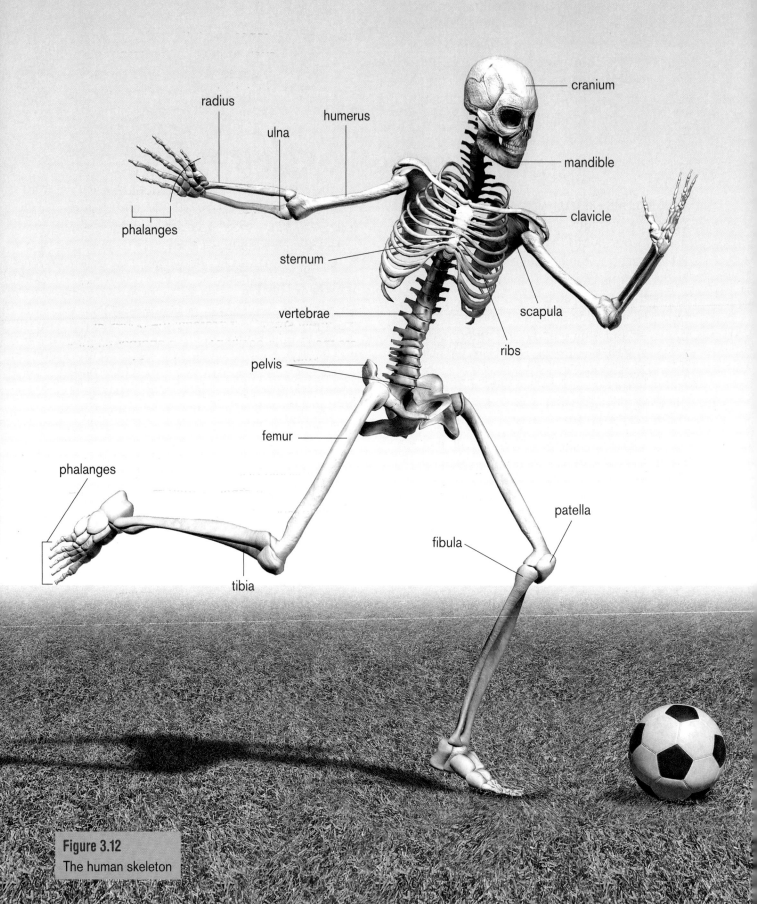

radius

humerus

ulna

cranium

mandible

clavicle

phalanges

sternum

scapula

vertebrae

ribs

pelvis

femur

phalanges

patella

fibula

tibia

Figure 3.12
The human skeleton

The rib cage is made of the sternum and the ribs. One end of the ribs attaches to the vertebrae, and curves forward; the other ends attaches to the sternum. (Try to feel where your ribs and sternum meet.) The rib cage has to be strong to protect your heart and lungs. However, it also has to be flexible. Inhale deeply, then exhale. What happens to your rib cage?

The pelvis is a broad and strong bone that supports the body. Archaeologists can often tell the sex of a skeleton by looking at the pelvis. A female's pelvis is usually broader than a male's. Why do you think this is?

The Limbs

The rest of the bones that make up a human skeleton are in the limbs, the arms and legs. Looking at their bones, arms and legs are very similar. The differences allow them to do different things—arms and hands are better for carrying objects or doing intricate tasks, and legs and feet are for support.

The humerus, radius, and ulna, as well as the femur, tibia, and fibula, are long bones. Because they are shaped like hollow cylinders, these bones can take enormous pressure. Even though they are very strong, they can break. If you have ever broken a bone, it was probably one of these.

Altogether there are more than one hundred little bones that make up your wrists, hands, ankles, and feet. Aren't you glad you don't have to memorize the names of all those bones? You might find it useful to know, however, that the long bones in your fingers and toes are all called phalanges.

The Joints

Wherever two bones meet, there is a joint. Some bones, such as those in the cranium, are so firmly connected that movement does not occur. Most joints, however, allow for some movement while still holding the bones in place.

Consider one of your movable joints—your elbow, knee, or wrist—any one of them will do. How many times do you think you bend that joint in an hour? In a day? In a month? You never oil it, and yet no matter how many times you use it, you expect it to last your entire life. Very few machines are that efficient!

Movable joints have three unique features that make smooth, easy movements possible: cartilage, ligaments, and fluid. Cartilage is extremely smooth. It protects the bones and lets them glide easily past one another as they move. Without cartilage, the bones would soon wear out. Ligaments are strong, fibrous bands that connect the bones that make up a joint. They hold the joint firmly together and, at the same time, allow for some movement. A thick, slippery fluid lubricates joints. Without fluid, cartilage would dry out and the bones would rub against each other as they move. Find the ligaments and cartilage in **Figure 3.13**.

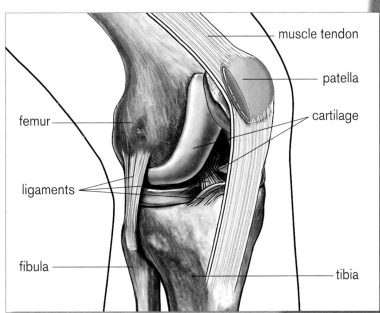

Figure 3.13
A knee joint

Together, the ligaments, cartilage, and bones make up the skeletal system. This system supports the body, protects internal organs, gives the body its shape, and provides places for the muscles to attach.

Analysis Questions

1. Which bone has a better shape for providing protection—the cranium or the humerus? Explain your answer.

2. How many phalanges do you have in your fingers?

3. List two bones that are important for protection. What do they protect?

4. List three bones that provide support.

5. Read each of these sentences. Write down the bone name you can use instead of the words in italics.

   a. You broke the *big bone* in your lower leg.

   b. He got punched in the *jaw*.

   c. The clown's hat was too big for her *head*.

   d. His posture is beautiful because he keeps his *back* so straight.

   e. She rested the *side of her lower arm* on her desk.

   f. The fighter pounded his *chest* and let out a yell.

   g. The baby is crawling on its *kneecaps*.

6. Compare the bones in an arm with the bones in a leg.

   a. List at least three ways they are alike.

   b. List at least three ways they are different.

Extension Activities

1. **Try It:** Long bones are shaped like cylinders. A hollow cylinder is surprisingly strong. Roll a sheet of paper loosely to make a cylinder, keeping it together with a rubber band. Stand it on end and see how much weight it will support. Will it support a pencil? a paperback book? a textbook? two textbooks? Using the same sheet of paper, try to form a shape that is stronger than the cylinder.

2. **Math Connection:** Practice your math skills and become more familiar with your own skeleton by completing the handout, *Your Own Bones*.

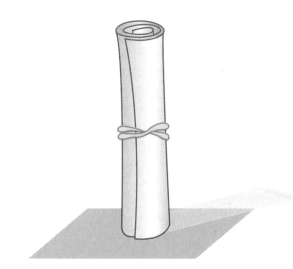

Vertebrates

Humans are vertebrates. So are all other animals that have a bony backbone. This includes the amphibians, reptiles, birds, mammals, and fish. In fact, the first vertebrates were fish that lived in the oceans over 500 million years ago.

Includes sciLINKS® NSTA

Topic: Vertebrates and Invertebrates
Go to: www.scilinks.org
Code: MSLS3e49

Fresh Bones

If you were describing a bone, you might say that it is "hard, white, and dry with tiny cracks all over its surface." While this description fits an old, dry bone, it would not fit a fresh one. We are all much more familiar with "dead" bones than we are with living ones such as those that make up our own skeletons.

How do fresh bones differ from dry ones?

What does a bone cell look like?

Inspect a section of fresh bone, and a microscope slide of bone to help you answer these questions.

- paper towel
- probe or toothpick
- beef bone, cut

- microscope
- prepared slide: bone
- Handout, *Fresh Bones*

Procedure

(Work with a partner. Record your findings on your handout.)

Part A: Fresh Bone Sections

1. Obtain a section of fresh bone from your teacher. Imagine what it looked like before it was cut. On your handout, record what bone you think it came from.
2. Record your first impression of living bone by completing the sentences on the handout.
3. Sketch your bone in the space provided on the handout.
4. Try to find the two types of bone tissue on your piece of bone—compact bone and spongy bone. (Use **Figure 3.14** to help you identify the two types. Not all bones have spongy bone. You may need to look at it on someone else's bone.) Record your observations on the handout.

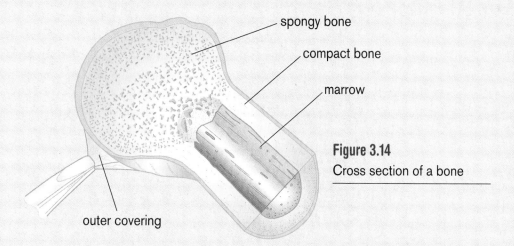

spongy bone

compact bone

marrow

Figure 3.14
Cross section of a bone

outer covering

5. Scoop out a little of the marrow and put it on the paper towel. Record your observations of its characteristics.

6. Pull away part of the outer covering of the bone. Notice how it attaches to the bone. Record your observations.

7. Since all living tissues need a blood supply, look for blood vessels in each of the four main parts of the bone. (Even tiny vessels count.) Record your findings.

8. Clean up your work area.

9. Look at the sections of bone that your teacher has set out for you to inspect. Compare them with the section you had. How are they alike? How are they different? Can you identify the bones?

Part B: Bone Cells

1. Look at **Figure 3.15**. It shows a magnified view of bone. Find the bone cells. How many do you think are shown in this sketch? (The answer is lots!)

2. Look at the microscope slide of bone that has been set up for you. Focus carefully so you can find the bone cells.

3. Sketch what you see on the handout. Label the parts you can identify.

cell

space

minerals

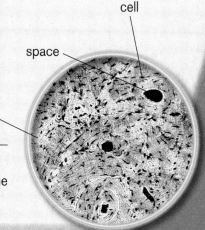

Figure 3.15
Microscopic view of bone

Analysis Questions

1. Did the bone you inspected have more spongy bone or compact bone?

2. Which parts of the bone—compact bone, spongy bone, outer covering, marrow—had a blood supply?

3. All of the things listed here can be found in a fresh bone. Copy the list and circle the ones that are also found in a dry bone.

   outer covering   marrow
   bone cells   spongy bone
   tiny blood vessels   compact bone
   minerals   water

4. Do you think a fresh bone is mostly living cells or minerals? Explain.

Conclusion

Using complete sentences, answer the focusing questions.

Extension Activities

1. **Try It:** Acid removes the minerals from bone. See what happens when you clean an uncooked bone, such as a rib or chicken leg, and soak it for a couple of weeks in vinegar. (Vinegar is a weak acid.)

2. **Going Further:** Ask a person who works in the meat department of a grocery store or meat market to help you get sections of bones from several kinds of animals (fish, turkey, hog, sheep, and cow). Look for ways that the bones are alike and ways they differ.

Bones and Blood

Are you thinking about structure and function? You just looked at the structure of a living bone, so now you should be thinking about functions. You should be asking questions such as, "why do we need marrow?" and "why are there two types of bone?" If you are really thinking like a scientist, you may wonder, "what do bone cells do?"

Parts of Bone

A bone has four main parts: the outer covering, compact bone, spongy bone, and marrow. (See **Figure 3.16**.) The outer covering is thin and tough. It provides a place for muscles to attach and contains cells that produce new bone.

Directly under the outer covering is a layer of compact bone. Compact bone looks quite solid. It is called compact because it contains few spaces. By contrast, spongy bone is full of spaces. It is easiest to see in the ends of long bones, but it can also be seen in ribs and other flat bones. Both types of bone are very strong.

The spaces inside spongy bone are not hollow. They are filled with cells—fat cells and cells that produce red blood cells. This **marrow** also fills the center of the long bones in the arms and legs, as well as the inside of many other bones. People are often surprised to learn that most blood cells are formed inside bones; they are not made in the heart or in any of the blood vessels. The newly formed blood cells enter the tiny blood vessels that are found throughout a bone.

Bone Cells

Think about what you've learned so far—the skeletal system is made up of bones, ligaments, and cartilage; and the bones themselves are made up of four main parts. But everything living is made up of cells, so where can you find the cells? They are in all four parts of a bone. A bone is so solid that sometimes it is hard to believe that it contains cells. Look at **Figure 3.17**. This shows what a very thin slice of compact bone looks like when it is viewed under the microscope. The bone cells are enclosed by minerals and located in tiny spaces within the hard part of the bone. Notice the skinny projections that extend from the cell out into the hard part of the bone. These projections are reaching out towards the microscopic blood vessels that are inside the bone.

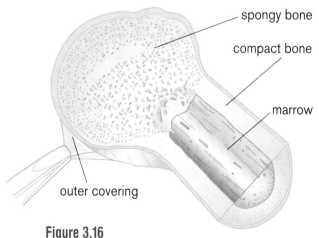

Figure 3.16
Cross section of a bone

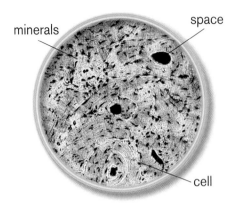

minerals   space

cell

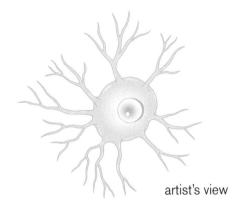

artist's view

Figure 3.17

This is a microscopic view of bone as it looks when magnified about 200 times. The artist's view is a drawing of what a bone cell might look like if we could take away the minerals in the bone without touching the cells.

Even though bone cells are trapped within bone, they are very active. They constantly remove the old minerals in bone and replace them with new minerals. In fact, over a ten year time span your bone cells gradually tear down and rebuild your complete skeleton. Even as you read this page, your bone cells are remodeling your skeleton!

A Blood Supply

Like all cells, bone cells need nutrients and oxygen. All of the work they do takes energy, and energy comes from nutrients. In order to get the nutrients, the cells must have a blood supply. Very tiny blood vessels wind their way through the bony tissue, delivering oxygen and nutrients to every single cell.

As the blood flows by, bone cells absorb calcium from the blood and deposit it around themselves to build new bone. Calcium is one of the minerals that makes bones hard. Without calcium, bones would not be very strong. Calcium is also what makes bones show up well on x-rays.

The blood supply is also important for healing a broken bone. Without blood, a broken bone would not mend itself. New bone cells need blood in order to grow. In a healthy young person's body, who is eating nutritious foods, it may be impossible to tell that a bone was ever broken once it has healed.

Timeline for Healing a Break	
immediately	blood collects around the break
1 week	young bone cells are produced
2 weeks	bone cells start to deposit minerals
6 weeks	bone has regained strength
1+ years	bone is completely healed

Figure 3.18

Try to identify the bones and locate the break in this x-ray.

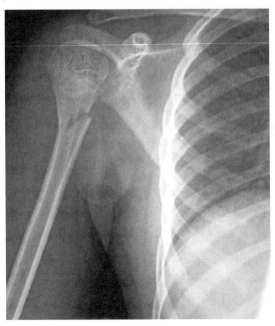

Growing Bones

Bones grow from growth centers located at their ends. The red lines in **Figure 3.19** show the growth centers on a femur. Growth centers are soft; when they harden, the bones stop growing. All growth centers do not shut down at the same time, but they do harden in a predictable order.

Because you are still growing, most of your growth centers are still soft. The growth centers in the bones that form the elbow are among the first to harden. They usually do so between 13 and 16 years of age. (The age range means that these centers may close down in one person at 13 while they may continue to grow in another person until age 14, 15, or 16.) Each person's bones grow and harden at his or her own rate. Among the last bones to harden are those that form the wrist. Most bones are fully hardened by the time a person is 25 years old.

Growth centers are softer than the rest of the bone so they appear fuzzy on an x-ray. This allows a doctor to use an x-ray to check a child's bone growth. The four x-rays in **Figure 3.20** show the development of bones inside a hand and a wrist.

A bone is not a simple thing! Some people spend their entire careers trying to understand how bones work.

Figure 3.19

The red lines show the growth centers on a femur. Which ones do you think could be damaged if a teen injures a hip?

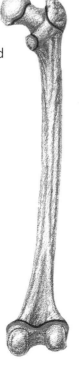

18 months

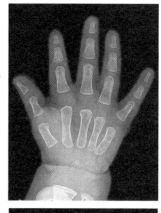

5 years

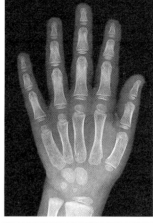

12 years

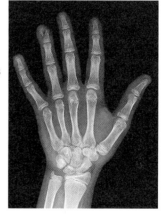

17 years

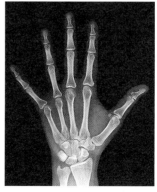

Figure 3.20

X-rays of the hands can be used to tell a person's age. From top to bottom, the following ages are shown: 18 months, 5 years, 12 years, 17 years.

Analysis Questions

1. List the four main parts of bone and one function of each part.

2. List at least three reasons why bones need a blood supply.

3. Think about growth centers for a minute.

   a. Where are some of the growth centers located, that are the first to stop growing?

   b. Where are the last ones to stop growing?

4. Compare the x-rays shown in **Figure 3.21** with those shown in **Figure 3.20** to estimate the ages of these two people.

5. HB Connections:

   a. What does the skeletal system do for other body systems?

   b. Why does the skeletal system need other body systems?

Extension Activities

1. **Going Further:** Read the labels on foods to find out which ones are low in calcium and which ones are high in it. Make a chart to display your data.

2. **Technology:** As scientists learned more about radiation, they thought of ways to use their knowledge. As a result, x-ray machines were invented to "see" inside a person's body. Find out what kinds of structures can be seen with x-rays. If your teacher has any x-ray films, or if you can get some discarded ones from a clinic, try to "read" them. Identify all of the bones and joints that you can see.

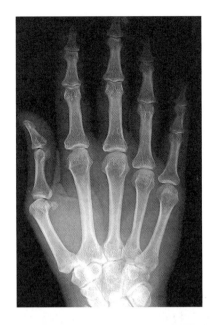

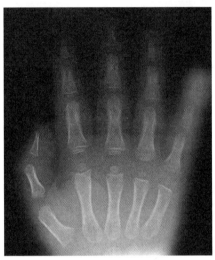

Figure 3.21

An Experiment with Rats

Are you feeling like a scientist? Let's hope so, because in this activity, you need to think like a scientist. Imagine that you have a friend, Scientist Senob, who is studying bone problems, particularly problems that cause weak bones. Weak bones bend under pressure. Senob has been reading all about this condition and has decided that it could be caused by lack of sunlight, sweets in the diet, or lack of exercise. Here are the questions Senob wants to answer:

Does lack of sunlight cause bones to be weak?

Do sweets in the diet cause bones to be weak?

Does lack of exercise cause bones to be weak?

Senob decided to use rats to see which of these things cause bones to weaken. Senob put four rats in each cage and made sure the rats had similar living conditions. In order to test an idea, Senob would change one of the conditions. Of course, whenever this happened, Senob recorded the differences.

Two of the experiments are now done and Senob wants you to read the results and reach your own conclusions. Here they are, see what you think.

Experiment I

Question: Does lack of sunlight cause bones to be weak?

Prediction: Rats will have weak bones when they do not get any sunlight.

	Rat Group I A	Rat Group I B
Sunlight:	12 hours each day	0 hours each day
Diet:	rat pellets	rat pellets
Exercise:	wheel	wheel
Results:	healthy bones	weak bones

Based on this experiment, how would you answer the question?

Experiment II

Question: Do sweets in the diet cause bones to be weak?

Prediction: Rats will have weak bones when they have sweets in their diet.

	Rat Group II A	Rat Group II B
Sunlight:	12 hours each day	12 hours each day
Diet:	rat pellets	rat pellets & sweets
Exercise:	wheel	wheel
Results:	healthy bones	healthy bones

Based on this experiment, how would you answer the question?

Well, do you think Senob is thinking like a scientist? Why? Here are three things that scientists think about when they work. Go through the list and decide which of these things you think Senob did correctly.

1. Ask a question that can be answered by doing experiments.

Senob started by asking "What causes weak bones?" This question was too hard to answer so Senob did some research to learn more about weak bones. Senob ended up with three questions to answer.

2. Design an experiment. Make sure it tests only one thing at a time.

Senob set up three experiments—one experiment to answer each one of the questions. Each experiment should test only one variable. (A variable is something that can change from experiment to experiment.)

3. Collect data. Organize it so someone else can read it. Senob made data tables and labeled them so other people could understand the results of the experiment.

Analysis Questions

1. What is a variable?
2. What are the three variables that Senob wanted to test?
3. What variable did Senob test in
   a. the first experiment?
   b. the second experiment?
4. Think about the third experiment that Senob should do.
   a. What question will Senob try to answer?
   b. Set up a chart like the others that describes the light, diet, and exercise each group of rats will receive.
   c. On the line for results, write down what you predict will happen to the rats' bones.

5. Referring to the data, predict whether or not each of these conditions would cause rats to have weak bones. Be sure to explain your answers.
   a. 12 hours sun, rat pellets, exercise wheels
   b. 0 hours sun, rat pellets with sweets, exercise wheels
   c. 12 hours sun, sweets only, exercise wheels
6. Why do you think it is important to test only one variable at a time?
7. Now Senob thinks that vitamins might prevent weak bones. Set up a chart that shows what experiment you would do to test whether or not this is true.
8. Senob wants to understand more about the causes of weak bones in rats and ways to prevent weak bones. What are some other questions that Senob could try to answer by doing experiments?

Did you realize that "Senob" is bones spelled backward?

Extension Activities

1. **Design and Do:** What strengthens hand muscles—vitamin C, drinking extra milk, or squeezing a tennis ball? Outline an experiment that you could do to answer this question. If your teacher approves, go ahead and try your experiment.
2. **History Connection:** Do a research project on rickets to find out how scientists discovered what caused this bone condition.

Rickets:
A look back in time.

During the 17th and 18th centuries, many children in western Europe developed rickets. Their bones were so weak that sometimes they bent under the weight of the body and became permanently misshapen. This condition is known as rickets.

People were worried and perplexed by the increase in the number of children who developed rickets. When they gathered the facts, they realized that children who lived in the city had rickets far more often than children who lived in the country. City life had been changing.

This was the era of the Industrial Revolution. For the first time, large factories produced clouds of black smoke that hung in the air for days. Many children no longer played outside; instead, they worked in factories and lived nearby in homes with small windows. Eventually, scientists realized that sunlight was important for the development of healthy bones. The children who lived and worked in these crowded and polluted cities were not getting enough sunlight to build strong bones.

Chapter 4

Muscular System

Which muscles do you use when you play soccer?

4-1

Chicken Wings

Using your right hand, grasp the upper part of your left arm and then make a fist with your left hand and bend your elbow. Do you feel your muscles move? The muscle on the front of your arm, your biceps, probably bulged the most.

Now try several more arm motions. Which movements do not involve your biceps at all? Which ones use the biceps the most? What's going on inside your arm as you do all of these motions? What do muscles look like? And what else is inside of your arm besides muscles?

It is difficult to look at structures inside your own body, but you can learn about the structures and functions of the human body by studying other animals. In this lesson, you will learn about the design of your arm by dissecting a chicken wing.

What makes up a chicken wing?

What can you learn about the structure of a human arm by taking apart a chicken wing?

Keep these two focusing questions in mind as you dissect the chicken wing.

- chicken wing
- dissecting pan
- scissors
- Handout, *Chicken Wings*
- soap
- paper towels
- plastic bag

Procedure   *(Work in a group with one or two other students.)*

1. Choose one person in your group to be the reader/recorder. The reader/recorder should read the dissection guide aloud and record the group's observations in the spaces provided on the guide.

2. Work quickly and carefully. Make sure that each person in your group can see what is going on.

3. If your group does not finish, save your wing in a plastic bag. Write your names and class period on the bag and give it to your teacher so it can be refrigerated.

4. Think safety! Thoroughly clean your hands, desk, and tools with soap and water.

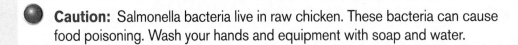 **Caution:** Salmonella bacteria live in raw chicken. These bacteria can cause food poisoning. Wash your hands and equipment with soap and water.

Conclusion

Use complete sentences to answer each of the focusing questions.

Extension Activities

1. **Dissection Skills:** You've dissected a chicken wing; now dissect a chicken leg. Compare the two structures. How are they alike? How are they different?

2. **Try It:** Tendons connect skeletal muscles to bone. Wiggle your fingers while you look at the tendons on the back of your hand. How many tendons can you count? The largest tendon in your body is called the Achilles tendon. You can feel it by pressing on your lower leg just above your heel. What is the purpose of this tendon? Find at least two other places in your body where you can feel a tendon.

Wings and Arms

Humans have arms, lizards have legs, and chickens have wings. All of these structures are similar, and yet they are not identical. In order to understand the differences in structures, think about their different functions.

Tissues

When you dissected the chicken wing, you observed several different kinds of tissue. A **tissue** is a group of similar cells that have a similar function. For example, many bone cells that are grouped together form bone tissue, and bone tissue, along with blood vessels and cartilage, forms the structures we call bones.

Did you notice the glistening white tissue on the ends of the bones? That was the cartilage. You may have noticed that there weren't any blood vessels running through the cartilage. Since cartilage does not have a very good blood supply, that's a clue to you that it may not heal very well if it is damaged. You'll find that this is true. When people damage the cartilage in their knees or other joints, they usually require surgery to remove the cartilage.

Figure 4.1

Remember, ligaments attach bone to bone. They strengthen the joint and keep bones from slipping out of place. Ligaments are extremely strong. Look at your notes—do ligaments have a blood supply?

Bone, cartilage, and ligaments are three tissues that you observed in the chicken wing. You observed at least six others—which ones can you recall?

Wing Bones, Arm Bones

Wings and arms have very similar bone structures. The upper section of each structure has one long bone, the humerus. They both have two long bones, the ulna and the radius, that form their lower sections. The "wrists," however, are less similar. Humans have many small, rounded bones in each wrist while most of these are fused together in birds. Ah-ha! You should be thinking, "different structures, different functions." Think about it. A human's wrist is very flexible and can move in many directions. A bird's wrist does not need to be very flexible. In fact, the entire wing is probably stronger for flight because the wrist does not move in many directions.

What about the rest of the bones in arms and wings? In humans, they form the fingers. Humans have five fingers, each of which is made up of two or three phalanges, but a bird has three "fingers," each of which contains only one or two phalanges. Find all of these bones in **Figure 4.2**.

In general, birds have fewer bones than humans, and they also have more "fused" bones. This gives their skeletons the most strength with the least weight.

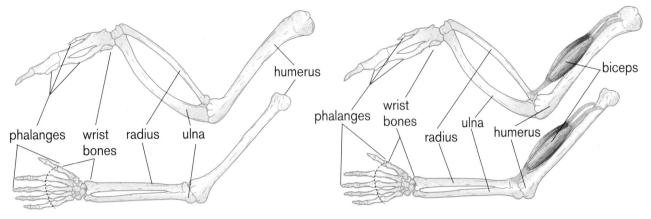

Figure 4.2

Compare the bones in a chicken wing (top) with the bones in a human arm (bottom).

Figure 4.3

The muscle along the front of the humerus in both humans and chickens is called the biceps.

Wing Muscles, Arm Muscles

The muscles you observed in the chicken wing are all called **skeletal muscles** because they attach to the bones that make up a skeleton. You have probably not had much experience observing muscles, so you may have had a harder time finding and counting the individual muscles than you did the bones.

However, you should have noticed that muscles do not attach directly to the bones; instead they are attached by **tendons**. (Did the tendons have a blood supply?) As with the bones, the arrangement of muscles in a chicken wing is very similar to the arrangement of muscles in a human arm. In this chapter you will learn more about your muscles and how they work.

Analysis Questions

1. What is a tissue?
2. List at least seven kinds of tissue that you observed in the chicken wing.
3. Compare tendons and ligaments. (Think about what you observed as well as what you have read.)
   a. List at least two ways they are similar.
   b. List at least two ways they are different.

4. Based on your observations, how would you describe:
   a. muscle tissue?
   b. a muscle?
5. Compare a chicken wing and your arm.
   a. List at least three ways they are similar.
   b. List at least three ways they are different.
6. Challenge: Look at the human and bird skeletons in **Figure 3.8**.
   a. Start just below the pelvis and try to name the bones in the bird's leg.
   b. Which way does a bird's "knee" point when it is bent?
7. All animals move during at least one stage of their life cycle. Why do you think animals need to move?

Extension Activities

1. **Animal Kingdom:** Find out how animals without bony skeletons move. Start by finding out about earthworms, sponges, jellyfish, and clams.
2. **Going Further:** Try to identify the biceps and triceps on a chicken wing (look at **Figure 4.3**). Pull gently on each one and watch what happens.

Animal Muscles

One of the most obvious differences between plants and animals is movement. Plants stay in one spot while most animals can move— they may swim, walk, or fly. Movement usually requires muscles of some kind. If animals have an internal skeleton, their muscles attach to their bones. Animals with external skeletons have muscles that attach to the inside of their "shells." The muscles that move an insect's wings must be incredibly "fit." A physiologist from Finland calculated that butterflies move their wings 5 cycles per second. Mosquitoes were faster, moving at 587 cycles per second, and the tiny midge beats its wings at the impressive rate of 1,046 cycles per second!

Animals such as squids, that have neither external nor internal skeletons, have evolved other systems for movement. Squids use a jet stream of water to push them forward. The water pressure may be strong enough to propel them out of the water and through the air for 30 meters (90 feet) or more.

Why do you think animals need to move, but plants don't?

Major Muscles

You now know the names of quite a few bones, but what about muscles? How many of them can you name? The human body contains more than 600 skeletal muscles and each of them has a name. Again, it isn't important to learn the names of all 600 muscles, but it does help to know the names of a few of them, and to understand their functions.

Muscle Names

The locations of eleven of the major muscles are shown in **Figure 4.4**. How many of these do you already know by name? Many of these muscles are mentioned in physical education classes and in fitness programs.

The two muscles that people are most likely to call by their names are **biceps** and **triceps**. The largest and heaviest muscle of the body is the **gluteus maximus**. You are probably sitting on your gluteus right now. The **quadriceps** is

actually a group of four muscles that all work together. You can see three of these muscles in **Figure 4.4**. You cannot see the fourth one because it is underneath the other three. This muscle group is very powerful. On the back of the thigh is another group of powerful muscles. Together, they are referred to as the **hamstring**. The big tendon, called the Achilles tendon, attaches the **gastrocnemius** to the heel. Look at **Figure 4.4** and find the five other muscles that are labeled but not mentioned here. Make one observation about either the size or location of each of these muscles.

Muscle Action

Skeletal muscles cause movement when they shorten or **contract**. You can tell what movements a muscle controls if you know where its tendons attach. Therefore, when you study the structure of a skeletal muscle, you should find out which bones it attaches to. For instance, **Figure 4.5** shows where the biceps attaches to the radius and the scapula. When the biceps contracts, it pulls on the radius and causes the elbow to bend.

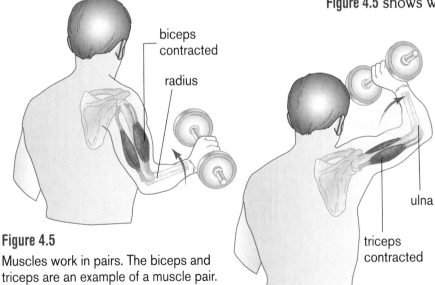

biceps
contracted

radius

ulna

triceps
contracted

Figure 4.5
Muscles work in pairs. The biceps and triceps are an example of a muscle pair.

Includes *SciLINKS*
NSTA

Topic: Muscle and Bone Interaction
Go to: www.scilinks.org
Code: MSLS3e66

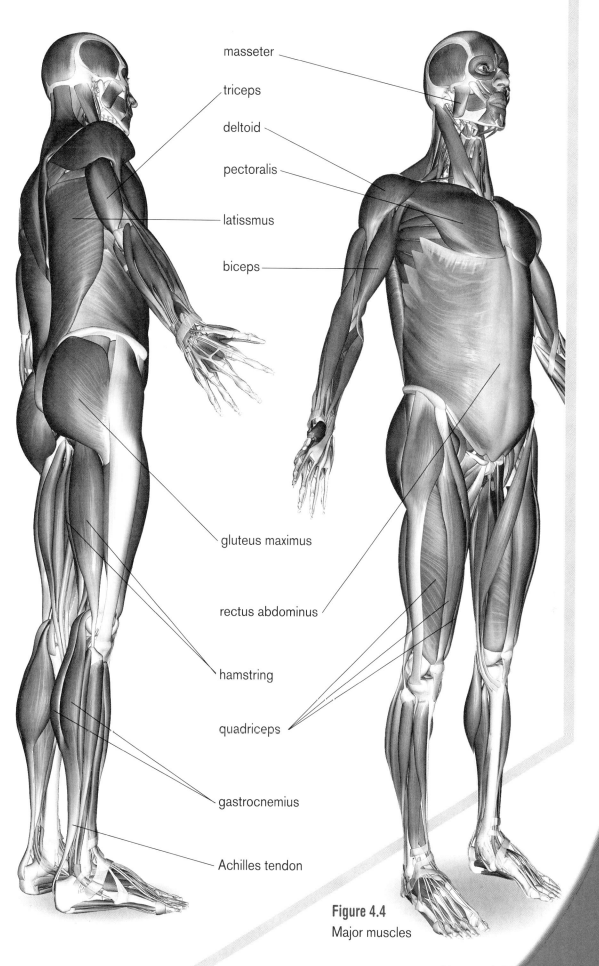

masseter

triceps

deltoid

pectoralis

latissmus

biceps

gluteus maximus

rectus abdominus

hamstring

quadriceps

gastrocnemius

Achilles tendon

Figure 4.4
Major muscles

In order to straighten the arm, other muscles are needed. The biceps cannot straighten the arm because a muscle cannot "push," all it can do is pull. Here is where the triceps is needed. At one end, it attaches to the scapula and humerus; the other end attaches to the ulna by the elbow. When it contracts, it pulls on the ulna and straightens the arm (see **Figure 4.5**).

Now it's your turn to predict the action of a muscle. The upper end of the gluteus attaches to the back of the pelvis; the lower end attaches to the back of the femur. Imagine what happens when the gluteus contracts. It pulls the femur backward, which straightens the upper part of your leg. You use your gluteus every time you walk, run, or climb. You also use your gluteus when you stand up from a sitting position. (Why?)

Muscular System

All of the muscles, along with tendons and blood vessels, make up the **muscular system**. Muscles alone, however, can't do much. Every movement you make is because muscles are pulling on your bones. By working together the muscular system and the skeletal system allow us to move.

Analysis Questions

1. Imagine that you tripped down several stairs and fell, landing on your arm and jarring your shoulder. Your doctor examined the injury and decided you had injured your deltoid. You are supposed to put an ice pack against the muscle and let it rest for several days.

   The doctor told you that one end of the deltoid is a broad tendon that attaches along the clavicle. This tendon also crosses over the top of the shoulder and attaches to the upper part of the scapula. (Point to these bones on your own body and imagine the muscle that attaches to them.) The other end of the deltoid attaches to the outer surface of the humerus. (Put your hand over this area of your arm.)

   a. Imagine that the deltoid is contracting. What motion does it cause?

   b. What movements should you avoid in order to rest your deltoid?

2. On your own body, locate each of the muscles labeled in **Figure 4.4**. Try to contract and relax each muscle so you can feel it move. Memorize the names and locations of at least 5 of these muscles.

3. Why do we say that the muscular system is a system?

Extension Activities

1. **Going Further:** Find a chart that shows the names of other major muscles. Choose several muscles that are not identified in this activity and learn their names and what movements they control.

2. **Models:** Construct a model to demonstrate the action of one particular muscle. Include the tendons and bones. Use your model to illustrate what happens when the muscle contracts.

Face Muscles

A network of small muscles in your face controls your expressions and communicates your mood. When you try to move one part of your face, such as by wiggling your ears, you often find yourself using many other face muscles at the same time.

Do the facial muscle exercises shown here. You may look silly but that's okay. While you do them, try to figure out which facial muscles you are using.

Think about your facial expressions. What parts of your face do you use when you look angry? happy? in pain? scared? Facial movements are unique to humans. Most animals' faces are immobile. Have you ever seen a giraffe laugh or a snake smile?

stretch top lip down

open nostrils

move mouth corners up and down

shut nostrils

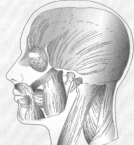

wink one eye, then the other

pull scalp back

pull scalp down

How Long?
How Many?

You probably know what it feels like to have a "tired" muscle; that's what happens when a muscle has worked too long. **Endurance** refers to how long a muscle can work before it tires. There are two ways to measure muscle endurance. You can measure how long the muscle can work or you can count how many movements it can make before it tires.

> How do you test the endurance of specific muscles?

After testing the endurance of your muscles, you will be able to answer the focusing question.

- old tennis ball
- paper

Procedure

Part A: Tennis Ball Grips

1. Make a data table to record your data.
2. **Tennis Ball Grips:** Read the four steps for this endurance test before you start.
   a. Predict how many you will be able to do. Record your predictions.
   b. With your right hand, squeeze a tennis ball as many times as you can. Use the same amount of pressure each time you squeeze. Count the squeezes.
   c. Record the number of squeezes you actually did on your data table.
   d. Repeat the three steps of this procedure, but this time use your left hand.
3. Add your data to the class data table.
4. Answer analysis questions 1 and 2.

Part B: Class Histogram

A **histogram** is a type of a graph that shows how often (frequency) data falls in ranges or intervals (see **Figure 4.6**). Your teacher will show you a histogram that represents all the data that was collected from each class doing the tennis ball grips. It shows the number of students (frequency) who were able to grip the tennis ball within a certain range.

Frequency Table	
Range of Test Scores	Frequency (Number of Students)
0–9	0
10–19	2
20–29	2
30–39	3
40–49	2
50–59	3
60–69	5
70–79	8
80–89	7
90–100	5

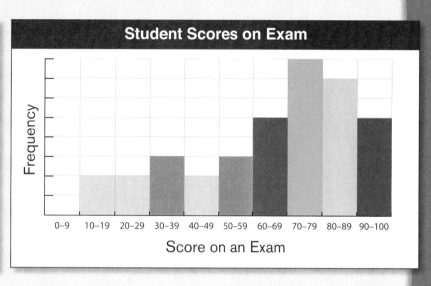

Figure 4.6

With your class or a partner, analyze the graph by answering the following questions.

a. What does the graph tell you about the endurance of muscles? What can you *infer* (conclude) from this information?
b. Why do you think class data was used rather than individual data?
c. What are some things that could cause this data to be inaccurate?
d. Is it fair to look at this graph in order to decide whose grip muscles have the best endurance? Why or why not?

Part C: Planning an Investigation

After completing the tennis ball grip activity, a Life Science team at another school wondered whether the girls or the boys in their team had the highest endurance. They developed a scientific investigation to test their question. They started planning the scientific investigation, but need help finishing it.

1. The first step in a scientific investigation is to develop a question that can be answered by doing an experiment, so the students came up with the following **testable question**.

Testable Question:

Is the muscle endurance of females stronger than the muscle endurance of males?

On your paper, under the heading "**Testable Question,**" write the testable question.

2. The students want to learn more about muscle endurance and what scientists have already discovered about their question before they did any more planning. So that you can have the same background knowledge as the class, complete the following:

   a. Read *Strength* and *Endurance*, found later in this chapter. Answer analysis questions 3 through 6.

   b. Read *Women Beat Men on Muscle Endurance*, page 78. Answer analysis questions 7 through 12.

   c. Optional: Conduct a web search on the gender differences in muscle endurance and muscle strength. What do scientific studies show about the difference between males and females in muscle strength and muscle endurance?

3. Before conducting an experiment, scientists usually have an idea, based on prior observations or research, about what they think will happen. Think about the research you just conducted. Based on that information, what do you *think* will be the answer to the testable question you wrote on your paper? This is called a **hypothesis**. A hypothesis is a possible explanation about an event that can be proved or disproved through an experiment or observations. Sometimes a hypothesis is written in an "***If…, Then…***" format. In other words, "***If*** ____ [what is manipulated in the experiment] ____, ***then*** ____ [How you think it will respond] ____."

For example, "***If*** *age affects the endurance of muscles,* ***then*** *a group of 20-year-old males should have more muscle endurance than a group of 65-year-old males.*"

Using the "*If…, then…*" format, write your hypothesis on your paper, under the heading "**Hypothesis.**"

4. The class needs help in designing its experiment. The students would like to model their experiment after the one completed at the University of Colorado, by having each student hold out their arm rigidly to the side while holding a *weight*. They will time how long each student can hold the position before tiring out. Help the class design their experiment. It is important that the experiment is a **fair test**. You conduct a fair test by making sure it

Includes *sci*LINKS
NSTA

Topic: Scientific Inquiry
Go to: www.scilinks.org
Code: MSLS3e72

tests only one thing (**variable**) at a time. A variable is something that can change from one experiment to another. Think about how you are going to control all the variables so you only test one. In this case, since the class is testing the difference between males and females, you will need to control all the variables except boys and girls.

a. What are all the variables you can think of in this experiment? Write them down on your paper under the heading, "**Variables**."

b. Except for the difference of boys and girls, how would you recommend controlling the rest of the variables? Write down your recommendation for each of the variables you listed above.

5. Read the procedure in **Figure 4.7** that the students wrote for this experiment. You will be conducting the same experiment so read it over and check to make sure you understand what needs to be done, step by step. If you have any questions, first ask the other members of your team to see if they can help you. If they can't, ask another team for help, and finally, ask your teacher for clarification.

Procedure: Muscle Endurance Test

Materials

- *Middle School Life Science* textbook
- stopwatch or clock with second hand

1. Make a data table to use when you record your data.
2. Work with a partner.
3. **Muscle Endurance Test:** Read the steps for this endurance test before you start.

a. With your arm out to your side, perpendicular to your body, hold a *Middle School Life Science* textbook for as long as possible. Be sure to keep your arm rigid and perpendicular to your body. Your partner should time how long you can hold up your arm. *Don't forget to keep all your variables constant.*

b. Record the time on your data table.

c. Repeat, only this time your partner will hold the book.

d. Add your data to the class data table.

Figure 4.7

6. Conduct the experiment by following the procedure, collecting data, and keeping all your variables constant except the one you are testing.

    a. Use a data table to record your data.

    b. When you are finished, add your data to the class data table.

7. In order to summarize and present data, an appropriate graph needs to be constructed, using the data collected.

    Using the class data, construct a **bar graph** on your paper. Your bar graph should compare the average time that boys and girls in your class can hold the textbook (a test for muscle endurance).

8. On your paper, write a *conclusion* by answering the following questions (first write down the heading, "**Conclusion**"):

    a. What does the class data tell you? Does it or does it not support your hypothesis? Why or why not?

    b. What other explanations could explain your results? What errors could affect your results?

    c. What other questions about muscle endurance do you have now that you've completed your experiment?

Analysis Questions ⎯ ⎯ ⎯ ⎯ ⎯ ⎯ ⎯ ⎯ ⎯

1. Which of your muscles did you test to the point of tiring when you did the tennis ball grip?

2. Possible sources of error: What are some things that could cause your data to be inaccurate?

3. Do you think it is possible for a person to have a lot of muscle strength without much muscle endurance or a lot of muscle endurance without much muscle strength? Explain your answer.

4. What is an example of a sport where muscle strength is more important than muscle endurance? What is an example of a sport where muscle endurance is more important?

5. Put these structures in the order that they are used when you bend your arm at the elbow: biceps, brain, nerve, radius.

6. In order for you to move, your muscular system must interact with other body systems. List at least 3 other systems that allow your muscles to work and describe how each of these systems interact with your muscles.

7. What university conducted the study, *Women Beat Men on Muscle Endurance*? What year were the results of the study reported at the American Physiological Society meeting?

8. What were the two exercises that the participants were asked to do? Do you think these exercises tested muscle strength or muscle endurance? Explain your answer.

9. What results were founded from the testing?

10. The author of the article, Abbie Thomas, writes, "But rather than some kind of motivational effect, the study found the difference was due to some feature of the muscles." What do you think might be a *motivational effect* in this study? Why?

11. Abbie Thomas continued to write, "The authors (of the study) suggest that, given that women are weaker than men…" Do you agree with this statement? Why or why not?

12. What three things do the authors of the study suggest could be the reason(s) for the differences in muscle endurance between men and women?

Conclusion

Use your data and experiences from this activity to answer the focusing question.

Extension Activities

1. **Design and Do:** Prepare a plan to improve your endurance. Have your teacher check your plan, and then try it. Report your results to the class.

2. **P.E. Connection:** Talk to a P.E. teacher about activities that improve strength and endurance. How does he or she decide which activities to use in P.E. classes? Does the teacher consider activities that focus on both types of muscle fitness—strength and endurance?

Strength and Endurance

Strength and endurance are important qualities for athletes, and they are important components of every fitness program. The more you understand about muscle fitness and how your muscles work, the more likely you are to make wise decisions and not do things that could harm your body.

Muscle Strength and Size

Strength is the amount of work that a muscle can do in a single try. You need a certain amount of strength to do all the things you have to do each day. If you start doing a little more than usual, you will be exercising your muscles a little more than usual and the muscles will get stronger and may begin to "grow." A muscle grows because the individual cells that make up the muscle grow larger and more small blood vessels form within it. The blood vessels form because the more a muscle works, the more oxygen it needs, and the only way muscle cells can get oxygen is from the blood. As a muscle gets stronger, the tendons, ligaments, and bones get stronger too. They have to, or else they would break when the muscles pull on them.

Exercise is the only safe way to increase the strength and size of a muscle. No special diet or magical food can turn a scrawny body into a muscular one. The "Popeye Diet" will not do it. Neither spinach, protein powders, nor any other food will develop muscle strength.

Sometimes people try drugs called steroids to help them "bulk out." This is dangerous since steroids can ruin a person's health. Steroids may cause liver and kidney damage, shrinkage of the testes, severe mood swings, heart disease, and cancer. They can also stunt growth by closing off the growth centers at the ends of bones. Steroids are not a good choice when it comes to developing muscles. So what can you do if you want strong muscles? Exercise and eat a balanced diet!

Figure 4.8
Many articles warn people about the dangers of steroid use.

When muscles are used, they grow, and when they aren't used, they shrink. Anyone who has worn a cast to support a broken arm or leg has experienced this phenomenon; it is called "muscle wastage." When it is not used, a muscle becomes smaller because the individual cells become smaller. Fortunately, once the muscle is used again, the cells enlarge and the entire muscle returns to its normal size.

Endurance and Tired Muscles

Another way to think about muscle fitness is to consider **endurance** or how long a muscle can work before it tires. A muscle tires when it runs out of oxygen and waste products build up in its cells. Eventually, these wastes reach such high levels that the muscle stops working. Luckily, the muscle can work again once it has rested and had time to get rid of the waste products and rebuild its oxygen supply.

The best way to improve a muscle's endurance is by exercise, particularly by aerobic exercise. An **aerobic exercise** is one where the heart and muscles are kept working at a steady pace. In this way, the muscle cells never run out of oxygen so they can work longer. (Aerobic means "with oxygen.") Aerobic exercise is recommended for losing weight, keeping the heart healthy, and improving fitness.

Not all exercises are aerobic. For example, a walk provides some exercise but it does not contribute much to a healthy body. A brisk walk or jog raises the heart rate to an aerobic level. A fast sprint is probably too fast to keep oxygen supplied to the muscles, so it is not aerobic. Aerobic exercises do the most benefit if they are continued for at least twenty or thirty minutes.

Figure 4.9

Muscle wastage is a problem for astronauts since they do not experience the normal pull of gravity on their bones and muscles. After a long stay in space, their muscle size decreases and their bones lose calcium unless they exercise while in space.

Figure 4.10

How could you decide whether or not these students are getting aerobic exercise?

Muscle Control: A Nerve Connection

Unless you have worked your muscles beyond their endurance, they are under your control. They do not move by themselves; they only work when you "tell" them to do so. Most of the time you don't think about which muscles are working because your brain controls them automatically. Nerves carry messages from your brain to specific muscles. When the message reaches the end of the nerve, chemicals are released that cause the muscle to contract. After contracting, the muscle quickly relaxes until it receives another message to contract. **Figure 4.11** illustrates this nerve-muscle interaction.

A lot of action goes on inside a muscle every time it moves! Wouldn't it be interesting to watch all this activity?

Women Beat Men on Muscle Endurance

In 2000, scientists from the University of Colorado reported to the American Physiological Society that women were able to beat men in muscle endurance tests. Abbie Thomas, in a report for ABC News Online Science, wrote that the scientists conducted a study, during which participants were asked to undertake two tasks. During the first task, the participants were instructed to hold their arms rigid as long as possible. Participants were directed to hold their arms rigid as long as possible for the second task as well, but this time a weight was attached. The scientists found that the women were able to perform both of these tasks longer than the men by an average of 75%.

Thomas reported, "But rather than some kind of motivational effect, the study found the difference was due to some feature of the muscles. The authors [scientists] suggest that, given that women are weaker than men, the difference may be due to some interaction between muscle strength and blood flow within the muscle. But it may also be due to the different types of muscle fibres in women and men, or even their different hormones."

Thomas, Abbie. "Women beat men on muscle endurance." **ABC Online Science, News in Science,** September 26, 2000; www.abc.news.au. June 24, 2008.

Figure 4.11
A muscle contracts when it is triggered to do so by a nerve.

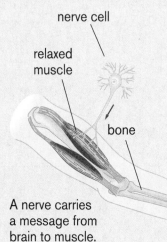

nerve cell

relaxed muscle

bone

A nerve carries a message from brain to muscle.

When the message reaches the muscle, the muscle contracts.

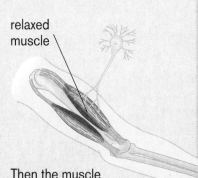

relaxed muscle

Then the muscle relaxes again.

Animal Olympics

If animals could compete in Olympic events, there would be some sure winners. When it comes to endurance, it would be hard to beat a sooty tern. This seabird can fly for as long as five years without stopping. It can even eat and sleep while in flight. Even though the cheetah is the fastest mammal, a pronghorn antelope could outrun it in a long-distance race. The antelope has much better endurance.

Humans should not compete against chimpanzees in events that test strength. Even though chimps are only about half as tall as humans, they are about three times stronger. In the insect world, ants are incredibly strong. Some ants can lift fifty times their own weight with their jaws!

Analyzing Activities and Injuries

You can apply what you have been learning about muscles and fitness by analyzing an activity. For instance, you could ask, "Is this activity aerobic?" or "Does it require strength?" If you are trying to improve your physical fitness, this analysis will be helpful to you. At the same time, you should be aware of the injuries that occur the most frequently during various types of exercise.

What are some of the most common sports-related injuries?

What should you do if you have one of these injuries?

Keep these questions in mind while you analyze several activities.

Materials

- Handout, *Analyzing Activities*
- Reading, *First Aid Guide*, page 84

Procedure
(Work with a partner.)

1. Talk with your partner and agree on three activities to add to the first column of the handout.

2. Together, decide how important strength, flexibility, and endurance are for performing each activity on the handout. Use this scale to rate your decisions.

    3-very important
    2-important
    1-may help a little
    0-not important

3. Different activities use different muscles. For each activity listed, indicate which muscle would tire the fastest; that is, which muscle would reach the "stop" point first? Be as specific as you can. For example, write "biceps and triceps" instead of "upper arm muscles."

Figure 4.12

Dancers must be flexible so they can move easily. They need to have the endurance to dance the length of a performance. Depending on their routine, they need strength in specific muscle groups.

4. Some activities are much more tiring than others. Rate each activity according to the amount of energy it requires. Assume a person does his or her best at the activity for 30 minutes. Use this rating scale.

   > 3-lots
   > 2-quite a bit
   > 1-some
   > 0-almost none

5. Decide if you think each activity is aerobic under normal circumstances. Answer *Yes* or *No* for each one. (Look back to the section, *Strength and Endurance*, if you need to review the meaning of aerobic.)

6. It is not unusual for people to injure muscles, bones, and joints when doing physical activities.

   a. Read "Common Aches, Pains, and Injuries" in the *First Aid Guide*.

   b. Decide which injury is most likely to occur for each activity on the handout. Write these in the "Common Injuries" column on the handout.

7. In the last column of the handout, initial the activities that you do at least one time each month.

8. Answer analysis questions 1 through 5.

9. Read *RICE: Treatment for Common Injuries* and then answer the rest of the analysis questions.

Analysis Questions _ _ _ _ _ _ _ _ _

1. Fill out the following information about one of your favorite activities.
   a. name of activity
   b. type(s) of muscle fitness it requires
   c. muscle(s) it exercises the most
   d. whether or not it is aerobic
   e. three common injuries that might occur when doing this activity

2. Referring to the activities on the handout, which activity do you think:
   a. requires the most strength?
   b. requires the most endurance?
   c. requires the greatest flexibility?
   d. takes the most energy?
   e. is the best example of an aerobic activity?

3. Give an example of an injury that affects each of the following:
   a. a bone
   b. a joint
   c. a muscle
   d. a ligament
   e. the skin

4. What is the difference between a sprain and a strain?

5. Which injury on the *First Aid Guide* seems to be the most serious? Why?

6. What does RICE stand for?

7. Why does it make sense to use RICE for an injury like a sprain?

8. Why aren't you supposed to use RICE for:
   a. a dislocation?
   b. a bruise?

9. RICE recommends using ice. In the past, people often used heat to treat an injury. What effect do you think heat would have on an injury?

Conclusion

Use complete sentences to answer the focusing questions.

Extension Activities

1. **Careers:** Following an injury, people may go to a physical therapist or an athletic trainer to help them regain strength, motion, flexibility, or coordination. Talk to one of these specialists and visit a clinic if you can. Find out what treatments they recommend for common injuries.

2. **Health Application:** Are you feeling ambitious? Improve your first aid skills! The American Red Cross, a local hospital, or your school may offer classes on what to do during medical emergencies. Check with them for schedules, topics, and times.

3. **Language Arts Connection:** Professional athletes are frequently injured. Read the sports pages in newspapers for two weeks and keep track of the incidents that involve injuries, the sport in which each athlete was involved, and the treatments that were used.

Figure 4.13

Athletic trainers work in schools as well as with professional sport teams.

 # First Aid Guide

Before treating any injury, always ask yourself these two questions:

1. What type of injury is it? *(For example, you do not want to treat a bone fracture the same way you would treat a muscle strain.)*
2. How severe is the injury? *(You can usually treat a mild injury if you use the treatment recommended here; seek medical help if the injury could be more serious!)*

Common Aches, Pains, and Injuries	
Bruise: bleeding under the skin	
Symptoms:	skin discolors (looks purplish) and, over days, lightens; may be tender to the touch
Cause:	a fall or a bump
Treatment:	if painful, apply ice or cold cloth with slight pressure
Cramp: a muscle contraction that doesn't relax immediately	
Symptoms:	sudden muscle pain, the muscle may feel hard
Cause:	may be unexpected; sometimes results from not drinking enough water
Treatment:	rest and gently massage the muscle, then gradually try to move it (not RICE)
Dislocation: a "separated" joint	
Symptoms:	pain, swelling, and skin discoloration at the joint; joint may look out of position
Cause:	stress at a joint
Treatment:	carefully apply ice, protect the joint, and get medical help
Fracture: broken bone (may be anything from a hairline crack to a total break)	
Symptoms:	extremely variable; tender to touch, swelling, bruising, slight to severe pain (may be mistaken for a sprain)
Cause:	stress on a bone
Treatment:	do not move if a break is likely; stop bleeding; apply ice; get medical help
Sprain: tearing of ligaments that hold a joint in position	
Symptoms:	varies from mild to extreme pain at a joint
Cause:	sideways pressure on a joint
Treatment:	RICE (use an elastic bandage)
Strain: a "pulled" muscle (torn muscle cells)	
Symptoms:	muscle feels tender; may swell; may stiffen overnight
Cause:	an overstretched muscle, bleeding within the muscle
Treatment:	RICE (use an elastic bandage)

RICE: Treatment for Common Injuries

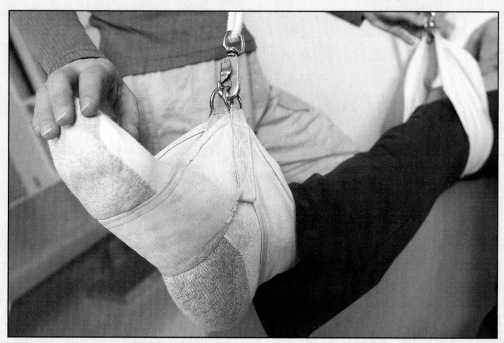

Figure 4.14
Using RICE to treat a minor injury

Aches and pains are the body's way of signaling that something is wrong. Many injuries will heal by themselves, but they will heal faster with proper care. RICE is a form of first aid that everyone should know. Here is what RICE stands for:

Rest: Rest the injured body part immediately! Continued use could make an injury worse.

Ice: Apply ice to prevent swelling. Cold causes blood vessels to contract or shrink so less blood flows to the injured area.

Compression: Compression means pressure. This also limits swelling. Swelling is painful and can make healing take longer.

Elevation: Elevate (raise) the injured part to a level above the heart, if possible. By doing this, gravity can help drain fluid from the injury so there will be less swelling.

Start RICE as soon as possible. It works for most injuries—muscle strains, sprains, and even some broken bones. Remember, ice is always safer than heat. However, if someone is seriously injured or very cold, do not make them colder by using ice.

Call a physician if any of the following occurs:

- immediate swelling
- a deformity (a bone could be broken or a joint dislocated)
- the pain lasts more than 24 hours
- you are not sure how to treat it

Smooth Muscles, Too

Any place in the body where there is movement there is probably a muscle. Your toes wiggle and forehead wrinkles because skeletal muscles contract. There are also muscles in arteries, intestines, and many other body organs. When these muscles contract, they cause blood to flow in arteries and food to travel through the intestines. This second kind of muscle is **smooth muscle**.

Differences Between Skeletal and Smooth Muscles

There are several differences between skeletal and smooth muscles. The differences can be illustrated by comparing the triceps (a skeletal muscle) with the wall of an intestine (a smooth muscle). Before you read on, think about the two kinds of muscles. How many differences can you think of?

Figure 4.15
Two types of muscles

Smooth Muscle

Skeletal Muscle

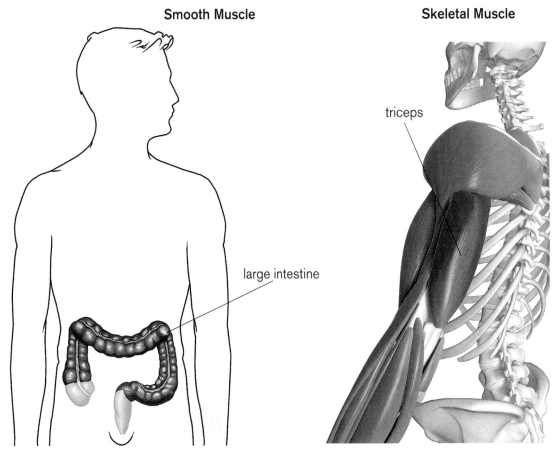

large intestine

triceps

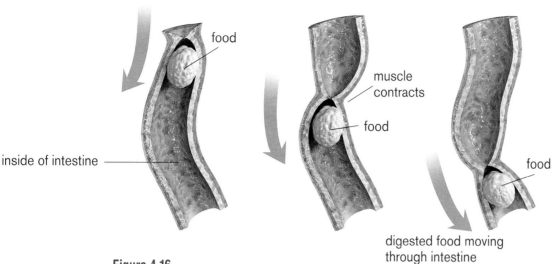

food

inside of intestine

muscle contracts

food

food

digested food moving through intestine

Figure 4.16

The contraction of smooth muscles moves digested food through the intestine. Can you explain why?

One obvious difference is the location of the muscles. Skeletal muscles attach to bones; smooth ones do not. They make up the walls of many hollow body structures.

This leads to a second difference between the two muscle types—their actions. When a skeletal muscle contracts, it pulls a body part, causing it to move. When a smooth muscle contracts, it squeezes. This squeezing is what moves food through an intestine or blood through an artery (see **Figure 4.16**).

A third difference is control. You control the movement of your skeletal muscles but smooth muscles contract automatically. They respond to changes inside the body. For instance, when the intestine fills, its walls stretch. This stretching causes the smooth muscles cells to contract and squeeze the food along its way.

Cell Differences

Like all tissues, skeletal and smooth muscles are made up of cells. Some of the differences between skeletal and smooth muscles involve the cells and must be seen through a microscope. Look closely at the two microscope views shown in **Figure 4.17**. An artist made these drawings while looking at thin sections of muscle through a microscope. You can see that the two types of muscle do not look the same.

Figure 4.17

Smooth muscle

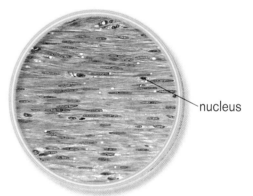

nucleus

fibers

nuclei

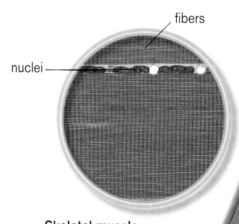

Skeletal muscle

If you are thinking like a scientist, you should be thinking, "Ah-ha! Different structure, different function!" That's right. Think about what would happen if these cells contracted. Can you see why skeletal muscles shorten and pull but smooth muscles squeeze?

You will see examples of smooth and skeletal muscles when you do dissections later on in the year. Try to remember to look for differences in the two types of muscle tissue.

Analysis Questions

1. Which of these structures contain smooth muscle? (Use the index to look up the structures you do not know.)

foot	stomach	hip	cheeks
lips	little finger	aorta	thumb
bladder	esophagus	iris	uterus

2. List three differences between skeletal and smooth muscles.

3. Write definitions for the terms skeletal muscle and smooth muscle.

4. Think of an activity that would exercise one of your smooth muscles.

5. Look at **Figure 4.17** while you compare smooth and skeletal muscle cells.

   a. List at least three ways they are alike.

   b. List at least three ways they are different.

6. In one or two sentences, describe the levels of organization of the muscular system. (Start with cells and end with system.)

Extension Activities

1. **Going Further:** In the meat department of your grocery store, buy something that contains smooth muscle tissue. Inspect the muscle tissue closely. How would you describe it? How does it compare with the skeletal muscle that you observed in the chicken wing?

2. **Going Further:** There is a third kind of muscle—cardiac muscle. Look at **Figure 4.18** and compare it with the drawings of smooth and skeletal muscle cells (**Figure 4.17**). What clues would help you recognize this type of muscle?

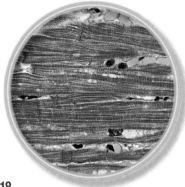

Figure 4.18
Cardiac muscle cells make up the heart.

Muscle Cells— Smooth or Skeletal?

Scientists must be good observers. As you have been doing the activities in this program, you have been practicing your own observational skills. Now you will be faced with a challenge. How good are your observational skills when you observe very small structures? And, how successful will you be when you compare two sets of your own observations?

What clues are most helpful when you are trying to identify muscle cells?

How does this activity help you understand more about muscles?

Use your microscope to look at two prepared microscope slides of muscle cells labeled *X* and *Y*. Your task is to decide which kind of muscle cell is on each slide.

- microscope
- prepared microscope slides: *X* and *Y*
- Handout, *Muscle Cells*

Procedure

1. Look back at **Figure 4.17** and fill in Part A on the handout. Be a careful observer! The better you are at describing the differences in the drawings, the better you will be at recognizing the differences through the microscope.

2. On the handout, circle the observations that you think will be the most useful when you try to decide what type of muscle tissue is on the microscope slide.

3. Start with either slide *X* or slide *Y*. Look at it first with low power and then with high power. Fill in the appropriate column in Part B on the handout.

4. Look at the other microscope slide, low power first. Fill in the rest of Part B on the handout while you look at the slide.

5. Compare what you have written in Parts A and B of your handout and then decide which type of tissue is on each microscope slide.

6. Read the information on the next page, *Muscle Cells at Work*.

7. Answer the analysis questions.

Analysis Questions

1. Which kind of muscle tissue is on slide *X*? Use evidence to support your answer.

2. Which kind of muscle tissue is on slide *Y*? Use evidence to support your answer.

3. What was the most difficult thing to complete in this activity?

4. Summarize the microscopic differences between skeletal muscle cells and smooth muscle cells.

5. What two things do muscle cells need when they work?

6. Why do you think your breathing rate goes up when you exercise? (Use evidence from the reading, *Muscle Cells at Work*, to support your answer.)

7. What three things are produced when a muscle cell breaks down food?

8. What does your body do with the three things that are produced when muscle cells break down food?

Conclusion

Write complete answers to the focusing questions. You will have to decide how many sentences you need to answer each question.

Extension Activity

Going Further: Put a very small piece of hamburger on a slide. Then cover it with a cover slip and observe it through the microscope. You may need to add one drop of diluted food coloring so the cells show up better. Can you find muscle cells?

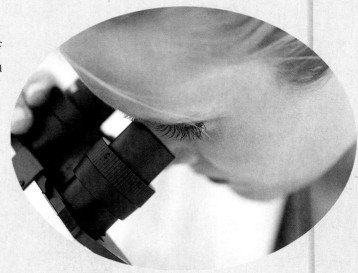

Muscle Cells at Work

Whenever you do something, you are putting your cells to work. You move because your muscles contract and your muscles contract because each muscle cell contracts. No other cells contract, only muscle cells. Think about it. Muscle cells have a highly specialized function: they contract. That's a clue that they must have a highly specialized structure, too. You observed the shape of a muscle cell—it is long and narrow, which is a logical shape for something that pulls.

Contracting is a lot of work for a cell, and doing work takes energy. Cells get their energy from food, and when they use food they need oxygen. When the nutrients from food break down during a cellular process called **respiration**, heat, carbon dioxide, and energy are produced. Figure 4.19 summarizes what happens inside a working muscle cell when plenty of oxygen is available.

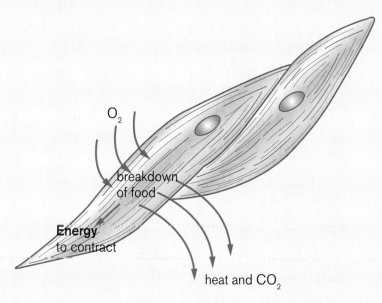

O_2

breakdown of food

Energy to contract

heat and CO_2

Figure 4.19

What does a working muscle cell produce when plenty of oxygen is available?

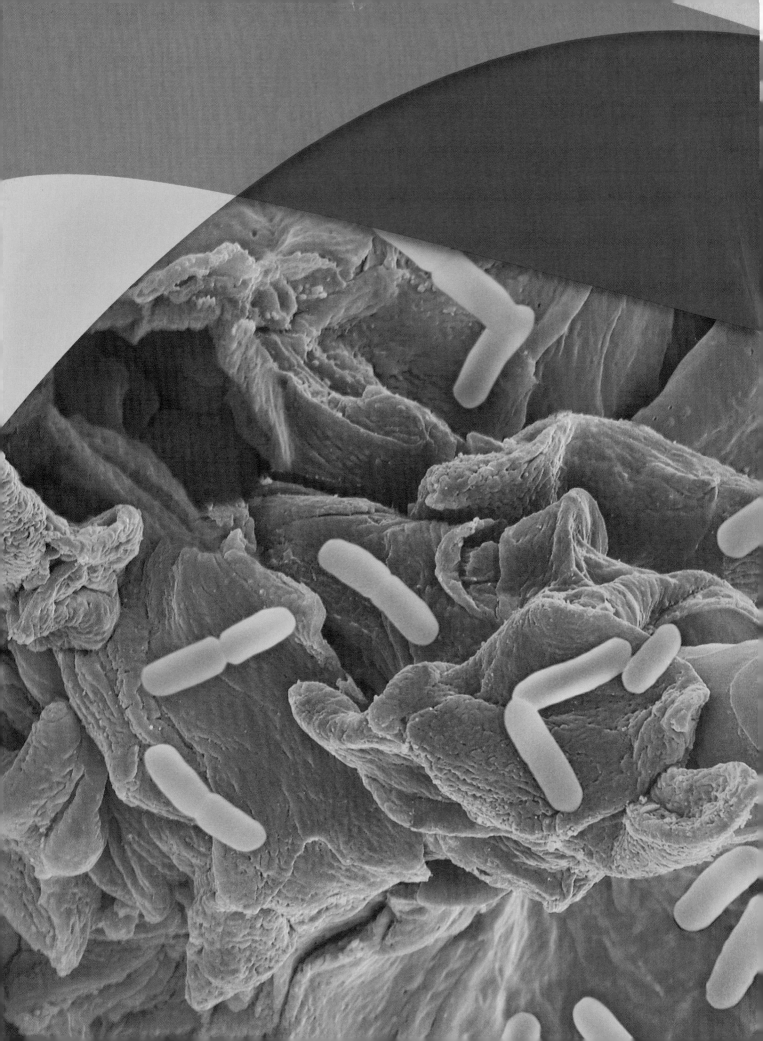

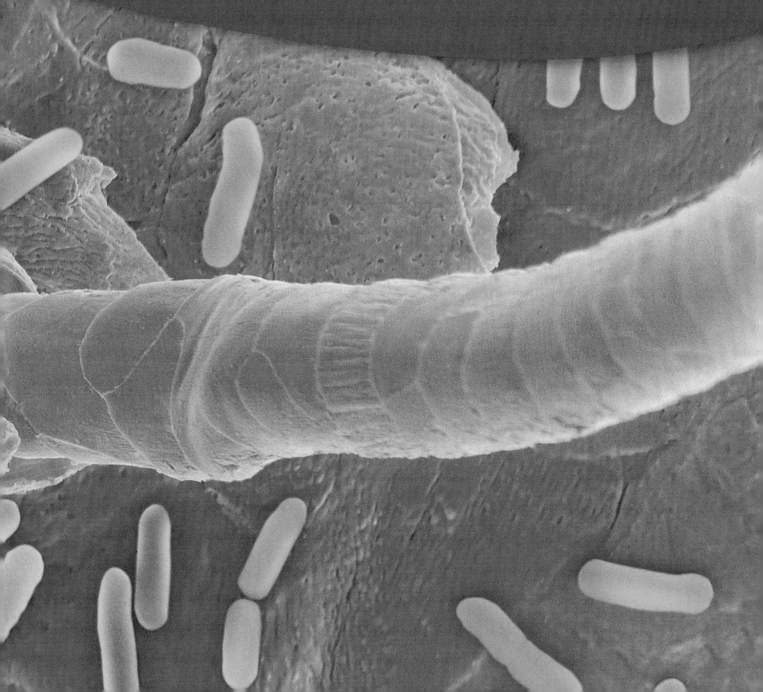

UNIT III

Body Protection

Chapter 5

Skin

Do any two people in your class have exactly the same skin color?

Who's Living on My Skin?

Millions of tiny organisms are living in your classroom. They are in the pencil sharpener, floating in the air, and on the surfaces of everything in the room. They can live just about anywhere because they can use just about anything for food. You don't know they are around because they're too small to see; they are *micro*-organisms. These microorganisms often settle down on your skin and try to "move in." Right now, your skin could be swarming with microorganisms and you wouldn't even know it.

How can you find out where microorganisms are living?

Where on your skin will you find the most microorganisms?

Find the answers to these questions by testing your own skin for microorganisms. You will be given a petri dish with agar (See **Figure 5.1**) to grow these tiny forms of life. Agar contains moisture and nutrients that allow microorganisms to grow.

Materials

- one plastic petri dish with nutrient agar
- tape
- wax pencil or marker
- two cotton swabs
- Handout, *Who's Living on My Skin?*

Procedure

(Work with a partner.)

Part A: Setting Up

1. Get your petri dish, but *do not* open it. You don't want to contaminate the agar.
   a. Put a piece of tape on one side of the dish so it acts like a hinge (See **Figure 5.1**).

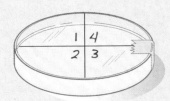

Figure 5.1
Labeled petri dish.

Figure 5.2
Barely lifting the lid.

   b. Write your names, date, and class period on the tape.

   c. Use the wax pencil or marker to mark the *bottom* of the petri dish in four sections. Number the sections 1, 2, 3, and 4, as shown in **Figure 5.1.**

2. Plan what you are going to do.

   a. Decide on three areas of skin to test. Choose places such as a foot, hand, neck, leg, or under a fingernail.

   b. On the handout, record the three skin areas you plan to test, on sections 1, 2, and 3 of the petri dish. You will leave section 4 blank so write "none" on your handout for this section.

3. Follow these three steps to test the first skin area you chose.

   a. Rub a clean cotton swab on the skin area.

   b. Lift the lid off the petri dish just high enough to insert the cotton swab as shown in **Figure 5.2. Do not lift the lid completely off.**

   c. *Gently* touch the cotton swab to the agar surface in section 1. Move it around a little but try not to break the surface of the agar.

4. Use a new cotton swab for each skin area you test. Follow the same steps for sections 2 and 3. Remember: Do not put anything on section 4.

5. Tape the other side of the dish so it will stay closed. Store it upside down where your teacher indicates. Plan to look at the dish in about four days.

6. Answer analysis questions 1 through 3.

Part B: What Grew?

1. Get your petri dish but do not open it. Leave it taped closed while you record your observations.

2. Fill in the data table on your handout. Your teacher may have you observe the microorganisms with a binocular microscope or magnifying lens in order to make your sketches. If so, remember to record the magnification.

3. Clean up: Take your dish to the sink, open the lid slightly, and carefully squirt some bleach solution into it. This will kill the microorganisms. Tape the lid closed again and put it where your teacher tells you.

Caution: Most of these microorganisms are not harmful. However, some of them could cause diseases such as staph and pneumonia, so do not take the lid completely off the dish.

When you use bleach as a disinfectant, be careful! It can burn skin, damage eyes, and take the color out of your clothing. ***Remember, safety counts!***

4. Add your results to the class data transparencies. (See **Figure 5.3**.)
   a. On the class data transparency for bacteria, draw a blue line to each part of the body that you tested. At the end of the blue line, near the body, put dots to show the number of bacterial colonies you counted. At the other end of the line, write the number of bacterial colonies.
   b. On the class data transparency for fungi, draw a red line to each part of the body that you tested. At the end of the line nearest the body, put dots to show the number of fungal colonies that grew. At the other end of the line, write the number of fungal colonies.

5. Answer analysis questions 4 through 12.

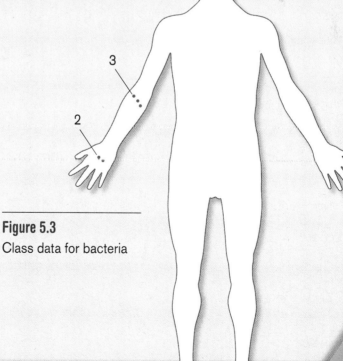

Figure 5.3
Class data for bacteria

Analysis Questions _ _ _ _ _ _ _ _ _

1. Why is it important not to expose the agar to air?
2. Why did the instructions say not to put anything in section 4?
3. Which section do you predict will show the most growth? Why?
4. How many kinds of bacteria grew in your petri dish? How many kinds of fungi?
5. How can you tell the difference between bacteria and fungi when they are growing on agar?
6. How did your results compare with your prediction? (See your answer to question 3.) How can you explain any differences?
7. List at least three reasons why some parts of the skin might have more microorganisms than others.
8. Look at the class data.
   a. Which parts of the skin resulted in the most growth?
   b. Which parts had the least growth?
9. Possible sources of error: Explain at least three reasons why someone could get wrong results when doing this experiment.
10. Based on the class results, which area of the skin do you think usually has the most microorganisms living on it?
11. What could you do to find out more about the microorganisms that grew on your petri dish?
12. Think of questions about microorganisms and how they grow.
    a. Write down at least two questions.
    b. Put stars by the ones you think you could answer by doing an experiment.

Conclusion

Write two short paragraphs, one to answer each focusing question. Remember to use your findings as evidence.

Extension Activities

1. **Design and Do:** Plan an experiment to show whether or not washing with soap affects the number of microorganisms that live on skin. If your teacher approves it, go ahead and try your plan.

2. **Careers:** Microbiologists, virologists, and bacteriologists all work with microorganisms. They work in places such as industrial plants, hospitals, health departments, and food packaging houses. Contact and/or visit one of these people. Find out about their work by asking questions such as these:

   What microorganisms do you check for?

   Why do you test for them?

   How do you test for them?

   What do you do if you find them?

3. **History of Science:** In the 1800s, Dr. Ignaz Semmelweiss tried to convince other doctors that they should wash their hands before doing surgery or delivering babies. Find out how Dr. Semmelweiss discovered the importance of cleanliness and what other doctors thought of his ideas.

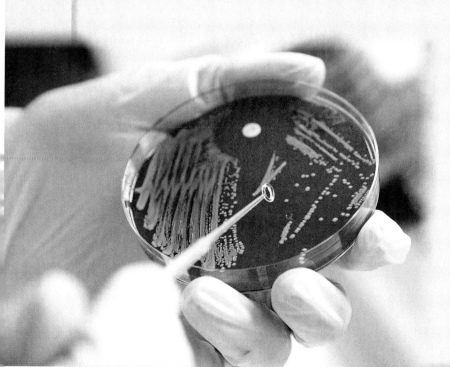

Figure 5.4

This bacteriologist is inspecting a petri dish in a hospital lab.

Microorganisms

Microorganisms can be hard to study because they are too small to see or hold. One way to learn more about microorganisms is to grow them on agar and then study the things that grow. Not all microorganisms, however, grow on agar.

What clues can you use to recognize the three main types of microorganisms?

You will be given microscope slides of four different microorganisms. Your task will be to decide what type of microorganism is on each slide. In order to answer the focusing question and complete this task, you will first need to learn more about microorganisms.

- prepared microscope slides of four microorganisms
- microscope
- Reading: *Three Types of Microorganisms*, page 102
- Handout, *Microorganisms*

Procedure

1. Read *Three Types of Microorganisms*.
2. Answer the analysis questions. Make sure you can answer all of them before you go on to the next step.
3. Get a microscope slide that shows one of the microorganisms.
   a. Focus it—this may be challenging!
   b. Record the magnification.
   c. Sketch it in the space that is provided on the handout.
4. Repeat step 3 for the other three microscope slides.

5. Use the evidence you have recorded on your handout and what you learned from the reading to decide what type of microorganism is on each microscope slide. Explain your decision on the handout.

Analysis Questions _ _ _ _ _ _ _ _ _ _

1. List the three types of microorganisms in order from smallest to largest.
2. When you are using a microscope, what clue is most helpful when you are deciding if you are looking at virus, bacteria, or a fungus?
3. How many cells—0, 1, several, or thousands—would you expect to find in each of the following structures?
   a. yeast
   b. sternum
   c. virus
   d. bacterium
   e. gluteus
   f. fungus fiber
4. Which types of microorganisms can grow on agar?
5. Some people think viruses are living organisms; other people disagree.

   a. Give one reason why a virus could be considered a life-form.
   b. Give one reason why a virus could be considered nonliving.

Conclusion

Write a paragraph that answers the focusing question.

Extension Activity

Research Project: Each of the four microorganisms you observed plays an important role in our lives. Write a report or make a poster about one of these four microorganisms.

Includes *sci*LINKS®
NSTA

Topic: Bacteria
Go to: www.scilinks.org
Code: MSLS3e101

Three Types of Microorganisms

There are more microorganisms on Earth than any other type of living thing. The three types of microorganisms mentioned the most often are fungi, bacteria, and viruses.

Fungi

Fungi are usually made up of many cells that form long, skinny strands. You don't usually see the strands because they are underground or in a rotting object where they have moisture. Some fungi send up stalks when they reproduce. Mushrooms are examples of these reproductive stalks. Molds and yeast are examples of microscopic fungi. Unlike other fungi, yeast are single cells that do not form long strands.

   Many types of mold will grow on agar. When they do, they look "furry" because of the long strands. (See **Figure 5.5**.) Different types of molds form different types of "furry blobs."

   Most fungi are decomposers; that is, they live on dead plant and animal tissues or on animal waste products. They are very important in recycling matter. Some types of fungi can be found on the human body.

Bacteria

Bacteria are some of the simplest forms of living things. They are single cells that may be round, rod shaped, or shaped like a spiral. Sometimes, bacteria seem to "stick together" to form strands. These microorganisms are so small that you can't see one without a microscope. To give you a better idea of just how small they are,

Figure 5.5

A mushroom is a fungus. The part you see produces spores, but most of the fungus is underground.

Yeast are fungi; these were magnified at 1,200 times.

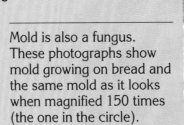

Mold is also a fungus. These photographs show mold growing on bread and the same mold as it looks when magnified 150 times (the one in the circle).

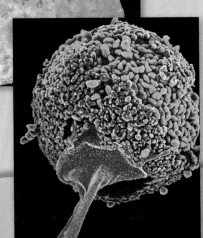

Figure 5.6

guess how many average-sized bacteria could fit in a single drop of water. (The answer is at the end of this reading.) Two types of bacteria are shown in **Figure 5.6**. Would you describe them as being round, rod shaped, or spiral?

Some types of bacteria will grow on agar. If a single bacterium lands on the agar, it may divide, resulting in two bacteria. These two will divide again, giving rise to four bacteria, and this will continue until there is a mass of millions (or more) of these tiny microorganisms. This mass is called a **colony**. You can see a colony without a microscope. Different kinds of bacteria form different kinds of colonies. For instance, one type of bacteria may form bright yellow, shiny colonies that are round in shape. Another type might form white colonies that appear rough and have an irregular shape.

Like many fungi, most bacteria are decomposers. Some bacteria can digest only one type of food, and others can eat a variety of foods. There are many bacteria that live on or in the human body. (What do you think they eat?) We need some of them to stay healthy, yet others can cause infections.

Viruses

Viruses are extremely small. They are so small that they can only be seen with an electron microscope. You cannot see them with the microscopes you have at school. One example of a virus is shown in **Figure 5.7**; notice the magnification of the photograph.

Unlike fungi and bacteria, viruses are not cells. In fact, they cannot function or reproduce unless they are inside a living cell. For this reason, they cannot grow on agar. The common cold, flu, measles, and mumps are all examples of diseases caused by viruses that can live in human cells. Other types of viruses live only in plant cells, and some viruses live only in bacteria.

Answer: *Two billion bacteria could fit in a single drop of water!*

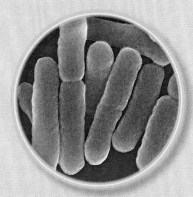

These bacteria are *Salmonella;* they cause food poisoning. (magnification: 5,530 times)

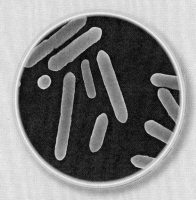

These are the bacteria that cause tetanus. (magnification: 2,000 times)

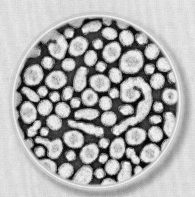

Figure 5.7

This is the virus that causes influenza. (magnification: 23,785 times)

5-3

Skin Microorganisms

The surface of the skin is a good environment for many microorganisms since it is warm, slightly moist, and provides places for them to "hide." All three types of microorganisms can live on skin. Some of them cause familiar skin problems.

Skin Fungi

Have you ever had athlete's foot or ringworm? Both of these skin problems are caused by fungi. Skin cells provide the nourishment the fungi need to survive. Fungi must absorb food from their environment because they have no chlorophyll and cannot make their own food like green plants do.

Red, itchy, flaky skin between the toes is a sign of **athlete's foot**. It is slightly contagious.

People can catch athlete's foot by walking barefoot where the fungus is located. Locker room floors at a gym, fitness center, or swimming pool are a common source of infection because this fungus likes moisture. Keep your feet dry if you want to avoid athlete's foot!

Ringworm is not a worm; it is another skin problem caused by a fungus. Scaly, round, itchy patches form on the skin where this fungus is living. It, too, is contagious. You can catch ringworm by touch, by exchanging combs or brushes, or from a pet dog or cat.

Ringworm and athlete's foot can both be treated with fungicides, which kills fungus. They are not serious conditions, but they are unsightly. The earlier they are treated, the easier they are to get rid of.

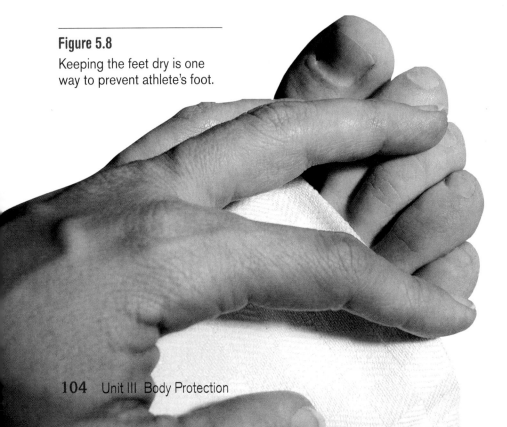

Figure 5.8
Keeping the feet dry is one way to prevent athlete's foot.

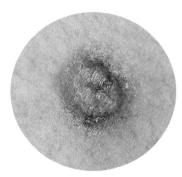

Figure 5.9
Ringworm starts as a scaly patch and grows to form a ring.

Skin Bacteria

Millions of bacteria live on your skin. Even the most thorough cleansing cannot clear them all. You usually are not aware they are there, but bacteria can become embarrassing when they cause body odor. The odor results when skin bacteria break down some of the chemicals that are normally found in perspiration. Some of these new chemicals may be smelly. Antiperspirants reduce the amount of sweat that a person produces, and some deodorants kill bacteria.

Skin Viruses

Viruses can also infect skin cells. Viral infections are difficult to treat because they live inside cells. Two examples of skin problems caused by viruses are cold sores and warts.

Cold sores look like blisters that form around the mouth. They start out as spots on the gums, lips, or inside of the mouth. These spots swell, redden, and hurt. When a cold sore appears, the virus is multiplying. When the infection clears up, the virus is "resting." It can be reactivated by stress, a cold, sunlight, or many other things. Luckily, most cold sores clear up on their own. A physician may suggest a painkiller or mouthwash to make the person feel better.

Warts are common in teenagers. There are several different kinds of warts, each caused by a different virus. Some people have a low resistance to these viruses so they may feel as if they are always getting warts. Viruses form warts by causing skin cells to divide rapidly. Warts usually disappear within a few months but some take years to clear up. Several medications may be purchased in drugstores to help treat warts. They act by destroying abnormal skin cells. These medications should never be applied to the face—facial skin is too sensitive for such strong medications.

These are just a few examples of the many skin problems that are caused by microorganisms. The next time you have a skin problem, you might want to ask, "Could it be caused by a fungus, virus, or bacteria?"

Analysis Questions

1. What should you look for in tennis shoes if you want to avoid athlete's foot? Explain your answer.

2. List two things a person could do to prevent body odor and explain why you think these things would work.

3. Why is a viral infection usually more difficult to treat than a bacterial infection?

4. Imagine that you have a contagious skin condition. List at least three precautions you would take to make sure no one else catches it.

5. A new skin disease has just been reported. What is one way to determine if it is caused by a microorganism?

Extension Activity

Health Application: Visit a store, and read the labels for at least two medications that can be used to treat either athlete's foot or warts. List the active ingredients. Decide which medication sounds best to you, and then ask the pharmacist about the advantages and disadvantages of using it.

Figure 5.10
Cold sores

Bacterial Growth

You won the lottery! The lottery commission offers winners two choices to receive their money. You can take a million dollars right now, or you can get one penny the first day, two pennies the second day, four pennies the third day, and so on. Each day, for one month, the amount of pennies you receive will double from the day before. Which method of payment would you choose? Why?

The second method of payment—receiving the pennies for a month—is comparable to cell division. Bacteria colonies grow by cell division. A single bacterium divides, resulting in two bacteria. These two divide, resulting in four bacteria, and so on. The time it takes for a population to double is called the **doubling time**.

Why does a colony of microscopic bacteria growing on agar become visible without a microscope?

To see how fast the number of cells increase when cells divide repeatedly, complete the following activity.

- piece of paper
- data table: *Cell Division*, page 107
- graph paper

Procedure

1. Copy the *Cell Division* data table on page 107 onto your own paper.
2. Use a piece of paper to represent a single bacterium cell.
3. Fold the paper in half. This represents the bacterium cell dividing. Each section denotes one daughter bacterium cell. Fill in your *Cell Division* data table with the number of resulting "bacteria cells."

4. Continue to fold the paper as many times as you can. For each fold, count the resulting number of daughter cells and record the number of "bacteria" on your data table.

5. Answer analysis questions 1 and 2.

6. Complete your data table for 10 doublings by calculating the number of resulting "bacteria cells" for each doubling left on your data table.

7. Use the information from your data table to complete a line graph. Don't forget to title and label your graph.

8. Answer analysis questions 3 through 8.

Number of Doublings	Number of Bacteria Cells
0	1
1	
2	
3	
4	
5	
6	
7	
8	
9	
10	

Analysis Questions

1. What pattern do you see in the number of bacteria for each division?

2. How many bacteria do you think will result from the next division?

3. Estimate how many bacteria you think there would be after:
   a. 12 doublings?
   b. 15 doublings?

4. Estimate how many doublings it would take before there are over one million bacteria in the colony. Explain your answer.

5. Look at your line graph. Why do you think the line forms a curve that gets increasingly steeper?

6. Read the quote from the The Andromeda Strain by Michael Crichton. Crichton wrote, "In this way it can be shown that in a single day, one cell of E. coli could produce a super-colony equal in size and weight to the entire planet Earth."

   Why do you think a single E. coli bacterium cell has never produced enough cells to equal the size and weight of Earth?

7. Perhaps billions of cells in your body die and are replaced every minute. How do you think your body replaces all the cells that die?

8. After completing this activity, would you still choose the same method for receiving your lottery winnings? Why or why not?

Challenge Questions

1. Calculate how much money you would have at the end of a month if you chose to receive your lottery winnings by starting with one penny, and then doubling the amount every day, for 30 days.

2. Calculate how many *E. coli* bacteria cells would be produced in one day (24 hours), if you start with one bacterium that divides every 20 minutes.

3. Calculate the weight of a colony of bacteria produced in one day (24 hours), if you start with one bacterium that weighs approximately 1×10^{-12} g and divides every 20 minutes. If the Earth weighs approximately 5.9763×10^{24} kg, was Michael Crichton correct when he wrote, "*In this way it can be shown that in a single day, one cell of* E. coli *could produce a super-colony equal in size and weight to the entire planet Earth.*"? (Hint: use your answer to question 2 to find the answer.)

Conclusion

Write a brief paragraph that answers the focusing question.

Extension Activity

Health Application: *E. coli* is a bacterium that is commonly found in the lower intestine of warm-blooded animals. Most *E. coli* strains are harmless, but some can cause serious food poisoning in humans, and are occasionally responsible for costly product recalls. Ask your doctor and your parents what you can do to reduce your risk of contracting an *E. coli* infection.

> "*The mathematics of uncontrolled growth are frightening. A single cell of the bacterium* E. coli *would, under ideal circumstances, divide every twenty minutes. That is not particularly disturbing until you think about it, but the fact is that bacteria multiply geometrically: one becomes two, two become four, four become eight, and so on. In this way it can be shown that in a single day, one cell of* E. coli *could produce a super-colony equal in size and weight to the entire planet Earth.*"
>
> —Michael Crichton (1969) *The Andromeda Strain*, Dell, N.Y. p. 247

Design an Experiment

Are you thinking like a scientist? If so, the microorganisms that grew on your agar should have left you with more questions than answers. For example, you may wonder if the same kinds of microorganisms are always living on your skin. Or you could ask, "what kills the microorganisms that live on skin?" There are dozens, maybe hundreds, of questions that you could ask. Choose one of your questions and design an experiment that will help you answer it.

What are some important things to think about when designing an experiment?

In order to find the answer to this question, you need to think like a scientist. With a partner, choose one question that you'd like to explore. Predict an answer, plan a procedure, collect data, and analyze your data.

- one plastic petri dish with nutrient agar
- tape
- other materials that you decide you need

Procedure

(Work with a partner.)

1. Think of a testable question you can answer by doing an experiment.

   a. Testable questions are questions which can be investigated through experiments or observations. They connect with scientific concepts, not opinions or beliefs. For example, "Which tastes better—coke or root beer?" is not a good testable question since it involves opinions rather than numerical measurements.

   b. A good testable question can be answered by a **fair test**. A "fair test" is when you only test one **variable** and keep all the rest the same. "Does the amount of fertilizer affect plant growth?" is a good testable question because the investigator can test different amounts of fertilizer and keep the rest of the variables—light, water, type of fertilizer, type of plant—the same.

2. Write a **hypothesis** for your testable question, using the "*If..., then...*" format.

   a. Think about the results you got when you grew bacteria and fungi on agar. Also, think about what you know after reading about bacteria and fungi. Based on that information, what do you think will be the answer to your testable question?

   b. Don't forget to write your hypothesis in an "*If..., then...*" format.

   "***If*** ____ *[what is manipulated in the experiment]* ____, ***then*** ____ *[How you think it will respond]* ____."

3. Plan what you are going to do. Be sure to design a **fair test** by making sure it tests only one variable at a time. Think about how you are going to control all the variables so you test only one.

   a. Write down your procedure, step by step. If you have written a strong procedure, someone else should be able to follow it exactly as you will.

   b. Ask another team to check your procedure. If they don't understand exactly what you plan to do, you may have to rewrite some of your steps.

   c. Have your teacher check to make sure that you have designed a safe experiment.

4. Complete your experiment. Follow the steps exactly. If you change something, make a change in your procedure.

5. Record your data on a data table.

6. Write a **conclusion** by answering the following questions:

   a. What does your data tell you? Does it support your hypothesis? Why or why not?

   b. What other explanations could explain your results? What errors could affect your results?

   c. What other questions do you have now that you've completed your experiment?

7. Answer the analysis questions.

Analysis Questions

1. What, if anything, surprised you about the results of your experiment? (Sometimes scientists learn more from an experiment when something unexpected happens than when the experiment goes exactly as they expected it to.)
2. If you were to do this experiment again, do you think you would get the same results? Why or why not?
3. What are some additional questions that you could answer about microorganisms by doing more experiments?

Conclusion

Write a paragraph that answers the focusing question.

Extension Activity

Design and Do: Design another experiment that will help you understand more about microorganisms. Start by looking at the questions you listed in answer to analysis question 3. Choose one question and design an experiment that would allow you to answer that question. If your teacher approves, go ahead and try your experiment.

Figure 5.11
Scientists are careful to record everything they do.

Eight Pounds of Skin

Skin is the largest and most visible organ of the body. An average adult's skin weighs 8 to 10 pounds (3.5 to 4.5 kg). Most of us take our skin for granted, seldom thinking about all of the things that it does. But skin is so important that if one-third of it is lost—from burns, for instance—a person will die.

Skin Functions

Skin is our tough but flexible outer covering. Its main function is protection. Skin protects our insides from the many microorganisms that live on the surface of our bodies. It also protects us from radiation, chemicals, and other harmful things in the environment.

Skin also helps keep body temperature within a normal range, and keeps moisture inside our bodies so our internal organs do not dry out. It is also a sense organ that keeps us aware of our surroundings.

Obviously, skin is important for our survival. If you think about it, it is an amazing covering. How can skin do all of these things? Look at its structure for answers.

Skin Structure

Skin has two main layers—the epidermis and the dermis. The **epidermis** is the surface of the skin; it is what you see. The top layer of the epidermis is dead cells. You wash off some of these

Figure 5.12
Structure of the skin

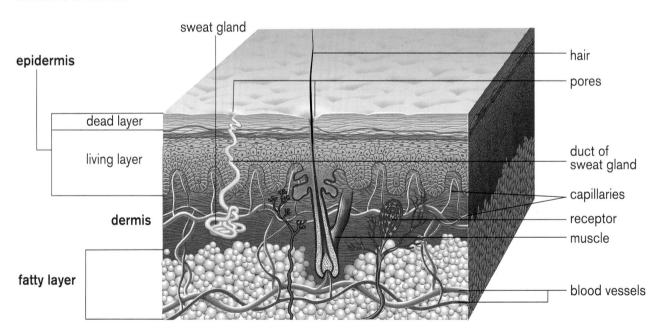

cells every time you take a bath or shower. They also brush off on your clothes. The cells that fall off your head and land on your shoulders are called dandruff. Most of these dead epidermal cells become part of the dust in the house. This constant shedding of cells keeps many microorganisms from entering the skin. A bacterium that attaches itself to a skin cell will be brushed off as soon as that cell is shed.

The bottom layer of cells in the epidermis is alive and active. These cells divide to keep producing new epidermal cells to replace the ones that die and fall off. An epidermal cell lives about one month. It forms in the bottom layer of the epidermis, gets pushed toward the surface of the skin as new cells form, dies and becomes part of the top layer of the epidermis, and finally dries and is brushed away. Your hair and nails also form from epidermal cells. Notice in **Figure 5.12** that the epidermis does not contain any blood vessels.

Beneath the epidermis is the **dermis**. It is much thicker than the epidermis and it does contain blood vessels. It also contains glands, nerves, some smooth muscle fibers, and the "roots" of hairs. You never see your dermis but you know it is there. Oil moves out of the oil glands that are in the dermis up onto your epidermis. You can feel the oil, especially by your nose. Sweat comes up out of the sweat glands to keep your skin cool. When you cut yourself, blood may stream out of the vessels that are in your dermis. Every time you feel something, you are using the nerves in your dermis. The ends of these specialized nerves receive messages and so are called **receptors**. Count how many receptors are shown in **Figure 5.12**.

Beneath the dermis is a layer of **fat** that keeps the body warm. It varies in thickness from one part of the body to another. For instance, it is usually thick over the abdomen but it is missing from the eyelids. Premature babies have not developed this insulating fat layer so they need help to stay warm. Incubators provide the environment they need until they gain weight and their fat layer develops.

Skin is held in place by **connective tissue**. Elastic fibers (these are made up of specialized cells) in the connective tissue allow your skin to stretch and spring back into place. Watch what happens when you gently lift some skin on the back of your hand and then let go. (Remember what the connective tissue looked like when you dissected the chicken wing?)

A Break in the Barrier

Usually, skin is an excellent protective barrier that keeps out dirt and bacteria. However, if this barrier is scratched, cut, or punctured, dirt and bacteria can get inside the body and cause infection. If the injury is not too serious, the skin can repair itself. For example, following an injury such as a small cut, slight bleeding is normal and can help "clean" the wound. The bleeding usually stops after a few minutes because blood contains substances that form clots. When the clot dries it becomes a scab. Scabs protect the damaged tissues and prevent dirt and bacteria from entering the body until new cells can grow to repair the cut. Read the first aid steps described in *Caring for Scratches and Small Cuts*.

Caring for Scratches and Small Cuts	
1.	Wash around the cut with hand soap.
2.	Rinse with clean running water.
3.	Blot dry with a clean cloth.
4.	Leave a small cut uncovered, unless it is on a finger or other place that may get dirty.

What happens if a person loses a large area of skin because of an injury such as a serious burn? Obviously, much of the protective barrier is gone. The person is at risk for developing all kinds of infections. However, one of the biggest problems is "drying out." A burn patient can lose gallons of fluid from their raw flesh in one day. The patient also will feel extremely cold without their skin to keep in the heat. Healthy skin is important for a healthy body!

Analysis Questions

1. What are four functions of skin?
2. Write down which layer of skin—the epidermis, dermis, or fat—relates to each of these characteristics.
   a. does not contain any blood vessels
   b. insulates the body
   c. has a layer of dead cells
   d. produces hair and nails
   e. contains glands and smooth muscle fibers
3. The reading says that skin is "tough." Give at least two examples of why this is true.
4. Use facts about the structure of skin to answer these questions.
   a. If a scratch does not bleed, why do you know that only the epidermis, not the dermis, has been cut?
   b. Do you think that cells in the epidermis could survive if we did not have a dermis? Explain your answer.
   c. Do you think that cells in the dermis could survive if we did not have an epidermis? Explain your answer.
5. Name at least five types of cells that you would expect to find in (or very near) the skin.

Extension Activities

1. **Design and Do:** About how many square centimeters of skin cover you own body? Think of a safe way to figure it out. Check your plan with your teacher, and then try it.
2. **Health Application:** Look up skin conditions such as blisters, scars, and pimples. Use what you know about skin structure to explain how these things develop.

Animal Skin

All animals have some kind of protective covering. In a tiny one-celled organism, this covering is simply the membrane that encloses its body. Large, multicelled animals have more complex skins that are made up of several layers. Each type of animal has specialized skin structures that improve its chances of survival. Feathers, quills, horns, and claws are all examples of structures that grow from specialized cells in the epidermis. Antlers grow from specialized dermal cells and so do the scales on fish.

Reptiles have a tough, dry, scaly skin. Lizards and snakes shed their epidermis periodically. Why do you think this helps their survival?

Whales have a thick layer of fat under their skin. This "blubber" keeps them warm and helps them stay afloat. A large whale may have a layer of blubber that is 2 feet thick!

Skin Colors, Skin Cancer

One of the most obvious characteristics of skin is its color. Sometimes people want to change their skin color. They may want it lighter, darker, browner, pinker, or whiter. How can they do this? They can't. They may be able to change it temporarily, but people inherit their skin color. A person's basic skin color is an unchangeable, genetic trait.

The color of skin depends on its structure. Color is determined mainly by the amount of pigment that is present—the more there is, the darker the skin appears. The cells that produce the pigment are located in the epidermis. **Figure 5.13** shows how pigment can be distributed in the cells of the epidermis. Blood flow also affects skin color. The more blood present in the capillaries, the redder, or pinker, the skin will look. (Remember, capillaries are the small blood vessels in the dermis.) A third factor that influences skin color is the thickness of the skin layers. The top layer of the epidermis is made up of dead cells that give the skin a faintly yellow hue.

Suntans and Sunburns

The sun can temporarily change skin color by darkening it. The color change occurs in stages. These stages are shown in **Figure 5.14**.

Drawing A shows a section of skin before it has been exposed to the sun. *Drawing B* shows what happens when a person gets a sunburn. The ultraviolet light rays damage the outer skin cells and the body reacts by sending more blood to the damaged cells. Since there is more blood flowing through the capillaries, the skin looks redder.

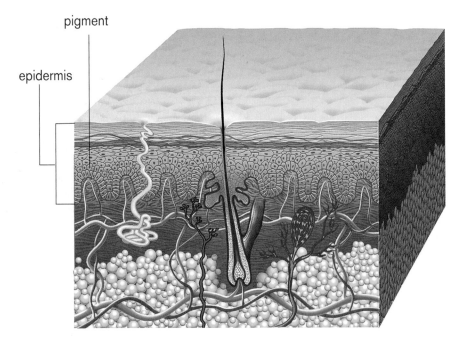

pigment

epidermis

Figure 5.13
Cells that produce pigment are located in the epidermis.

Includes sciLINKS®
NSTA

Topic: Skin Cancer
Go to: www.scilinks.org
Code: MSLS3e116

Figure 5.14
Changes that occur when a person burns and tans.

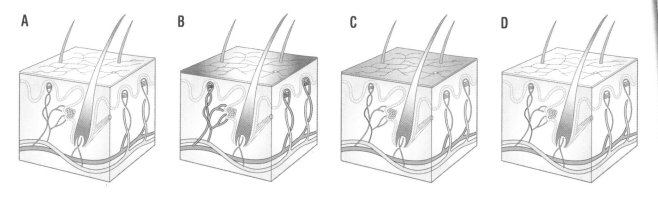

A   B   C   D

Next, the cells in the lower part of the epidermis divide and form new skin cells to replace the damaged ones. As this "new skin" is formed, the redness fades. The skin looks tan because more pigment is now present and the top layer of dead cells is thicker. Look for these changes in *Drawing C*.

After several weeks, the skin returns to its original color, as shown in *Drawing D*. The extra dead skin cells have washed off and the new skin cells contain the "normal" amount of pigment.

Tanning is a protective reaction. When the sun's rays burn skin cells, extra pigment forms to block the rays and protect the cells. This is true for everyone, whether they have inherited a very dark skin color or a very light one.

Sunlight and Skin Cancer

Sunlight is made up of many kinds of rays including infrared light (which is what makes sunlight feel warm), visible light rays (which allow us to see), and ultraviolet light rays (which contain very high energy levels).

Figure 5.15
The color of a person's skin tells something about his or her ancestry.

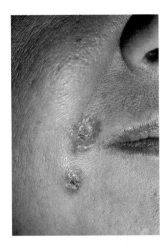 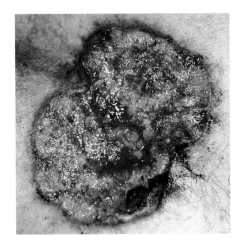

Figure 5.16
These can be signs of skin cancer.

Because they have such high energy levels, ultraviolet (UV) rays can be particularly harmful to skin. These are the rays that cause sunburns and skin cancer. Anyone can get skin cancer; it doesn't matter if your skin is very light or very dark. However, the lighter a person's skin is, the higher the risk is for cancer.

People do not think about this risk when they are young, but they should. Some experts believe that most sun damage occurs before a person is 20 years old. Cancer develops slowly but irreversibly; once it starts, it continues to develop.

Know These Signs of Skin Cancer
A mole, birthmark, or beauty mark that changes color or texture, increases in size or thickness, or is irregular in shape.
A skin growth that gets larger.
A spot or growth that continues to itch, hurt, crust, scab, or bleed.
A sore that does not heal within four weeks, or heals and then reopens.
If you ever notice any of these symptoms, see your doctor!

Everyone should know the signs of skin cancer. It usually appears on areas that are not covered by clothing such as the face, neck, ears, forearms, and hands. If caught early, skin cancer can be curable. The signs of skin cancer are listed in the chart and pictured in **Figure 5.16**.

Protection

Think about protecting your own skin from sun damage. Dermatologists (skin doctors) suggest these four "sunwise" steps, especially for light-skinned people:

- Use sunscreen. Put more on every few hours, especially if you are perspiring or swimming.
- Cover up when you will be in the sun for a long time. Wear a hat, long sleeves, and long pants.
- Time your outdoor activities for early morning and late afternoon. Avoid the most intense sun rays around noon.
- Avoid tanning salons. The UV light rays in tanning booths can damage your skin.

Analysis Questions

1. Why do you think skin contains pigment? What is its function?
2. The cells that produce pigment are in the epidermis, not in the dermis. Why does this make sense?
3. Look back through this section and list examples of skin changes that are caused by the sun.

Extension Activities

1. **Health Application:** Sunscreens are rated according to their SPF values. SPF stands for "sun protection factor." Find out the difference between a sunscreen that has an SPF rating of 4 and one that has a rating of 40.

2. **Community Resources:** Contact a dermatologist, a cancer clinic, or the American Cancer Society to request information about skin cancer. Read the information and make your own list of what you can do to prevent skin cancer.

Figure 5.17

According to the American Cancer Society, one out of every seven Americans will develop skin cancer. What could these people do to lessen their chances of skin cancer?

Chapter 6

Immune System

Why do you think a person gets "shots"?

6-1

Health Problems

An organism is healthy when all of its systems are functioning properly. If even one part of a system isn't working right, problems result. There are all kinds of problems; they could last a lifetime, or only a few seconds. Some are typical of babies, others of teenagers, and yet others of adults.

What types of health problems are most familiar to students your age?

Brainstorm a list of health problems and then sort the problems into groups. Think about what you learned about sorting in previous activities. Sorting helps you identify patterns, and helps you become aware of what you do and do not know.

- scissors
- red colored pencil
- transparent tape or glue

Procedure

(Work with a partner.)

Part A: Sorting Health Problems

1. Cut a half sheet of notebook paper into 16 to 20 equal-sized pieces.
2. Neatly write the name of one health problem on each small piece of paper. (You will use these papers again later.)
3. Sort the problems into groups based on cause. If you don't know the cause of a problem, put it into a group called "unknown."

Cold

Virus

4. In the lower right corner of each piece of paper, write the cause of that particular problem.

5. With your partner, answer analysis questions 1 through 3.

Part B: Thinking Like a Biologist

1. You have been studying systems, and for each system you studied, you learned about certain health problems associated with it.

   a. Sort the 16 to 20 health problems into groups based on which system is most directly affected by the problem.

   b. If you do not know where to put a health problem, put it into a category called "unknown."

2. Organize your 16 pieces of paper by system as described here.

   a. On a separate sheet of paper, use a pencil to write down headings for each body system. (Write lightly in case you need to erase and move a heading.)

   b. Place the health problems associated with each system under the appropriate heading. You may need more room for some systems and less room for others.

   c. When you have placed all the health problems where you want them, use tape or glue to attach the paper slips.

3. Of the health problems you listed, which ones are caused by micro-organisms (or germs) and could be passed from one person to another? Using a red pencil, circle the problems you think are caused by microorganisms.

4. Put stars next to six health problems on your list that you think are most likely on other students' lists, too.

5. Answer analysis questions 4 and 5.

Analysis Questions _ _ _ _ _ _ _ _ _

1. You were asked to think of 16 to 20 health problems. Do you think you could have named 100?

2. Comment on how familiar you are with causes of health problems.

   a. Is it easy or difficult for you to identify the causes? Why do you think this is so?

   b. List the causes you used to group your health problems.

3. Besides their causes, list at least two other characteristics you could use to sort health problems.
4. Look at the systems that were affected by the health problems you named.
   a. Which systems were affected by at least one health problem?
   b. Which body systems were not affected by any health problems?
   c. Based on the health problems you named, which system was affected the most often?
5. Look at the causes of the health problems.
   a. How many of the health problems you named are caused by accidents?
   b. How many of the health problems you named are caused by microorganisms?
   c. How did you decide whether or not a problem is caused by a microorganism? (What criteria did you use?)

Conclusion

After a class discussion, answer the focusing question.

Extension Activity

Health Application: Use references to learn about the health problems that were least familiar to you. Find the cause of each problem and decide what system it primarily affects.

Protection from Germs

Infectious (or contagious) diseases are the most common cause of health problems. These are the diseases you can catch from other people, animals, the food you eat, the air you breathe, or the things you touch. Some infectious diseases are mild, such as the common cold, whereas others, such as AIDS, can result in death. What causes infectious disease and how does the body protect itself when an infection develops?

Causes of Infectious Disease

Many infectious diseases are caused by microorganisms, but some are caused by organisms that can be seen without a microscope. Viruses, the smallest type of microorganism, cause all kinds of health problems. In addition to the common cold and AIDS, viruses cause influenza (the "flu"), chicken pox, shingles, rabies, and mononucleosis (also known as "mono"). Bacteria cause diseases such as tuberculosis, tetanus, strep throat, and some forms of pneumonia. Ringworm and athlete's foot are examples of diseases caused by fungi. Parasites such as lice, tapeworms, ticks, chiggers, and fleas can also cause health problems. The relative sizes of some of these organisms are shown in **Figure 6.1**.

Topic: Communicable Diseases
Go to: www.scilinks.org
Code: MSLS3e124

Figure 6.1

This drawing is to help you compare the sizes (but not the appearances) of disease-causing organisms. Everything is shown 10,000 times larger than normal. Large parasites can't fit on this page. A parasite such as a tapeworm can reach a length of 6 meters. If it were enlarged 10,000 times, it would be 60,000 meters long, which is much too long to show here.

small parasite   bacteria   virus

Lines of Defense

The skin blocks the entry of many microorganisms. It is a tough, outer covering without any entrances, unless, of course, there is an injury. Microorganisms can enter through any break in the skin, which is why it is important to clean cuts and scratches.

Bacteria that are still alive when food is swallowed are usually killed by stomach acids. If organisms are inhaled, they may be trapped in the tiny hairs in the nose, or caught in the mucus that coats the respiratory tract. Then the cilia move the mucus, along with trapped microbes, up and out of the lungs.

Not all bacteria are bad. Populations of "good" bacteria live on the skin, numbering millions per square centimeter. They also live in places such as the mouth and large intestine, between your toes, in your ears, and under your fingernails. When "foreign" bacteria try to settle in, they have to compete with the local residents. So, think of your good bacteria as friends, helping to keep you healthy by destroying microbes that could make you sick.

Natural Protection: White Blood Cells

Sometimes disease-causing microorganisms get past the first-line defenses. They may enter the body through a wound or through the respiratory or digestive systems. When this happens, they face destruction by white blood cells. White blood cells do not stay in the blood vessels. They squeeze out and roam through all the tissues of the body.

There are several kinds of white blood cells. Each kind is specialized to do a particular job. One kind, called macrophages ("big eaters"), works like a clean-up crew. When they find something foreign they engulf and destroy it. **Figure 6.2** is a photograph of a macrophage engulfing a bacterium. Once inside the cell, the bacterium is destroyed by enzymes.

As macrophages move through the body, they pass by normal, healthy cells and molecules the body needs. In other words, they ignore anything that is a normal part of you. Each day macrophages destroy dead cells, including billions of worn-out red blood cells. They also clear the surfaces of the air sacs of dust, lint, and other particles that make it into the lungs. And, of course, they are always on the watch for microorganisms that could make you sick.

Figure 6.2

Macrophages can engulf and destroy microorganisms.

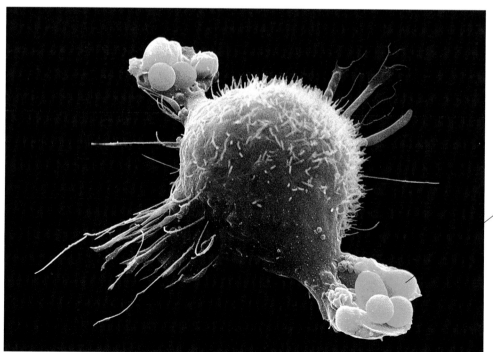

microorganism

Other kinds of white blood cells do not destroy foreign substances. Instead they make specialized proteins called **antibodies** that can "recognize" unique parts of particular foreign substances. There are millions of different antibodies. Each antibody recognizes only one kind of microorganism. Antibodies leave the white blood cells and move with the blood throughout the body. When they bump into something they recognize, such as a bacterium, they attach to it. This "labels" the bacterium for destruction. The antibody-bacterium combination is then engulfed and destroyed by macrophages and other types of white blood cells. This is illustrated in **Figure 6.3**. Antibodies are particularly important when one kind of microorganism builds up in your system. Antibodies help the body get rid of many microorganisms of the same type quickly.

When you catch a disease, your body makes lots of antibodies against that one kind of microorganism. For example, if you catch measles, your white blood cells make antibodies against measles. Measles antibodies can attack only the measles virus. They cannot fight the chicken pox virus, a flu virus, or any other kind of virus. Later, if you "catch" the same microorganism again, you already have some of the needed antibodies and your body can quickly make more. When this happens, you are said to be **immune** to that disease. That is why you get some diseases, like measles, only once in your lifetime.

What about the common cold? Most people catch many colds over a lifetime, so they don't seem to become immune. That's because there are over 200 viruses that cause the symptoms we call a "cold." You never catch the same cold twice, but there are over 200 colds that you can catch.

In some people, substances such as pollen or cat hair can set off the immune system in the same way that dangerous microbes do. When this happens, the person is said to have an allergy. There are times when people feel pretty miserable because of their allergies, but at least they know their immune system is working!

The Immune System

So what is the immune system? It's the system that fights off infections. It's made up of all the **white blood cells** that cruise throughout the body, as well as the organs

Figure 6.3
Some white cells produce antibodies. Each kind of antibody is made to "recognize" one particular microorganism.

antibodies

microorganisms

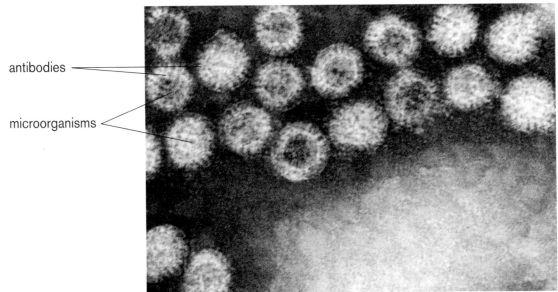

where white blood cells are made and stored. Unlike other systems, the immune system is not well defined.

Like all blood cells, white blood cells are produced in the **bone marrow**. Many of them are delivered to organs such as the lymph nodes or spleen. Here the white blood cells grow and specialize, and are stored until they are needed.

Lymph nodes are small, bean-shaped structures. They are located throughout the body, but clusters of them are found in the neck, armpits, abdomen, and groin. Lymph nodes store white blood cells and release them into the bloodstream when they are needed. They may become enlarged and tender when they are active. The lymph nodes in the back

of the throat are called **tonsils**. They are in a good position to trap the infectious organisms that enter the mouth. Find these organs in **Figure 6.4**.

The **spleen** is on the left side of the body along the upper edge of the stomach. It is below the diaphragm and behind the ribs so you cannot feel it. The spleen filters blood, removing wastes and old red blood cells. The white blood cells mop up the remains of the old red blood cells and destroy any microorganisms that are in the blood.

As you can see, all the parts of this system are designed to make sure white blood cells will be ready to fight microorganisms as quickly as possible.

Figure 6.4

It's hard to draw a picture of the immune system. A few of its major organs are shown in this diagram.

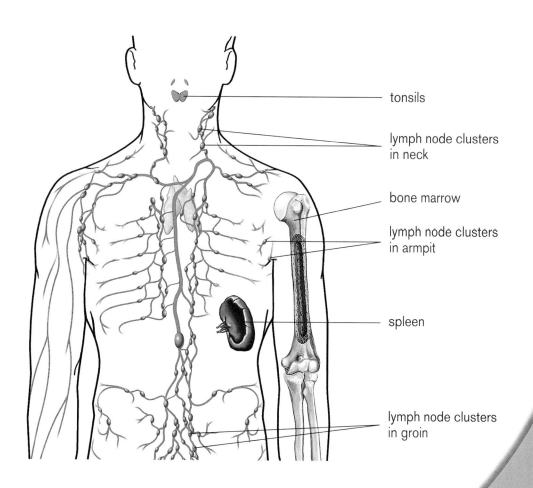

tonsils

lymph node clusters in neck

bone marrow

lymph node clusters in armpit

spleen

lymph node clusters in groin

Extra Help: Antibiotics

Sometimes disease-causing microorganisms reproduce faster than they can be destroyed by white blood cells. Thousands of them may invade your cells and make you sick. When this happens, your body may need help to fight the infection. Fortunately, there are drugs to help in this battle. These powerful drugs, called antibiotics, block the growth of bacteria; they do not affect viruses. The effects of some frequently prescribed antibiotics are described in the table below.

Antibiotics must be used carefully. If used too often, antibiotics can kill the good microorganisms that help keep you healthy. If you take too much antibiotic, you could kill your own cells. That's why antibiotics require a doctor's prescription.

The immune system is very complicated, but hopefully you now have some idea of how it works. Can you imagine millions of white blood cells traveling throughout your entire body, killing microorganisms and breaking them down into "raw materials"? Can you see why some people call white blood cells scavengers?

Antibiotic	Action
Penicillin	prevent bacteria from forming cell walls
Bacitracin	
Streptomycin	prevent bacteria from making proteins
Erythromycin	
Tetracycline	
Sulfa drugs	prevent bacteria from making DNA

Analysis Questions

1. List the types of organisms that can cause infectious diseases.

2. List the ways your body tries to block the entry of microorganisms.

3. Describe two ways that white blood cells defend your body.

4. Which of the following kinds of antibodies have your white blood cells produced?

   anti-flu   anti-rabies
   anti-mumps   anti-strep throat
   anti-ringworm   anti-"pinkeye"

5. List the parts of the immune system.

6. What is an antibiotic?

7. Make a table that summarizes information about the following infectious diseases. Include information about the cause of each disease, how you think the organism usually gets into the body, and whether or not the disease can be helped by an antibiotic.

   flu   tetanus
   pneumonia   strep throat
   malaria   a cold
   hepatitis

8. Explain how the immune system interacts with at least three other systems.

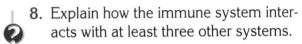

Extension Activity

Health Application: Learn about schistosomiasis, a disease caused by a flatworm, that is a health problem in many parts of the world.

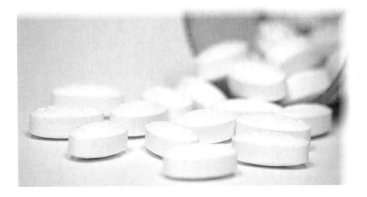

Defense Systems

Every animal is at risk of being destroyed by microorganisms. That's why all animals have a defense system of some kind. Their outer covering is the first line of defense. Sometimes this covering is tough, but some animals use unpleasant or poisonous chemicals to discourage potential invaders. All multicellular animals have cells that can destroy microorganisms, but only vertebrates produce antibodies.

A primate with alopecia (skin disease causing complete hair loss).

Preventing Infectious Disease

Infectious diseases have always been a major cause of death. Hepatitis, measles, tuberculosis, and numerous other diseases kill millions of children every year. Deaths from some diseases, however, have dropped dramatically in the last century. What has changed? Has the immune system become more efficient? No, as a result of medical research, we now have better ways to treat and prevent infectious diseases. A major breakthrough in our understanding took place in England in the late 1700s, as a result of one person's research on one disease—smallpox.

A Historic Experiment

The very word, smallpox, terrified people. Epidemics of the dreaded disease swept through towns, killing many of the residents. Those who recovered were often left with heavy "pox" scars or, worse yet, blinded. Children were often hit the hardest during the epidemics. It was not unusual for parents to lose their entire family to this deadly disease. Survivors could, however, take some comfort in knowing they would never get smallpox again.

A case of smallpox starts like many other infectious diseases, with headaches, fever, nausea, and backaches. Then the skin turns scarlet and a flat rash appears. The rash raises to form pustules over the surface of the body and inside the nose and mouth, making swallowing and breathing difficult. The face may swell so much that the person is not recognizable. If the pox enters the eyes, blindness can result. A high fever is common. Some historical reports indicate that one out of every five people who caught smallpox died from it. (See **Figure 6.5**.)

Figure 6.5
Victims of smallpox suffered terribly.

Smallpox was a truly frightening disease. People were willing to try almost anything to protect themselves from it. For example, it was a common practice for physicians to grind up dried smallpox scabs into a fine powder and have people inhale it. By doing so, people caught a less serious form of the disease. However, it still was severe, disfiguring, and contagious. The disease continued to spread.

An English country doctor named Edward Jenner noticed that milkmaids (the women who milked the cows) seldom caught smallpox. However, they often caught a milder form of pox called cowpox. Dr. Jenner wondered if cowpox protected people from catching smallpox. In 1796, he tested his idea by scraping some material from a cowpock off the hand of a young milkmaid and injecting it into the arm of a young boy. As expected, the boy developed cowpox. Six weeks later, after the boy had recovered from cowpox, Dr. Jenner inserted some pus from a fresh smallpock under the boy's skin. This was risky; the boy could have contacted smallpox and died. However, he did not get sick.

Dr. Jenner published a report about his experiment. He suggested that cowpox vaccines could stop the spread of smallpox. He pointed out that, even if people got sick, they would have only a mild case of cowpox which wasn't very contagious or deadly. At first, people ridiculed Dr. Jenner, but eventually his ideas were accepted.

We now know that smallpox and cowpox are both caused by viruses. The viruses are so similar that, once the body has become immune to one of the viruses, it can also attack the other virus. The last reported case of smallpox was in 1977.

Dr. Jenner's procedure of injecting harmless (or dead) microorganisms into a person or animal so it will not be infected by a more dangerous one is called **vaccination**. (This term comes from the Latin word vacca, which means cow, in honor of Dr. Jenner's early work with cowpox.)

Modern Day Vaccinations

Today, vaccines are widely used to protect people from infectious diseases. No one has had smallpox for over 30 years. Because of vaccinations, other major human diseases are coming under control, such as diphtheria, tetanus, whooping cough, polio, measles, and yellow fever. Children do not like to get "shots," but, if given the choice, do you think they would rather have the disease or the shot?

Most states have laws that provide guidelines for childhood vaccinations. What are the laws in your state? Which vaccinations have you received?

Research Continues

Even though deaths from infectious diseases have dropped considerably in the last century, they still kill millions of people each year. Some of the suffering and death could be prevented if vaccination programs were expanded. (See **Figure 6.7**.) Some programs, however, have to wait until researchers learn more about the disease and how it works.

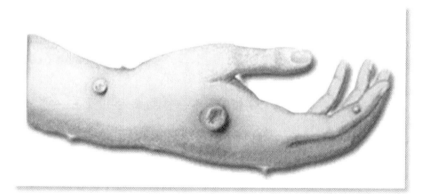

This drawing shows a cowpox lesion on the hand of Sarah Nelmes, the young milkmaid who participated in Dr. Jenner's experiment.

Includes *sciLINKS*®
NSTA

Topic:	Vaccines
Go to:	www.scilinks.org
Code:	MSLS3e131

Malaria, measles, polio, tuberculosis, and HIV all take their toll on people. Maybe some day, they too will become diseases of the past. Is this the kind of work that would interest you? Would you like to study an infectious disease and discover how to prevent it?

Analysis Questions

1. What is a vaccination?

2. Why do you think a young boy was allowed to be part of Dr. Jenner's experiment?

3. Why did people become immune to smallpox if they had cowpox?

4. Look at the graph in **Figure 6.7**.

   a. What three diseases caused the most deaths among children under 5 years of age due to diseases that could have been prevented by routine vaccination?

   b. What are two reasons children might not get vaccinated?

5. Look back at the 16 health problems you thought of when you completed the activity in *Chapter 6-1, Health Problems*.

   a. List the health problems that are caused by microorganisms. (You may need to look up some of the health problems in a dictionary or other resource.)

   b. Circle the infectious disease on your list that you think is the most serious.

6. If you could develop a vaccination that would eliminate one infectious disease, which one would you choose? Why?

Extension Activities

1. **History of Science:** Learn more about Edward Jenner and his work with smallpox, or read about Louis Pasteur and his work with rabies.

2. **Health Application:** Each state has its own laws about vaccinations. Call your local health department and ask for a copy of the immunization requirements. Ask your parents or your doctor to help you fill in your own immunizations record. Check to see if your vaccinations are up to date.

Figure 6.6

A doctor vaccinating a young child.

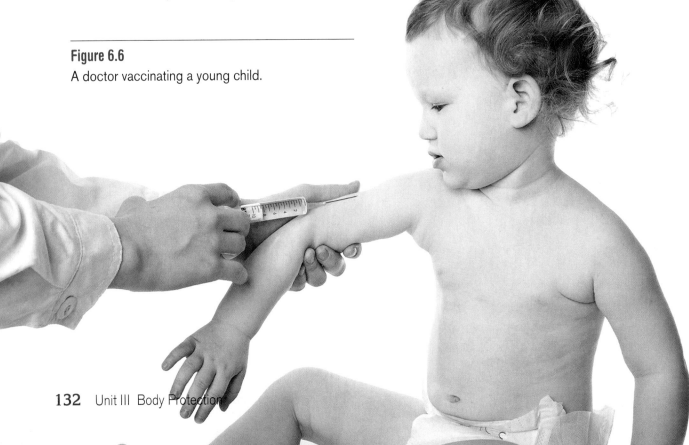

Figure 6.7

In 2002, the World Health Organization (WHO) estimated that 1.4 million deaths among children under 5 years of age were due to diseases that could have been prevented by routine vaccination. This represents 14% of global total mortality in children under 5 years of age.

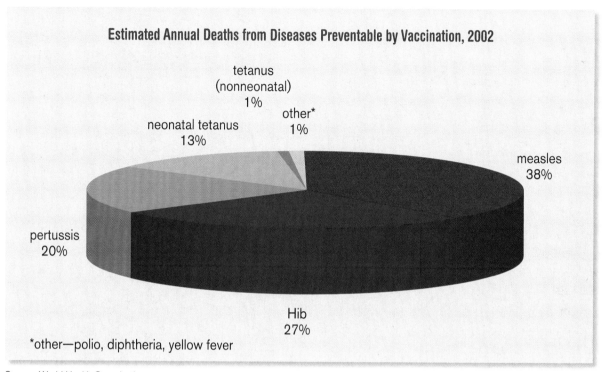

Estimated Annual Deaths from Diseases Preventable by Vaccination, 2002

tetanus
(nonneonatal)
1%

other*
1%

neonatal tetanus
13%

measles
38%

pertussis
20%

Hib
27%

*other—polio, diphtheria, yellow fever

Source: World Health Organization

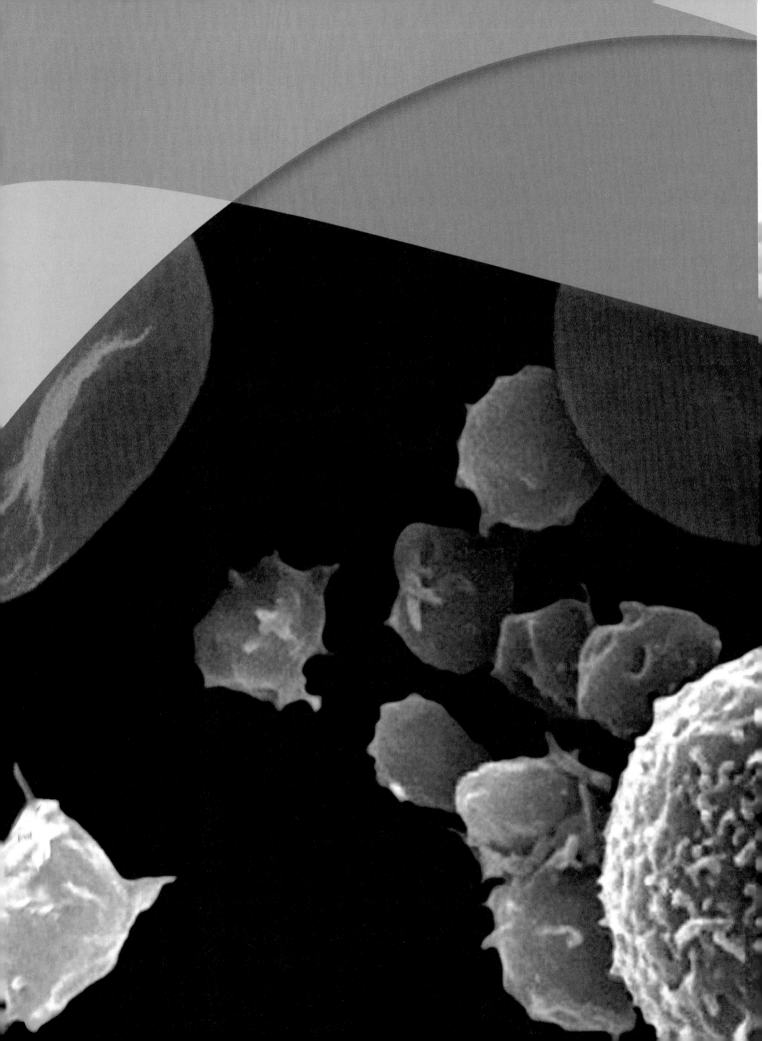

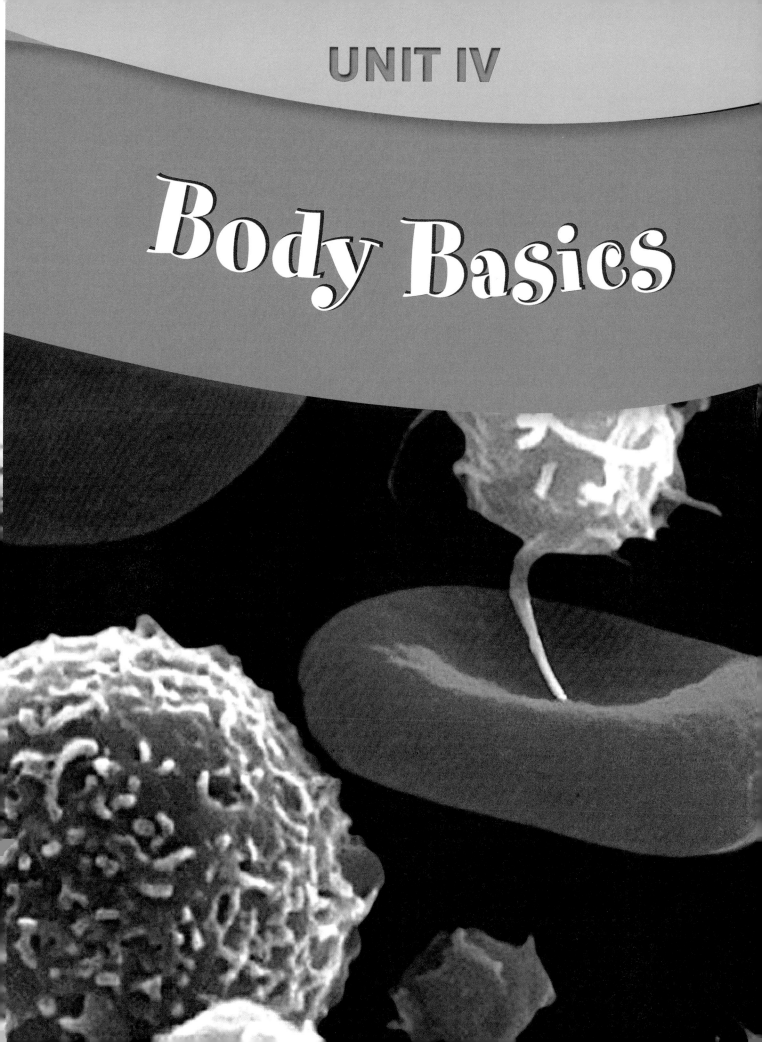

UNIT IV

Body Basics

Chapter 7

Foods

If you were shopping at this market, which foods would you choose? What makes something a food?

Thirty-Two Foods

An automobile can't run without gasoline and a body doesn't work without food. Food provides the energy your muscle cells need whether you're emptying trash, climbing stairs, or writing answers to the analysis questions. It even takes energy to breathe. Some animals eat very specific foods, but humans eat a wide variety of foods. What comes to mind when you hear the word food?

What is a food?

What categories help you see patterns to the foods you eat?

Think about the answers to these two focusing questions while you name 32 foods and then design a way to sort the foods into groups. After you've sorted your foods using one set of characteristics, sort the foods again using another set of characteristics.

- six sheets of notebook paper
- colored pencils: green, orange
- transparent tape

Procedure *(Work with a partner.)*

Part A: Sorting Foods

1. Fold and tear two sheets of notebook paper into 16 squares each so you end up with 32 pieces of paper that are all about the same size.
2. Neatly write the name of one kind of food on each piece of paper. (You will use these papers again later.)
3. With your partner, decide on groups (or categories) you want to use to sort the foods. Use these groups to sort the foods.

4. On a clean sheet of paper, write the names of your groups. Under each group heading, list several examples of your 32 foods which fit into that group.

5. Answer analysis questions 1 through 3.

Part B: A Second Sorting

1. After a class discussion, sort the foods to show where they come from. Use these categories:
   a. Foods that come from plants (examples: lettuce and bread)
   b. Foods that come from animals (examples: eggs and cheese)
   c. Foods that come from both plants and animals (examples: beef burrito, cheese pizza)
   d. Foods that you don't know where they come from

2. Use the following color codes to underline the food names:
   a. Foods that come from plants—underline with green
   b. Foods that come from animals—underline with orange
   c. Foods that come from both plants and animals—underline with green and orange
   d. Foods that you don't know where they come from—don't underline

3. Answer analysis question 4.

Part C: A Third Sorting System

1. Sort the foods one more time. Start with three new sheets of paper. Put one of these headings at the top of each sheet:
   a. Foods that are good for you
   b. Foods that are not good for you
   c. Foods that we don't know if they're good or bad

2. Put each food on the page where it belongs. When you have sorted all 32 foods, tape them down.

3. Write your conclusion.

4. Read *Sorting Foods* at the end of the lesson.

5. Save your three pages of sorted foods. You will use them again in a later activity.

Analysis Questions

1. What problems did you have putting the foods into groups?
2. Why might someone want to sort foods using the groups you used?
3. You used one set of characteristics to sort 32 foods. Think of another set of characteristics you could use. List the names of the new groups.
4. Use the second sorting method to look for a pattern in the sources of foods you named. How many foods fit into each of the following categories?
    a. comes from animals
    b. comes from plants
    c. comes from both plants and animals
    d. don't know where it comes from

Conclusion

Using complete sentences, answer the focusing questions.

Homework: Food Diary

Keep a food diary. For 24 hours, list every food you eat and the amount you eat. **Figure 7.1** shows an example of some entries from a student's food diary.

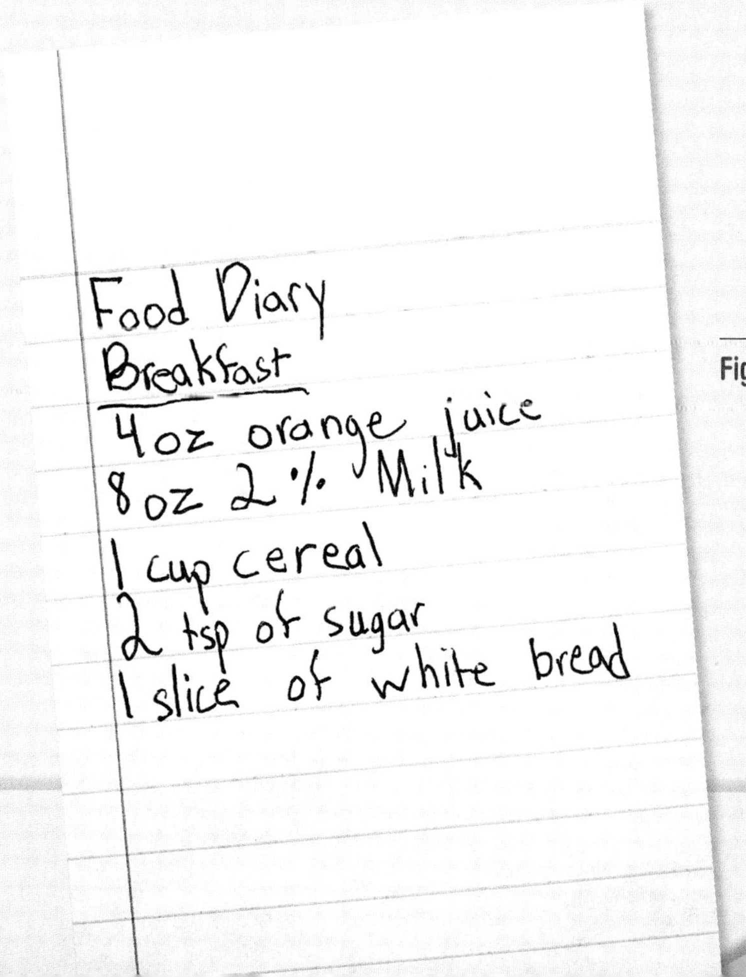

Figure 7.1

Food Diary
Breakfast
4oz orange juice
8oz 2% Milk
1 cup cereal
2 tsp of sugar
1 slice of white bread

Sorting Foods

How many different ways did your class think of for sorting foods? Probably a lot. There are many ways to think about foods, and the method you use depends on your purpose. Dietitians and food scientists often find it useful to group foods based on their fat, protein, and carbohydrate contents. These are the groups you will use as you study foods in this chapter.

Nutrients are the substances in food that are needed to keep the body healthy. Fats, proteins, and carbohydrates are sometimes called the **major nutrients**. Along with water, they make up the bulk of most foods. In addition to the major nutrients and water, food also contains vitamins and minerals.

All the nutrients are needed for growth, repairing damaged tissue, and regulating body functions. Fats and carbohydrates are used for energy. Protein is necessary for cells to function. All animals need the major nutrients in their diets, but the amounts they need vary.

Includes SC*LINKS*
NSTA

Topic: Nutrients
Go to: www.scilinks.org
Code: MSLS3e140

Fat Test

Fat is one of the three major nutrients. Our bodies need some fat in order to function normally. And yet, we are often told that Americans eat too much fat. In order to determine the amount of fat you eat, you need to know which foods contain fat.

How can you test for fat in foods?

What foods contain fat?

Think about these questions while you test at least ten foods for fat. Remember to look for patterns in your data when you answer the second focusing question.

Materials

- brown paper towel or a sheet of paper
- fat (shortening or oil)
- samples of ten foods
- Handout, *Food Test Results* (also used in 7-5, 7-7, and 7-10)

Procedure

(Work with a partner.)

Part A: Getting Ready

1. Make a data table.
   a. Read the entire procedure so you know what to do. Jot down the types of things you need to record.
   b. Make one row for each control and each food you will test.
   c. Make one column for each type of data you will record (in this case, predictions and results).

2. Mark your paper towel into 12 sections.
   a. Label one section fat and another section water, as shown in **Figure 7.2**.
   b. In the other sections, write the names of the 10 foods you will test.

3. Make your predictions. With your partner, discuss each of the foods listed on your data table. In the prediction column, write "yes" by the foods you think contain fat and "no" by the foods you think do not contain fat.

Figure 7.2

Part B: Testing for Fat

Twelve stations, each with a different food, have been set up around the room. Make sure you know which station you should go to first. Read the first three steps before you start.

1. At each station, take a small amount of the food to be tested. The amount you take should be smaller than a raisin.
   a. Firmly smear the food on the correct section of your paper towel.
   b. Brush off the remains of the food into a waste container.
   c. To test liquid, place one drop in the correct section.
2. Take turns placing the foods on the paper towel. Wipe your fingers between foods or use different fingers so you don't mix the foods accidentally.
3. When you have tested the 12 foods, return to your seat, set the paper towel aside, and answer analysis questions 1 through 3.

Part C: Analyzing the Results

Figure 7.3

1. Watch the water spot on your paper. When it is dry, you can analyze the results of the food tests.
2. First, look at the section you used to test fat. Hold the paper towel up to the light as shown in **Figure 7.3**. Look at it from both front and back. This shows what a very positive (+ +) fat test looks like.
3. Look at the water section from both front and back. Since water does not contain fat, this shows what a negative (−) fat test looks like.

4. Look at the food tests on the paper and record the results on your data table. Use a double plus (+ +) if a food contains a lot of fat, a single plus (+) for a food with some fat, and a minus (−) for foods with no fat.

5. Answer analysis questions 4 through 7 and write your conclusion.

Analysis Questions — — — — — — — —

1. In your own words, write down what you should learn by doing this activity.

2. List at least four "facts" you know about fat in foods.

3. Think about the predictions you made on the data table. How did you decide whether or not you thought something contained fat?

4. Why did you have to wait for the water spot to dry before you analyzed the results?

5. Based on your results, which of the foods contained fat?

6. Which foods contained fat that you did not expect to contain fat? (Look at your predictions.)

7. Copy this list of foods; then circle the ones you think contain fat.

margarine	corn
carrot	sausage
bread	lettuce
doughnut	crackers
banana	sunflower seeds

Conclusion

When you answer the first focusing question, explain how to tell the difference between foods that contain water and foods that contain fat.

When you answer the second focusing question, look at your data. Look at the types of foods that contain fat and the kinds that do not. Remember that you're watching for patterns.

Extension Activity

Going Further: If you think you've found a pattern and can recognize a type of food that contains fat, test more foods to see if you're right. For example, you may decide that all fruits or all cookies contain fat. If so, test more fruits or more cookies to find out if that's true.

7-3

Fats and Oils

All organisms contain fat, but some organisms (and some parts of organisms) contain more fat than others. For example, pumpkin seeds are much higher in fat than the pumpkin itself, and a chicken heart contains more fat than chicken muscles. Because fat is found in every organism, that's a clue there's something very important about fat—but what? Why is fat important? Where does it come from?

Body Fat

Fat serves many important functions in the human body. A few of these functions are listed below.

- Fat molecules make up the *structure of cells*. Every single cell, and therefore every part of the body, contains some fat.

- Fat is a *source of energy*. When the body stores fat, it is storing energy to use later.

- Fat *provides insulation*. The layer of fat beneath the skin serves this important function.

- Fat acts as *padding*. It surrounds the heart, kidneys, and other body organs. It helps hold them in place and protects them from blows.

- Fat *transports some vitamins* through the body.

- Fats are used in the *production of several hormones*.

As you can see, some body fat is important for survival. One of the most important functions of fat is that it can be used for energy when we are not getting enough food. If we eat

extra fat, the body stores it in specialized fat cells. Masses of fat cells form fat tissue. Any extra protein or carbohydrate in the body gets changed to fat, too. This is true not only for humans but for all vertebrates.

The ability to store fat allowed people to survive when food was hard to find. For example, before the arrival of Europeans, Native Americans relied on their environment to meet their food requirements. They hunted and gathered edible plants. Some tribes planted crops. In the summer and fall months, people ate well, but during winter and spring, food became sparse. People without adequate fat reserves often became sick or died. Today, most people live much more inactive lives and have access to a steady supply of food. Some people would say that the ability to store fat has more disadvantages than advantages. What do you think?

Sources of Fat

High-fat foods can come from either plant or animal sources. Think about the results you got when you tested foods for fats. Did you test enough foods to begin to see any patterns? For example, all nuts and seeds contain lots of fat. That means foods such as peanut butter and sunflower seeds are high in fat. The vegetable oils used in cooking come from the seeds of plants such as corn, soybeans, and olives. (Oils are fats that are liquid at room temperature.) Chocolate, avocados, and coconut are other examples of fatty foods from plants (see **Figure 7.4**). Why do you think seeds and nuts are high in fat?

Figure 7.4

All of these foods come from plants, and they are all high in fat. How many of these foods have you eaten?

Foods that come from animals also contain fats. Fats are found in egg yolks and in most dairy products such as cheese, butter, milk, and cream. All meats contain fat even if you cannot see it. As you may remember, meat is muscle and muscle cells use fat as an energy source.

Kinds of Fat

There are many kinds of fat. All fats, however, are made up of carbon, hydrogen, and oxygen. Different kinds of fat are made by putting carbon, hydrogen, and oxygen together in different ways (see **Figure 7.5**). The three kinds of fats mentioned in this reading are cholesterol, saturated fats, and unsaturated fats.

Our bodies produce some cholesterol, but we also get it from the foods we eat. **Cholesterol** is an animal fat; it is not found in foods that come from plants. Egg yolks, cream, cheese, lard, and red meats are all high in this fat. Examples of foods that contain cholesterol are shown in **Figure 7.6**.

Trans fats are made when vegetable oils have been "hardened" to make them solid. During the hardening process, unsaturated fats become saturated. **Saturated fats** are also found in animal fats. These fats are the white, greasy substance that you see on uncooked beef. In **Figure 7.5** you can see that saturated fat molecules are straight. The unsaturated molecules are similar, but they have "bends" in them.

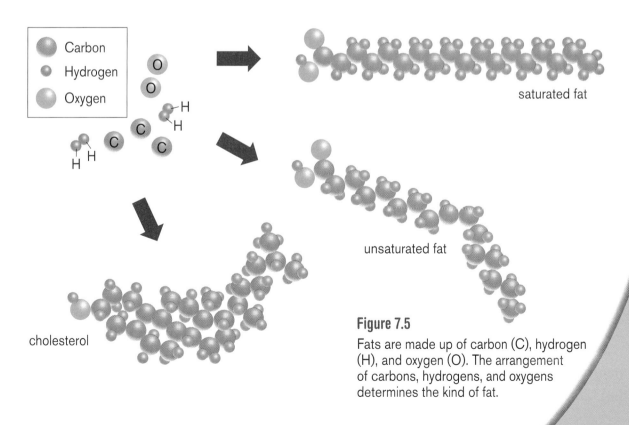

Figure 7.5

Fats are made up of carbon (C), hydrogen (H), and oxygen (O). The arrangement of carbons, hydrogens, and oxygens determines the kind of fat.

Figure 7.6

These foods are all high in cholesterol. Which ones do you eat regularly?

"Fat-wise" Decisions

If you are like most Americans, you may be eating too much fat. High-fat diets can lead to weight problems, diabetes, and heart disease. People who want to avoid heart trouble cut back on foods that contain cholesterol, trans fats and saturated fats. Studies indicate that teenagers and children who have high levels of cholesterol circulating in their blood have an increased chance of developing heart disease as adults.

Since plants do not produce cholesterol and most oils are unsaturated, most vegetable oils are "safe" to eat. However, coconut and palm oils are exceptions to this rule. Even though they come from plants, they are high in saturated fat.

Six recommendations for making "fat-wise" food decisions are given in **Figure 7.7**. Read the recommendations and then close your eyes and see how many of them you can remember.

As you can see, fats are important molecules. They are necessary for the survival of all living organisms. But too much of a "good thing" can cause problems, so start thinking about the fats in your own diet.

Analysis Questions

1. List six functions of fat in the body. Circle the three functions you think will be the hardest for you to remember.

2. List three foods that contain animal fat.

3. List three foods from plants that are high in fat.

4. Which of these fats should you avoid?

   cholesterol   animal fats
   saturated fats   unsaturated fats

5. Refer to the six recommendations for a "fat-wise" diet.

   a. Which is easiest for you to follow?

   b. Which is hardest for you to follow?

6. For each pair of foods listed, decide which is the "fat-wise" choice. Explain your decisions.

   a. a pat of butter or a pat of margarine

   b. corn oil or coconut oil

   c. baked ham or fried ham

7. Look at a microscope slide of fat cells.

   a. Draw what you see.

   b. Explain why the structure of a fat cell is appropriate for its function.

Extension Activity

Going Further: Observe and compare several different kinds of fats (raw beef, chicken fat, vegetable oils, shortening, etc.). What characteristics do fats share? How are they different?

Figure 7.7

Recommendations for Cutting Back on Cholesterol and Saturated Fat
1. Choose low fat versions of foods.
2. Trim off the fat when you eat beef, ham, or other meats.
3. Choose fish and poultry over beef or pork when you are given the choice.
4. Eat no more than 3 or 4 egg yolks per week. This includes eggs used in foods such as cookies, cakes, and casseroles.
5. Use unsaturated fats instead of saturated fats whenever possible.
6. Eat boiled, baked, or broiled foods instead of fried foods. Especially avoid foods that are deep-fat fried.

Fat Needed!

Many animals rely on stored fat for survival. Grizzly bears, for example, are always big eaters, but, as fall approaches, they may spend 20 hours a day in search of food. Each bear gains weight, accumulating large amounts of body fat. Like other hibernating animals, bears need the extra fat to survive the long winter.

During hibernation, bears maintain a normal body temperature, and their cells continue to function. The stored fats are slowly broken down to provide the energy the cells need. When bears emerge from their dens in the spring, they have used up their fat storage and are very hungry animals.

Fats in Your Diet

How can you find out how much fat you are eating? One way you can do so is by adding up the number of grams of fat in the foods you eat. (The dietary fat recommendations for 11- to 14-year-olds are shown in the chart.) You can find out how many grams of fat are in a serving of food by reading the package label or by using a nutrient chart. The chart in Appendix C, *Nutrients in Foods,* provides information on some of the most frequently eaten foods.

Daily Recommendations for 9- to 14-year-olds	
Nutrient	**Amount in Grams**
fat	62–95 grams

How does the fat in your own diet compare with the daily recommendations?

In general, how does the amount of fat eaten by students in your class compare with the recommendations for daily fat intake?

Analyze the foods in your food diary for their fat content to answer the first question. Then decide if you are making fat-wise food choices.

Materials

- your food diary
- orange colored pencil
- chart in Appendix C: *Nutrients in Foods*

Procedure

Part A: Individual Data

1. Using your lead pencil, put a check mark by the foods in your food diary you think contain fat.

2. Add a column to your food diary and title it "Grams of Fat."

3. Use the orange colored pencil to record the number of grams of fat found in each food. You can find this information in the chart, *Nutrients in Foods*.

4. Add up the number of grams of fat you ate on the day you kept the food diary. Write this number on a scrap of paper and put it in the Class Data Tub (your teacher will tell you where it is).

5. In your food diary, draw an orange star by any food that contains more than 7 grams of fat per serving. These are high-fat foods.

6. Put an orange circle around each food that contains cholesterol or saturated fat. Remember: This includes all foods that contain eggs.

7. Answer analysis questions 1 through 5.

Part B. Class Data

1. Look at the class data to answer the second focusing question. (Hint: Graphs are useful when you want to find a pattern in data.)

2. Answer analysis questions 6 and 7, then write your conclusion.

Analysis Questions _ _ _ _ _ _ _

1. Analyze your findings.
   a. How many grams of fat did you eat on the day you kept a food diary?
   b. In general, what is the suggested range for grams of fat for someone your age?
   c. How did the number of grams of fat you ate compare with the suggested range?

2. Look at the chart, *Nutrients in Foods,* in Appendix C. Find three foods that are high in fat. Write down the names of the foods and the number of grams of fat in a serving of each.

3. Look at the chart, *Nutrients in Foods*. Find three foods that have 0 grams of fat per serving. Write down the names of these foods.

4. List the five foods from your diary that are highest in fat. For each of these foods, list a low-fat food you could eat instead. Here is one example:

High Fat	Low-Fat Alternative
1/2 cup ice cream	1/2 cup frozen yogurt

5. Here is a lunch menu.

2 slices white bread	1 apple
2 tablespoons peanut butter	3 chocolate chip cookies
2 tablespoons grape jelly	1 cup whole milk

a. How many grams of fat are in this lunch? (Use the information in Appendix C, *Nutrients in Foods*.) Show your work.

b. Which foods contain cholesterol?

c. What could you change to make this a more "fat-wise" meal?

6. Look at the class data.

a. What is the range in amount of fat students ate? (least to most grams per day)

b. What would you estimate to be the average number of grams of fat that students in your class eat each day?

7. Here are questions from students who analyzed their own food diaries. Write an answer to each student.

a. Matt drinks two glasses of 2% milk for dinner. On a nutrient chart, he found that a glass of 2% contains 5 grams of fat. Now he wants to know, "Should I skip milk at dinner because it contains 10 grams of fat?"

b. Pam's mother is supposed to eat less fat so she is fixing low-fat foods for the family. When Pam analyzed her food diary, she found that the chicken she ate contained 9 grams of fat and the meat loaf contained 12 grams. Pam wants to know, "Is my mother choosing low-fat foods?"

c. Jennifer loves butter and uses about a teaspoon at a time on her breakfast muffin, lunch sandwich, and dinner vegetables and potatoes. She wants to know, "Since a teaspoon of butter only contains 4 grams of fat, is it okay to keep putting it on my food?"

Conclusion

Write a few sentences in answer to each of the focusing questions. Support your answers with evidence from your data.

Extension Activity

Health Application: Ask a fast-food restaurant for a copy of their menu. Go through each item and, using a red pencil, circle the items that are high in fat. Circle the low-fat choices with green.

How Much?

The recommended daily amounts of nutrients needed are given in ranges. You will have to decide how much is right for you. Here are examples of nutrient amounts for two students. These examples may help you choose values to fit your needs.

Mark loves to read. He spends his spare time reading mysteries. He has grown several centimeters in the last two years, but he hasn't started his growth spurt yet. For Mark, 65 grams of fat per day would be the recommendation.

Kendra is always on the go. She loves physical activity—riding bikes, roller blading, swimming. She is not the tallest girl in class, but she has been growing steadily for the last year. Kendra should choose 95 grams as the amount of fat that is appropriate for her.

Figure 7.8
When you make food choices, try to select low-fat items.

Protein Test

Of the three major nutrients, protein is the most critical because your body cannot create protein "from scratch." You must get your protein from foods. If you are aware of which foods provide protein, you will be able to put together a better diet. Hundreds of foods are good sources of protein. How many can you name?

How can you test for protein in foods?

Which foods are good sources of protein?

A distinctive chemical reaction takes place when a few drops of nitric acid are placed on protein. Egg white is very high in protein. Use it as a control so you will know what a positive protein test looks like. Then test five other foods to find out which ones contain protein.

- test tube rack
- egg white
- seven test tubes
- safety goggles
- paper towels
- Handout, *Food Test Results* (from Activity 7-2, *Fat Test*)

- masking tape
- samples of five foods
- water in dropper bottle
- nitric acid in dropper bottle

Procedure

(Work with a partner.)

Part A: Getting Ready

1. Read this procedure and think about what you will need to do. List the things you must record, and then design your data table.
2. On your data table, record the five foods that your teacher tells you to test.

3. Make your predictions. Discuss each of the foods with your partner. If you think a food contains protein, write "yes" in the appropriate cell of the data table. If you don't think it contains protein, write "no."

4. Get your test tube rack and two test tubes (see **Figure 7.9**). These test tubes are for your controls.

   a. Put nine or ten drops of water in one test tube.

   b. Put a small amount of egg white (about the size of half a raisin) in another tube.

Part B: Testing for Protein

1. **Put on your safety goggles.**

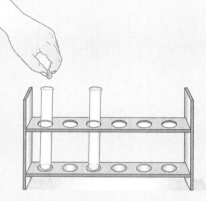

Figure 7.9

2. Carefully add ten drops of nitric acid to each of the two test tubes. Gently swirl the test tubes to mix the contents and then put the tubes back in the rack.

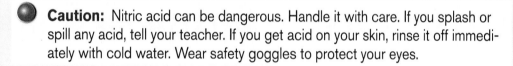

Caution: Nitric acid can be dangerous. Handle it with care. If you splash or spill any acid, tell your teacher. If you get acid on your skin, rinse it off immediately with cold water. Wear safety goggles to protect your eyes.

3. After three or four minutes, make your observations.

   a. Notice what happens when nitric acid is added to egg white. Most foods that contain protein will give a similar reaction.

   b. Record the results on the data table.

4. Prepare to do the protein test on your five foods.

   a. Using masking tape and a pen, label each test tube with the name of one of the five foods.

   b. Place small amounts of each food (about the size of half a raisin) in the proper test tube. You may need to crush or crumble the foods.

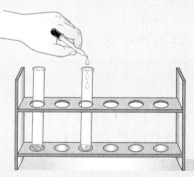

5. **Make sure you are wearing your safety goggles.**

   a. Take turns adding ten drops of nitric acid to each test tube (see **Figure 7.10**).

   b. Swirl to mix, then put the test tube back in the rack.

   c. Wait five minutes.

Figure 7.10

6. Record the results on the data table. Use a double plus (++) for foods with lots of protein. (They will react as the egg white did when nitric acid was added.) Use a single plus (+) for foods with some protein and a minus (−) for foods with no protein.

Part C: Finishing Up

1. Carefully dump the contents of each test tube where your teacher tells you. Rinse the test tubes with cold water, and then wash them with soap.
2. Clean up your work space.
3. Write the answers to the analysis questions and your conclusion on the back of your data table.
4. During the class discussion, fill in the handout, *Food Test Results.*

Analysis Questions

1. What happens when nitric acid is placed on protein?
2. If your lab partner spilled some nitric acid on his or her hand, what would you do?
3. Look at your protein test results. Which foods do not seem to have any protein?
4. Amy found that olives contain protein. Ted's results showed that olives did not contain protein. What are two possible reasons for their different results?
5. If you get nitric acid on your skin or fingernails, they turn yellow and your skin will burn. Explain why the color change occurs.

Conclusion

Write a brief paragraph in answer to each of the focusing questions. Be sure to refer to your data as evidence for your answers.

Extension Activity

Going Further: If you think you've found a pattern and can recognize a type of food that contains protein, test some more foods to see if you're right. For instance, you may decide that all vegetables contain protein. If so, test more vegetables to find out if that is true.

7-6

Protein: Beans or Beef?

Have you ever seen a child such as the one pictured in **Figure 7.11**? This boy is listless and irritable. He has barely enough energy to walk, let alone to play. If he were to get hurt, his wounds would take a long time to heal, and his immune system is so weakened that he quickly catches infectious diseases. The fever, vomiting, and diarrhea that go along with many childhood diseases will weaken him even more. What is wrong with this child?

Even though he may look well fed, this boy is undernourished. What appears to be fat is actually water; his system is holding fluid. He has not been getting enough protein in his diet so he has a condition known as kwashiorkor. Kwashiorkor is all too common among young children in countries where there is a food shortage. Without enough protein in their diets, people cannot be healthy.

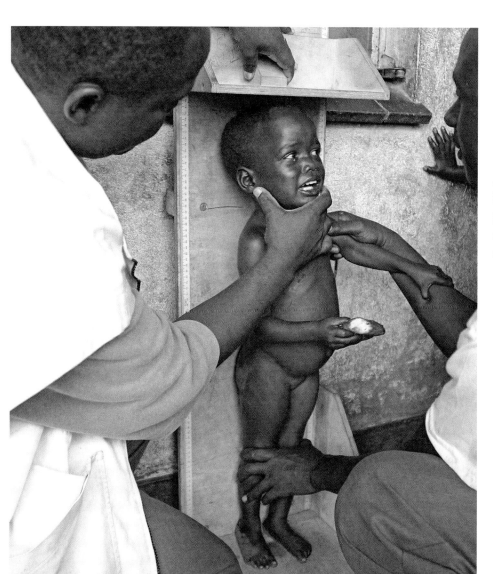

Figure 7.11
Kwashiorkor affects millions of young children in Latin America, Asia, and Africa.

Functions of Protein

Proteins are essential for the functioning of cells. There are many kinds of proteins, and each protein has its own function. A few of the many things proteins do are listed below. You will learn more about protein functions as you study other body systems.

- Protein molecules are part of the *structure of cells.* Every cell in your body contains protein. You need a continuous supply of protein so you can continue to produce new cells.
- Besides cells, protein *makes up other structures* such as bones, hair, and fingernails.
- *Enzymes* are proteins. They control all of the chemical reactions that take place in the body.
- The *antibodies* that fight infections are proteins.

- If you are low on fats and carbohydrates, your body can use protein *for energy.*

Proteins are not unique to humans. Microscopic bacteria and protists, fungi, and multicellular plants and animals all contain protein. Structures such as spider webs, wool, feather, and horns are also made of proteins. **Figure 7.12** describes other examples of protein function.

Protein Sources

Enzymes, antibodies, and other protein structures do not magically appear. They have to be made. Plant cells make the proteins they need. Animal cells produce the proteins they need by "rearranging" the proteins they get from food. All foods contain some protein, since plant and animal cells, like human cells, need protein. Some foods, however, contain more protein than others.

Figure 7.12

The Amanita is poisonous because of a protein it makes.

Red blood cells can carry oxygen because of the protein hemoglobin.

The white color of milk is primarily because of one kind of protein.

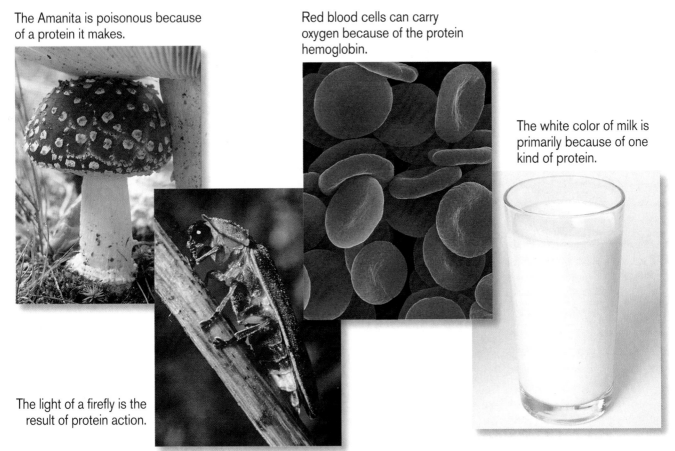

The light of a firefly is the result of protein action.

Grains and cereals (rice, wheat, oats, and rye) as well as nuts and seeds are all high in protein; so are seeds that grow in pods, such as peas, lentils, kidney beans, soybeans, and lima beans (see **Figure 7.13**). Most other vegetables and fruits are low in protein. People from other countries rely on plant products for more of their proteins than Americans do. For example, a Brazilian eats an average of 50 pounds of beans a year, whereas an average American eats about 7 pounds of beans a year.

Muscle cells contain a lot of protein, so all meats are high in protein. Egg whites, milk, cheese, and other dairy products are also high in protein (see **Figure 7.14**). Americans eat more animal protein—meats and dairy products—than do people from most other countries.

Amino Acids

Like fat, proteins are big molecules. If you could look at a protein molecule with a super microscope, you would see that it looks like a chain of smaller molecules, as shown in **Figure 7.15**. These smaller molecules are called **amino acids**. There are 20 different amino acids and they can be linked in millions of different combinations. Each kind of protein is one specific arrangement of amino acids. Like fats, proteins (and amino acids) are made up of carbon, hydrogen, and oxygen. In addition, proteins contain some nitrogen.

Our cells assemble the amino acids to make the proteins we need. For instance, if you eat pinto beans, your digestive system breaks down the protein into single amino acids and then your cells put the amino acids back together to make proteins such as those found in hair, hemoglobin, and connective tissue.

Figure 7.14

All of these are high-protein foods that come from animal sources. Which of them do you eat most often?

Figure 7.13

Many foods from plants are good sources of proteins. Which of these do you eat most often?

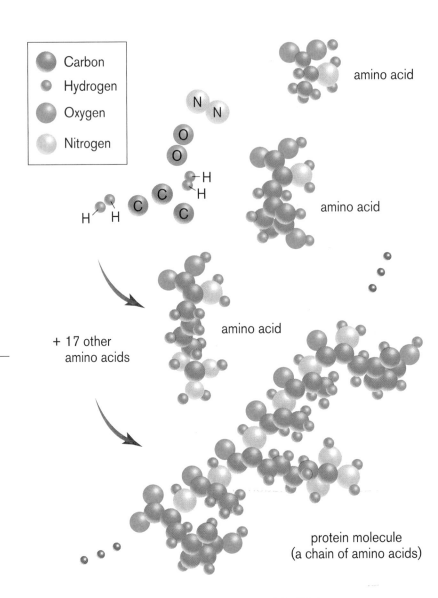

Figure 7.15

Carbon, hydrogen, oxygen, and nitrogen combine to make amino acids. Hundreds, maybe even thousands, of amino acids are linked together to form a protein.

Carbon
Hydrogen
Oxygen
Nitrogen

amino acid

amino acid

amino acid

+ 17 other amino acids

protein molecule (a chain of amino acids)

How Much?

You should eat some protein every day, but you probably don't have to worry about not getting enough—most Americans eat too much protein. Some people think high-protein diets will make them better athletes. Actually, if you eat too much protein, cells in your body just break it down and change it to fat. Your body does not store protein for later use.

The amount of protein a person needs depends on a lot of things. All people need some protein to replace old proteins. However, when a person is growing, or wounds are healing, new cells are being created and more protein is needed. A small person does not need to eat as much protein as a large person does. If you eat a variety of foods every day, including some fruits, vegetables, whole-grain breads, dairy products, and meats, you will probably get enough protein.

Daily Recommendations for 9- to 14-year-olds	
Nutrient	**Amount in Grams**
fat	62–95
protein	34–52

Analysis Questions

1. List five functions of proteins.
2. Give five examples of foods that are:
   a. good sources of protein
   b. poor sources of protein
3. Think about the protein in your diet.
   a. How many grams of protein are recommended for people your age?
   b. About how many grams of protein should you eat each day?
4. A woman needs to eat more protein if she gets pregnant. Why?
5. Here is a lunch menu.

   2 slices white bread
   2 tablespoons peanut butter
   2 tablespoons grape jelly
   1 apple
   3 chocolate chip cookies
   1 cup whole milk

   a. How many grams of protein are in this lunch? (Use the information in Appendix C, *Nutrients in Foods.*) Show your work.
   b. If you ate this lunch, how many grams of protein would you need to get from your other two meals?
6. Option: Analyze your food diary to see how many grams of protein you ate on the day you kept the diary.

Extension Activities

1. **Social Studies Connection:** Choose a country that you would like to learn more about and do some research. Find out what foods are eaten there and circle the ones that are high in protein.
2. **Health Application:** Read *Vegetarian Diets.* Copy the ingredients for two meatless main dishes. Put a star by each ingredient that provides protein. Would the combinations make complete proteins? If you feel ambitious, prepare one recipe and serve it to your family.

Vegetarian Diets

Vegetarians (people who do not eat meat) need to know about proteins when they choose their foods. If foods do not contain all 20 amino acids, the body can make most of the missing ones. However, there are some amino acids the body cannot make. These are the **essential amino acids.**

Proteins from animal sources contain all of the essential amino acids. For this reason, they are called *complete proteins*. However, plant proteins are not complete; they are missing some of the essential amino acids. If eaten together, plant foods can provide all the essential amino acids. Here are examples of good protein combinations:

> corn tortilla with refried beans   cereal with milk
>
> macaroni and cheese   pea soup and rye bread
>
> peanut butter on whole wheat bread

In general, grains and beans together make a complete protein; so do dairy products or eggs when they are eaten with any of the high-protein plant foods.

Carbohydrate Tests

A moose's diet is very high in carbohydrates. So is a grasshopper's, a beaver's, and a bee's. How about yours? What foods do you eat that contain carbohydrates? In order to answer this question, you need to know that carbohydrates are the foods that contain fiber, starch, and/or sugar.

How can you test for starches?

How can you test for sugars?

You will use two chemical tests. One tests for starch and one tests for sugar. (There is not a simple test for fiber.) You'll use an iodine solution to test for starch. The test for sugar is more complicated, but it's fun to do—just be careful, because you will use a heat source and a chemical called Benedict's solution.

Use cornstarch as a control for starch so you will know what a positive starch test looks like, and use corn syrup as the control for sugar. Then test five foods to see if they contain starch or sugar. When you are done, you should be able to answer the focusing questions.

Materials

For both tests:
- samples of five foods
- water in dropper bottle
- Handout, *Food Test Results* (from Activity 7-2, *Fat Test*)
- paper towels
- Handout, *Carbohydrate Tests*

For the starch test:
- cornstarch
- aluminum foil
- iodine in dropper bottle

For the sugar test:
- seven test tubes
- test tube rack
- masking tape
- corn syrup

At the sugar test station:
- hot plate
- beaker
- Benedict's solution in dropper bottle
- safety goggles
- test tube holder

Procedure

(Work with a partner.)

Part A: Getting Ready

1. On your handout, record the 5 foods that your teacher tells you to test.
2. Make your predictions. Discuss each food with your partner and decide if you think it contains starch and/or sugar. Use a plus (+) if you think the substance is present in the food, and a minus (−) if you think it is not in the food.
3. Your teacher will tell you which test to do first.

⬤ **Caution:** Iodine can stain your clothes. Be careful when you use it.

Part B: Testing For Starch

1. Get your aluminum foil and set it on the desk.
   a. Put three drops of water on the foil.
   b. Put about the same amount of cornstarch on a different place on the foil.
2. Add three or four drops of iodine to the water and three or four drops to the cornstarch (see **Figure 7.16**).

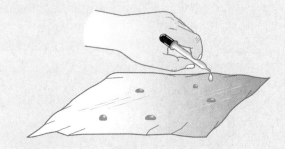

Figure 7.16

3. Notice what happens when cornstarch is mixed with iodine. Any food that contains starch will react in the same way. Answer question 1 on the handout.
4. On the foil, place a small amount of each of the five foods you will test.
   a. Test each food for starch.
   b. Record your results on the handout. Use a plus (+) if a food contains starch. Use a minus (−) if it does not.
5. Throw away the aluminum foil with the food samples.
6. If the sugar test stations are all in use, answer analysis questions 1 and 2.

Caution: You will be working with hot liquids. Use the test tube holder and wear safety goggles.

Part C: Testing for Sugar

1. Get seven test tubes and put them in the test tube rack.
   a. Put nine or ten drops of water in one test tube.
   b. Put the same amount of corn syrup in another test tube.
2. Add ten drops of Benedict's solution to each of the two test tubes. Gently swirl each test tube to mix the solutions (see **Figure 7.17**).
3. Put on your safety goggles. Place the two test tubes in a beaker of hot water and leave them there for about three minutes (see **Figure 7.18**).
4. Notice what happens when corn syrup is heated with Benedict's solution. Most foods that contain sugar will give a similar reaction. Answer question 2 on the handout.
5. Prepare to test your five foods for sugar.
   a. Label each test tube with the name of the food you will place in it. Put the tape high on the test tube so it doesn't fall off.

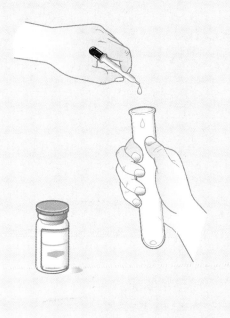

Figure 7.17

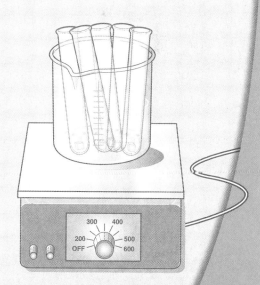

Figure 7.18

b. Put a small amount (about half the size of a raisin) of each food into the proper test tube.

c. Add nine or ten drops of water to each test tube.

6. Add ten drops of Benedict's solution to each test tube and gently swirl to mix. Heat the tubes as you did the others.

7. Record your results on the handout. Use a double plus (+ +) for lots of sugar when the results are similar to the corn syrup's reaction. Use a single plus (+) when there is a slight color change and a minus (−) for a negative test (no color change).

8. Wash your test tubes.

Part D. Finishing Up

1. Clean up your work space.

2. Write the answers to the analysis questions and your conclusion on the back of your handout.

3. During the class discussion, complete the sugar and starch columns on your handout, *Food Test Results*. Do not lose this handout; you will use it again in Activity 7-10, *Major Nutrients*.

Analysis Questions _ _ _ _ _ _ _ _ _

1. Look at the results of the starch test.

    a. Which foods contained starch that you predicted would contain starch?

    b. Which foods contained starch that you did not think would contain starch?

2. List at least six other foods that you predict would test positive for starch if you used iodine to test them.

3. Look at your data for testing water.

    a. Did water contain sugar or starch?

    b. Why did the directions say to test water for sugar and starch?

4. Look at the results of your sugar tests.

    a. Which foods contained sugar that you predicted would contain sugar?

    b. Which foods contained sugar that you did not think would contain sugar?

5. List at least six other foods that you predict would test positive for sugar if you used Benedict's solution to test them.

6. Tony and Pat both tested jelly for sugar. Tony got a yellow color change. Pat got a red color change. List three possible explanations for their different results.

7. Look at your data table.

   a. Which foods did not contain starch or sugar?

   b. What do you think these foods are made of?

8. Where do foods that contain carbohydrates come from—plants, animals, both, or neither? Use evidence from your data to support your answer.

Conclusion

Write two paragraphs, one to answer each of the focusing questions.

Extension Activities

1. **Going Further:** In answer to analysis questions 2 and 5 you made some predictions. Test your predictions and record your results.

2. **Going Further:** There are several kinds of sugars. Some of these can be detected with Benedict's solution; others cannot. Test several kinds of sugars. Read the ingredients on package labels to help you find different kinds. Which ones react with Benedict's? Which do not?

Sugar, Starch, and Fiber

Every bite of food you eat starts as a ray of sunshine. Plants capture the energy from the sun to make sugar. Their cells can then change the sugar to starch and fiber. Sugar, starch, and fiber all taste different, and have different properties. So why are they all called carbohydrates?

Carbohydrate Structure

All carbohydrates are made of carbon, hydrogen, and oxygen. That's where the word carbohydrate comes from.

carbo = carbon
hydrate = water (hydrogen and oxygen)

sugar molecules

starch molecule
(a chain of sugar molecules)

Carbon
Hydrogen
Oxygen

During photosynthesis, plants combine carbon dioxide (carbon and oxygens) and water (hydrogens and oxygen) to make sugar. **Sugar** is the smallest carbohydrate molecule. When many sugars are hooked together, they form either a **starch** or **fiber** molecule, depending on how they are connected. The sugar molecules in fiber are hooked together in such a way that human cells cannot break them apart (see **Figure 7.19**).

Carbohydrates in Foods

When you tested foods for sugar and starch, you may have noticed a pattern. Foods that come from plants usually contain either sugar or starch, and maybe both.

High-**starch** foods make up a major part of our diets. These foods are not the leafy parts of plants; instead, they come from roots and seeds. They include potatoes, rice, corn, and products made from grain, such as breads, rolls, and pasta.

Fiber is found in many parts of a plant. It provides support and structure for the plant. Raw vegetables, many fruits, and whole-grain breads and cereals are all high in fiber.

Figure 7.19
Sugar molecules are made up of carbon (C), hydrogen (H), and oxygen (O). When many sugar molecules are hooked together, they form either sugar or starch.

There are several kinds of **sugar**. The sugar that probably comes to your mind is table sugar, or sucrose, but there are many other natural sugars. For example, fruit sugar is fructose and the sugar used by muscles for energy is glucose. Note that the names of sugars usually end with "ose."

Figure 7.20 shows examples of foods that are high in each type of carbohydrate.

Many foods contain more than one kind of carbohydrate. As you can see from the chart, *Carbohydrates in Foods,* a carrot contains nine grams of carbohydrates—five grams of natural sugar, three grams of fiber, and one gram of starch.

Most of the carbohydrates we eat come from plant sources. Foods from animal sources, such as meats and eggs, contain little, if any, carbohydrates. Dairy products, however, contain a type of sugar called lactose.

Includes sci**LINKS**
NSTA

Topic: Carbohydrates
Go to: www.scilinks.org
Code: MSLS3e167

Figure 7.20

High-fiber foods

Starchy foods

High-sugar foods

Carbohydrates in Foods				
Food	Total Carbohydrates	Sugar	Starch	Fiber
apple, 1 medium	15 g	10 g	2 g	3 g
bread, white, 2 slices	20 g	2 g	17 g	1 g
carrot, 1 medium	9 g	5 g	1 g	3 g
cookie, chocolate chip	9 g	4 g	5 g	0 g
jelly, grape, 1 T*	13 g	9 g	4 g	0 g
milk, whole, 1 cup	11 g	10 g	1 g	0 g
peanut butter, 2 T*	6 g	2 g	2 g	2 g
potato, 1 medium	27 g	3 g	22 g	2 g
rice, cooked, 3/4 cup	35 g	0 g	34 g	1 g

* T = tablespoon

Carbohydrates in the Body

As with fats and proteins, carbohydrates are important for the functioning of our cells and our bodies.

- Sugar and starch are an important *energy source.* Fats and protein can also be used as energy but not as easily.
- Fiber *helps foods move* through the digestive tract. People who eat enough fiber are not likely to have problems with constipation.
- Foods that supply starches also *supply other nutrients*: vitamins, minerals, and some protein. (Sugary foods are seldom rich in other nutrients.)
- Like fats and proteins, carbohydrates also form part of the *structure of cells.*

In other organisms, carbohydrates have other functions. Wood is strong because of tough fiber molecules. The external skeletons of arthropods (insects, spiders, crustaceans, and other animals) are made up of another kind of carbohydrate.

Making Choices

In general, you will be eating adequate amounts of starch if you eat fruits and green vegetables each day and include potatoes, rice, or beans with your meals. When possible, choose foods that provide both fiber and starch, such as whole-grain bread instead of white bread. Of course, if you like to think in terms of grams, you can always add up the number of grams of carbohydrates you're eating (see the chart).

The recommendations for carbohydrates are based on starch and fiber intake. You really don't have to eat any sugar. The body can only use so much sugar; the rest it changes to fat. The main problem with most sugary foods is that they do not contain other nutrients. If you drink a soft drink and munch on a candy bar, you won't be hungry for dinner. That's why it is important to eat foods that supply the necessary nutrients first, and then, if you are still hungry, eat something sweet.

At first, it may be hard to remember which foods contain which nutrients. But after you have looked up the contents of quite a few foods in nutrient charts, you will begin to develop an "instinct" for what nutrients foods contain.

Daily Recommendations for 9- to 14-year-olds	
Nutrient	Amount in Grams
fat	62-95
protein	34-52
carbohydrate	300-500

Analysis Questions

1. What are the three types of carbohydrates?

2. Which of the three types of carbohydrate is made of the smallest molecule?

3. What is the difference between starch and fiber?

4. List four functions of carbohydrates in the body.

5. Here is a lunch menu:

   2 slices white bread
   2 tablespoons peanut butter
   2 tablespoons grape jelly
   1 apple
   3 chocolate chip cookies
   1 cup whole milk

   a. Make a table like the one here to analyze the carbohydrate content of this lunch. (Use the chart, *Carbohydrates in Foods*, to find the information you need to fill in the table.)

   b. If you ate this lunch, how many grams of carbohydrates would you be eating?

   c. If you ate this lunch, how many grams of carbohydrates would you need to get from your other two meals?

   d. Of the grams of carbohydrates in this lunch, do more come from sugar or starch?

   e. What's one change you could make to increase the fiber content of this meal?

6. Option: Analyze your food diary in terms of fiber and sugar. Did you eat enough fiber? Too much sugar?

Extension Activity

Going Further: Read the ingredient lists on food package labels. List as many kinds of sugar as you can find. Remember, the names of many sugars end with "ose."

Food	Grams Carbohydrate	Grams Sugar	Grams Starch	Grams Fiber
1 apple				

Hi-Lo Label Hunt

You don't have to use a nutrient chart every time you want to learn about a food. You may be able to find the information you need by reading a package label. Federal law requires that all packaged foods be clearly labeled because it is important that you, as the consumer, know exactly what you are buying.

What can you learn about a food product by reading a food label?

Compare four food labels. What information appears on all the labels? Look for information about the nutrients you have studied. When you are finished, you should be able to answer the focusing question.

- four food labels
- Handout, *Hi-Lo Hunt List*
- Reading, *Reading a Label*, page 173

Procedure

Part A: Completing the Hunt *(Work in teams of two or four students.)*

1. Divide the labels among the members of your team.
2. Choose a recorder who can write quickly and neatly. He or she will be responsible for filling in the *Hi-Lo Hunt List*.
3. You will have 12 minutes to fill in the handout. Help each other locate the items on the labels so your group can work quickly.
4. Everyone on your team should be prepared to share the information on the list with the rest of the class.

Part B: Reading About Labels   *(Work independently.)*

Read *Reading a Label* and then answer the analysis questions and write your conclusion.

Analysis Questions

1. What five items must appear on a food label?

2. In your own words, explain the difference between a nutrient and an ingredient.

3. Here is a list of 12 items. Some of them are nutrients; the others are ingredients.

flour	salt	rice	carbohydrate
vitamin C	protein	water	cornmeal
iron	egg	onions	baking soda

   a. List the ones you think are nutrients.

   b. List the ones that are used as ingredients.

4. Look at the ingredients shown here from a food label. Decide which product the label comes from—smoked beef jerky (a dried beef product), all-beef hot dogs, or beef-flavored dog food. Explain your decision.

   Beef, water, dextrose, salt, corn syrup, spices and flavorings, sodium phosphates, paprika, sodium nitrate

5. Use the nutrition information on the label in **Figure 7.21** to answer these questions.

   a. What is the main ingredient in Brand *A* cereal?

   b. How much cereal is considered to be one serving?

   c. How many grams of sugar does this cereal contain?

   d. What do you predict would be the results of a fat test on this cereal?

   e. What do you predict would be the results of a protein test?

   f. What do you predict would be the results of a starch test?

   g. What do you predict would be the results of a sugar test?

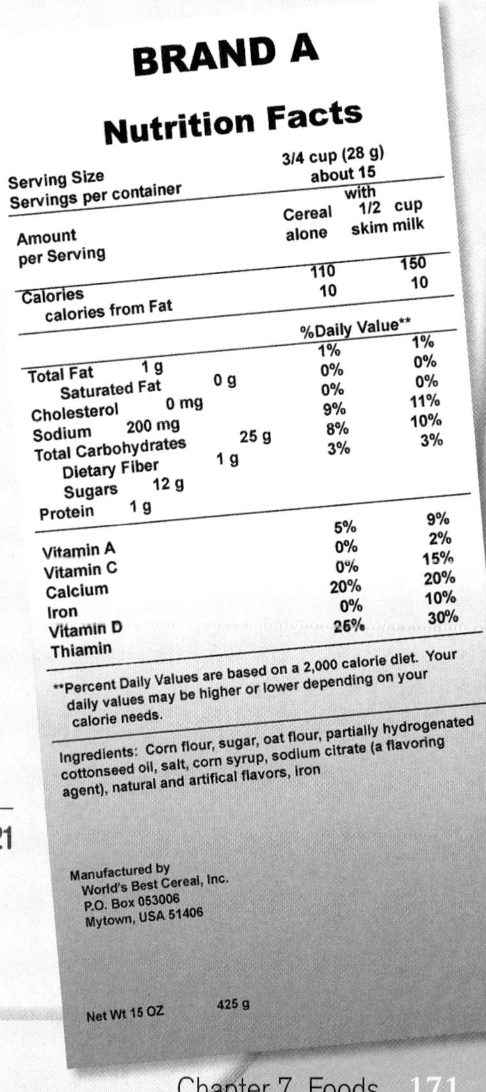

BRAND A

Nutrition Facts

Serving Size 3/4 cup (28 g)
Servings per container about 15

Amount per Serving	Cereal alone	with 1/2 cup skim milk
Calories	110	150
calories from Fat	10	10

	%Daily Value**	
Total Fat 1 g	1%	1%
Saturated Fat 0 g	0%	0%
Cholesterol 0 mg	0%	0%
Sodium 200 mg	9%	11%
Total Carbohydrates 25 g	8%	10%
Dietary Fiber 1 g	3%	3%
Sugars 12 g		
Protein 1 g		

Vitamin A	5%	9%
Vitamin C	0%	2%
Calcium	0%	15%
Iron	20%	20%
Vitamin D	0%	10%
Thiamin	26%	30%

**Percent Daily Values are based on a 2,000 calorie diet. Your daily values may be higher or lower depending on your calorie needs.

Ingredients: Corn flour, sugar, oat flour, partially hydrogenated cottonseed oil, salt, corn syrup, sodium citrate (a flavoring agent), natural and artifical flavors, iron

Manufactured by
World's Best Cereal, Inc.
P.O. Box 053006
Mytown, USA 51406

Net Wt 15 OZ   425 g

Figure 7.21

6. Which cereal do you think is most nutritious, Brand *A* (**Figure 7.21**) or Brand *B* (**Figure 7.22**)? Why?

7. Challenge: Here are the ingredients from a recipe for cookies. List the ingredients in the order they would appear on a food label.

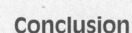

3 1/2 cups flour
1/2 teaspoon baking soda
1/4 teaspoon salt
1 cup shortening
1 cup sugar
1 teaspoon vanilla

Conclusion

Using complete sentences, answer the focusing question.

Extension Activity

Going Further: Look at the information printed on cat or dog food labels. In what ways is this information like that on "people" food labels? How is it different? Read the list of ingredients. Is there anything in pet food that is not in the food you eat?

BRAND B

Nutrition Facts

| Serving Size | 1/2 cup (29 g) | |
| Servings per container | about 17 | |

| Amount per Serving | with | |
| | Cereal alone | 1/2 cup skim milk |

| Calories | 80 | 120 |
| calories from Fat | 5 | 5 |

		%Daily Value**	
Total Fat	1 g	1%	1%
Saturated Fat	0 g	0%	0%
Cholesterol	0 mg	0%	0%
Sodium	200 mg	9%	11%
Total Carbohydrates	23 g	8%	10%
Dietary Fiber	9 g	36%	36%
Sugars	6 g		
Protein	4 g		

Vitamin A	15%	20%
Vitamin C	5%	5%
Calcium	5%	20%
Iron	30%	30%
Vitamin D	0%	10%
Thiamin	25%	30%

**Percent Daily Values are based on a 2,000 calorie diet. Your daily values may be higher or lower depending on your calorie needs.

Ingredients: Wheat bran, sugar, malt flavoring, salt, fig juice concentrate, iron, vitamin A palmitate

Manufactured by
Fine Foods, Inc.
P.O. Box 111399
Ourcity, USA 12246

Net WT 17 OZ   481 g

Figure 7.22

Reading a Label

We take food labeling for granted. If a label says "large pitted olives," that is exactly what we expect to find in the can. We can do so because the law requires that certain basic information appear on all labels.

Required by Law

All labels, including those on boxes, bottles, jars, cans, and bags, must include the following information:

1. the name of the product
2. the net weight (This is the weight of the food; it does not include the weight of the package.)
3. the manufacturer's name and address (This is included in case you want to contact the manufacturer.)
4. nutrition information (More information about this is given later.)
5. the ingredients (More information about this is given later.)

Look at the sample label in **Figure 7.22** on page 172. Find all five required pieces of information.

Ingredients

The first ingredient listed on a label is present in the greatest amount by weight; the last one is present in the smallest amount. Many of the ingredients listed on a label will be familiar to you because they are the foods used to make the product. However, you may not recognize the names of things such as the preservatives added to give the product a longer shelf life or the food colorings used to give the food more eye appeal. Unless you have an allergy to one of the additives, you will probably be most interested in the main ingredients anyway.

Nutrition Information

The information in the section titled "Nutrition Facts" tells about the nutrients in a product. Serving size is the first thing listed in this section. All the nutrition information is based on the nutrients in a single serving. If you eat more or less than the serving size, you will have to take that into consideration when you read the nutrition information.

Calories per serving and the number of calories that come from fat appear at the top of the nutrition information list. These figures are useful for people who think about their food in terms of calories.

The list of nutrients includes information that is important to your health. Here you can find the number of grams per serving of each of the major nutrients. You can also find the number of grams of saturated fat, trans fat, sugar, and fiber per serving.

Notice that the amounts of cholesterol and sodium (salt) are in milligrams (mg), not grams. That's because these nutrients are found in much smaller amounts in our foods. In general, most people should eat less than 300 mg (0.3 g) of cholesterol per day and less than 2,300 mg (1 teaspoon) of sodium per day.

This information is useful if you know something about the number of grams of each of these nutrients you should eat per day. People who don't think in terms of grams may find the *% Daily Value* to be more useful. This tells about how much of your recommended daily intake is met by one serving of the food.

Food labels may also contain information about vitamins and minerals, if the food contains them.

Other Information

In addition to the items required by law, you can find a variety of other things on the label. For example, it may tell you more about the contents or show you a picture of the product. What other items have you noticed on a label?

The Major Nutrients

You started this chapter by thinking about familiar foods. Then you studied each of the major nutrients separately—fats, proteins, and carbohydrates. You tested foods for the major nutrients and you read about each of these important substances. It's time to pull all of this information together.

What has been the most important thing you've learned in this chapter so far?

There are two parts to this activity. In the first part, you will finish filling in the handout, *Food Test Results.* In the second part, you will review and summarize what you have learned about fats, proteins, and carbohydrates. These two tasks will help you reflect on what you have learned about the major nutrients.

- Handout, *Food Test Results* (from Activity 7-2, *Fat Test*)
- chart in Appendix C: *Nutrients in Foods*
- Handout, *Fats, Proteins, and Carbohydrates*

Procedure

Part A: Food Test Results

The handout, *Food Test Results,* summarizes your data from the four food tests you did. In this part of the activity, you will check your results with the results scientists got when they analyzed the foods in chemistry labs.

1. Look up the nutrient content of an apple in the chart in Appendix C, *Nutrients in Foods.* Record the number of grams of fat found in an apple in the fat column on your summary table. Also record the number of grams of protein and carbohydrate in an apple.
2. Look up and record the nutrient content for each of the other foods you tested.

Figure 7.23

Carbohydrates, fats, and proteins are made up of the same substances—carbon, hydrogen, and oxygen—and yet they are very different nutrients. Look at these drawings. What do you think is the main difference that characterizes each type of nutrient?

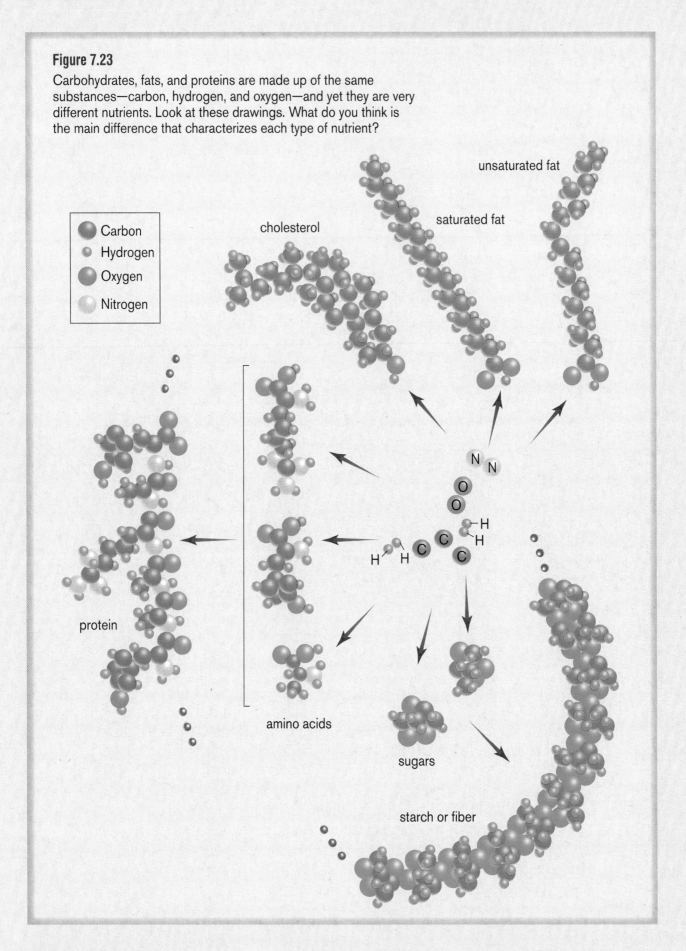

Carbon
Hydrogen
Oxygen
Nitrogen

cholesterol

unsaturated fat

saturated fat

protein

amino acids

sugars

starch or fiber

N N
O O
H
H
H H C C C H

3. Compare your test results with the nutrient analysis.
    a. Foods that have several grams of fat should have tested positive for fat. Did they test positive?
    b. Similarly, check your data for protein and carbohydrates.
    c. Circle any data that you think could be wrong.
4. Answer analysis questions 1 and 2.

Part B: Summary: Fats, Proteins, and Carbohydrates

Fill in the handout, *Fats, Proteins, and Carbohydrates,* to summarize what you have learned about these substances so far.
1. Refer to your handouts and the readings to find the information you need.
2. Do not fill in the last two rows yet. You will fill this in after you study digestion in the next chapter.

Analysis Questions — — — — — — — — —

1. Based on your analysis, which food test do you think gives the most accurate results? Explain your answer.
2. Based on your analysis, which food test do you think is most inaccurate? Explain your answer.

Conclusion

Write a half-page answer to the focusing question. Remember to explain what you learned and why you think it is important.

"You Eat Like a Bird"

All animals need protein, carbohydrate, and fat, but the amounts they need vary greatly from animal to animal. Some animals, such as many birds, need a lot of food to survive. That's because birds' bodies "work faster" than ours do. The smaller the bird, the faster its body works. A tiny hummingbird eats the equivalent of its body weight each day. Think about it. If you weighed 91 pounds, you would have to eat 91 pounds of food every day to eat like a hummingbird!

A Final Sort

You started this chapter by thinking about 32 familiar foods and sorting the foods into groups that made sense to you. Then you sorted the foods two more times. For one sort, you grouped the foods based on whether they came from a plant or animal source. For the last sort, you taped the foods onto one of three pages depending on whether you thought the food was or wasn't "good" for you.

What are the major nutrients and what do they have to do with living organisms?

In this activity, you will re-sort the 32 foods you named when you started this chapter. While you do so, you will be able to see evidence that you have learned about foods and nutrients.

- 32 foods (3 pages from Activity 7-1, *Thirty-Two Foods*)
- Handout, *32 Foods—Final Sort*

Procedure

Work with the partner you had when you did Activity 7-1, *Thirty-Two Foods*. You will need the three pages on which you taped your 32 foods after you did the third sort.

1. Check the foods you put on the page titled, "Foods that are good for you."

   a. If you think the food is in the right place, use the red pencil to write "ok" by the piece of paper with the name of the food on it.

   b. If you've changed your mind, use the red pencil to write where you think the food belongs.

2. Follow the same procedure to check the foods you put on the page titled, "Foods that are not good for you." By each food on this page, write a sentence explaining why the food belongs in this group.

3. Next to each food on the last page, "Foods that we don't know if they're good or bad," write down whether you now think the food is "good" or "bad" nutritionally.

   a. Write one sentence explaining each of your answers.

   b. If you still don't know whether a food is "good" or "bad," write a sentence explaining why you are having a hard time deciding about this food.

4. Sort your 32 foods one more time, but don't untape the papers. You will be sorting the foods based on their nutrient content.

   a. Look at the handout, *32 Foods—Final Sort*. Notice that there are three circles labeled fats, proteins, and carbohydrates.

   b. The space where the circles overlap indicates a combination of nutrients. For example, where the fat and protein circles overlap there is a space for foods that contain significant amounts of both these nutrients.

   c. Choose one of your 32 foods and decide which of the major nutrients it contains in significant amounts, and then write the name of the food in the appropriate space. (For example, potato chips would go in the fat and carbohydrate section; hard candy would go in the carbohydrate section.)

   d. Write the names of your other 31 foods in the correct sections of the handout. If you don't have any idea what a food contains, put it in the box with the question mark.

5. Answer the analysis questions and write your conclusion to this lesson.

Analysis Questions _ _ _ _ _ _ _ _

   1. What is the difference between a food and an organism?

   2. In addition to fats, proteins, and carbohydrates, what are organisms such as deer, octopuses, and grape vines made of?

   ❷ 3. Why do most consumers have a digestive system?

Conclusion

Write a half-page answer to the focusing question.

Chapter 8

Digestion

Which parts of the digestive system can you identify in this picture?

Food's Journey

Since you just studied food, you may have guessed that you would study the digestive system next. As with most other topics, you already know something about this one. You may have studied it in school, but most of your knowledge probably comes from experience. You have had lots of practice putting things into your digestive system. But have you been thinking like a scientist? Have you thought about what happens to your food once it disappears from sight? Have you looked for patterns to the ways your body reacts to different kinds and amounts of food?

Before you start studying this system, it's important to think about what you already know . . . and about what you don't know.

Why is the human digestive system made up of so many different parts?

Which parts of the digestive system do you think a person could live without? Explain.

Sketch a human digestive system. Make it as complete as possible and label the parts. While you work, talk about the focusing questions with your partner.

- blank paper
- pencil with eraser
- colored pencils or crayons

Procedure
(Work with a partner.)

1. Think about what parts make up the human digestive system. Make as complete a list as you can. (Work from memory. *Don't look through this chapter for ideas!*)

2. Use your pencil to sketch the digestive system. You should:
   a. fill the paper with your sketch
   b. enclose the sketch within an outline of the human body
   c. include all the parts you listed
   d. label the parts

3. Color your drawing to help identify the parts.

4. With your partner, decide how important each part is for the functioning of the system. Label each part to show your decision.

   **-very important
   *-kind of important
   x-not important

5. On the back of your sketch, list at least four questions you have about the digestive system.

6. Answer the analysis questions and write your conclusion.

Analysis Questions _ _ _ _ _ _ _ _ _ _

1. Do you think one part of the digestive system is more important than another? If so, which part do you think is most important? Why?

2. Which parts of the digestive system do you think you know the most about?

3. You had seen bones, muscles, and skin before you studied them. What about the digestive system? Which parts of it have you actually seen?

4. Why do you think you need a digestive system?

Conclusion

Answer the focusing questions and write a sentence or two that explains what you learned by doing this activity.

Saliva Action

During digestion, the three major nutrients—fats, proteins, and carbohydrates—have to be broken down so they can be used by the cells. Starch, one kind of carbohydrate, will be broken down into sugars, protein will become a mixture of amino acids, and fats will become individual fat molecules.

The teeth and tongue start this process by mangling, grinding, and crushing. This changes the appearance of food, but it does not affect the molecules. The saliva released in the mouth, however, starts the chemical reactions that will break down some big molecules.

What does saliva do?

In this activity, you will observe saliva, and then you will do some chemical tests to learn more about saliva. First, you will use the iodine test to test the following for starch: saliva, starch, and a mixture of saliva and starch. Then you will test the same three things for sugar using Benedict's solution.

Materials

- test tube rack
- six test tubes
- masking tape
- cornstarch
- iodine in dropper bottle

Clean-Up Station
- test tube brush
- soap
- bleach solution
- paper towels

Sugar Test Station
- Benedict's solution in dropper bottle
- hot plate
- beaker
- test tube holder
- safety goggles

Procedure

Part A: Getting Ready

1. Copy this data table onto your paper. (What does the data table tell you about what you will do?)

Substance Tested	Part B: Starch Test		Part C: Sugar Test	
	Prediction	Result	Prediction	Result
saliva				
starch				
saliva & starch				

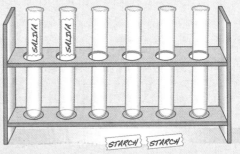

Figure 8.1

2. Get your test tube rack, six test tubes, and masking tape. Label the test tubes as shown. (See **Figure 8-1.**)

   a. Label two test tubes *saliva*.

   b. Label two test tubes *starch*.

   c. Label the other two test tubes *saliva and starch*.

3. Gently "chew" on your cheeks to build up a supply of saliva in your mouth. Then collect about 3 mL of saliva in each of the two test tubes labeled *saliva*. Collect another 3 mL of saliva in the two test tubes labeled *saliva and starch*. (Four of your test tubes should now contain small amounts of saliva.)

4. Observe the saliva and answer analysis question 1.

5. Place a pinch of cornstarch into each of the two test tubes labeled *starch* and into the two test tubes labeled *saliva and starch*.

6. Mix the test tubes containing saliva and starch by swirling gently.

7. Rearrange the test tubes in the rack so you have two sets. Each set should contain one test tube with saliva, one with starch, and one with both.

Part B: Testing for Starch

1. Get ready to test the contents of one set of test tubes for starch. Think about what will happen. Record your predictions on the data table.
2. Add a few drops of iodine to each of the three test tubes in a set.
3. Record your results.

Part C: Testing for Sugar

 Caution: You will work with hot liquids. Use the test tube holder and wear safety goggles.

1. Get ready to test the other set of test tubes for sugar. Think about what will happen. Record your predictions on the data table.
2. Go to the sugar test station.
   a. **Put on your safety goggles.**
   b. Add ten drops of Benedict's solution to each tube and swirl gently.
   c. Heat the test tubes in the water bath for about three minutes.
3. Record your results.

Part D: Cleaning Up

1. Dump your test tubes down the drain, and then clean them with soap and water. Finally, rinse them in the bleach solution. *Be careful with the bleach; it can ruin your clothes.*
2. Put away your equipment.
3. Answer the rest of the analysis questions and write your conclusion.

Analysis Questions _ _ _ _ _ _ _

1. Observe the saliva.
   a. List at least three observations that describe saliva.
   b. What do you think saliva is made of?
2. Compare your predictions with your actual results. What, if any, results surprised you? Explain.
3. How would you describe the action of saliva?
4. Why did you have to test saliva alone for both starch and sugar?
5. Possible sources of error: List several things that could cause your results to be wrong.

6. Which of these foods are at least partly digested by saliva?

bread	apple	hot dog
peanut butter	cookie	lettuce
jelly	butter	cracker

7. If you leave the test tubes out for two days, what do you think will happen to the starch in the saliva?

Conclusion

Write a paragraph that summarizes what you learned by doing this activity. Make sure that your paragraph includes an answer to the focusing question.

Extension Activity

Design and Do: In analysis question 6, you made some predictions about the action of saliva on nine foods. Design a procedure to check your predictions. With your teacher's permission, go ahead and do your experiment.

Digestion Begins

Food cannot be used by the body unless it can get into the cells. Nutrients on the outside of a cell cannot be used for growth, energy, or any other cell function. Somehow the food you eat has to be broken down into small enough particles for your cells to use. This process is called **digestion**.

Digestion starts in the mouth and continues throughout the approximately 9-meter length of the digestive system. Once food passes between the lips, it begins a series of changes that eventually breaks it down enough for cells.

Mash and Mix

Bite, chew, and chomp. The teeth tear and grind large food particles into smaller ones. The chewed food is also softened and moistened. The lips and tongue are very muscular and keep moving the food to position it for chewing. The cheeks and lips keep the food from spilling back out of the mouth. Given enough time, chewing can turn food into fairly small pieces.

Saliva Action

No matter how long you chew your food, you cannot break it down into small enough pieces for your cells to use. Even a tiny piece of food is not small enough to get through a cell membrane. Only very small molecules can get into cells. To break food into such small molecules, chemical changes must take place in the food. These changes start as soon as saliva mixes with the food. **Saliva** is produced in the salivary glands that are scattered throughout the mouth. (See **Figure 8.2**.) Saliva is mainly

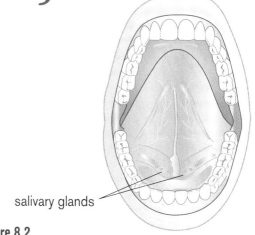

salivary glands

Figure 8.2
One pair of salivary glands is located in the membranous tissue under the tongue.

water, but it also contains enzymes that help digest food.

Salivary enzymes break down long starch molecules into shorter ones (See **Figure 8.3**). Starch molecules are too large to enter a cell, but sugar molecules are small enough to cross a cell membrane.

There is not enough time for starch to be completely broken down when it is in the mouth. It will not be completely digested until farther along in the digestive tract.

Down the Tube

When you swallow, your tongue forces the chewed food to the back of your mouth. A flap of tissue closes over the windpipe so

Includes *sci*LINKS.
NSTA

Topic: Digestive System
Go to: www.scilinks.org
Code: MSLS3e187

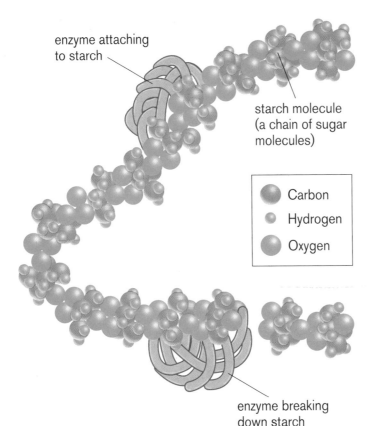

enzyme attaching
to starch

starch molecule
(a chain of sugar
molecules)

- Carbon
- Hydrogen
- Oxygen

enzyme breaking
down starch

Figure 8.3
Enzymes in saliva can break long starch molecules into shorter lengths. Given enough time, the enzymes can completely break down starch into sugar.

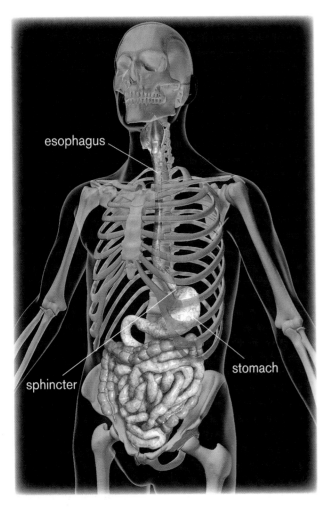

esophagus

sphincter

stomach

nothing gets into your lungs. The food itself is moved into the esophagus. The **esophagus** is a muscular tube that extends from the back of your mouth to your stomach. (See **Figure 8.4**.) When the smooth muscle fibers in the esophagus contract and relax, they "squeeze" food down to the stomach. This keeps food moving toward the stomach. It even makes it possible to swallow if you are lying down or standing on your head.

Food spends only about ten seconds traveling through the esophagus. At the bottom of the esophagus is a muscular ring called a **sphincter**, which opens to let food into the stomach. This sphincter is normally closed, so it keeps food from moving back up into the esophagus once the food is in the stomach. What do you think happens to food when it is in the stomach?

Figure 8.4
Measure from the top of your throat to the bottom of your sternum to estimate the length of your esophagus.

Enzymes

Chemical changes are always going on in your cells. Some molecules are being taken apart while others are being assembled. These changes do not happen by themselves. A special group of proteins, called **enzymes**, controls these reactions.

Each kind of enzyme controls a particular kind of chemical change. For instance, the enzymes in the digestive system break down food. The enzymes in saliva break down starch, but they cannot break down fat or protein. Other enzymes digest fat and still others digest protein.

Enzymes are not only involved in digestion. Some enzymes make new cell membranes; others control cell division. In fact, enzymes control every single body process.

Analysis Questions

1. Digestion begins in the mouth.
   a. Describe how the appearance of food is changed in the mouth.
   b. What chemical change takes place in the mouth?
2. Explain why food doesn't get in your windpipe when you swallow.
3. In your own words, explain what an enzyme is.
4. When you eat, how long do you think it takes food to get from your fork to your stomach?
5. Here is a lunch menu:

   peanut butter and jelly sandwich
   apple
   chocolate chip cookies
   milk

   a. Which of these foods will change appearance in the mouth?
   b. Which ones will be chemically changed in the mouth?

Extension Activity

Going Further: Learn more about your own mouth by completing the handout, *A "Tour" of the Mouth.*

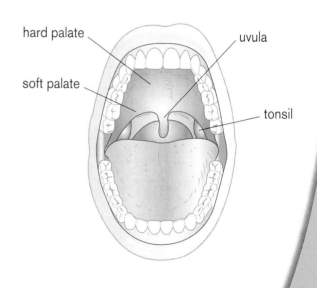

hard palate

soft palate

uvula

tonsil

St. Martin's Stomach

You may think of scientists as working in sterile laboratories, doing their investigations under carefully controlled conditions. This isn't always the case. Some informative experiments have been conducted under very unusual circumstances. For example, we learned a lot about how the stomach works as the result of an accident that took place in the early 1800s at Fort Mackinac, an outpost near the border between the United States and Canada. (Mackinac is pronounced "Mackinaw.") A man by the name of Alexis St. Martin was shot in the stomach, and as a result of the injury, his doctor spent the next eight years studying how the stomach works.

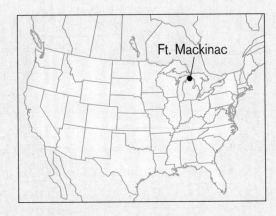

How did Alexis St. Martin's stomach injury help us understand the structure and function of the stomach?

Read about St. Martin's accident and some of Dr. Beaumont's early observations. Imagine that you are the doctor, and think about what you would have learned from these observations. Then, while working with a group of other students, read about another of Dr. Beaumont's experiments and plan a way to present the experiment to the class. Keep the focusing question in mind while you work.

- Reading, *A Historic Accident,* page 192
- Handout, *Beaumont's Notes from One Experiment*
- miscellaneous props

Procedure

Part A. Understanding the Situation

1. Read about St. Martin's injury in *A Historic Accident*. Imagine how you would have felt if you were St. Martin.
2. Read the experiment described in **Figure 8.7** (see page 194), and then answer analysis questions 1 through 4.

Part B. More Experiments
(Work in a group.)

1. Read Dr. Beaumont's notes for the experiment assigned to your group.
2. Answer the four analysis questions for this experiment. Make sure you all agree on the answers.
3. As a group, design a 3- or 4-minute presentation of this experiment to give to the rest of the class. Your group may use illustrations, skits, or another type of dramatization. Be creative, but remember, students should learn something about the stomach from your presentation.

Analysis Questions

1. What specifically was Dr. Beaumont trying to learn about the stomach? (Try to word this in the form of a question.)
2. What did Dr. Beaumont do to try to answer his question? (What was his procedure?)
3. What could Dr. Beaumont conclude based on the results of his experiment?
4. What might be a logical follow-up experiment for Dr. Beaumont to do?

Conclusion

Write your answer to the focusing question.

Extension Activity

Going Further: Learn more about Alexis St. Martin and his well-known stomach. While you read, keep asking yourself, "What did we learn about the stomach as a result of this man's accident?"

A Historic Accident

Fort Mackinac in the northern Michigan territory was a bustling trading post for the American Fur Company. Some people lived at the post, but most people came to trade and then returned to the forests and prairies. Indians would bring squash, wild rice, and game to exchange for American tobacco and wool blankets from England. Trappers brought fox and beaver pelts that they trapped during the winter. Some pelts they sold, receiving silver coins as payment. Other pelts they traded for liquor, tobacco, buckskins, coffee, and food. A few American soldiers mingled with the crowd, but there were few women or children.

June 6, 1822

On the morning of June 6, 1822, an 18-year-old French Canadian *voyageur*, Alexis St. Martin, was standing in the fur company's store. Just a few feet away, another man was holding a shotgun and it accidentally fired. St. Martin fell to the floor, blood pouring from his side. William Beaumont, a U.S. Army surgeon, elbowed his way through the crowd to examine the wounded man.

St. Martin looked ghastly. The shot had entered his left side, breaking his ribs

> *voyageur* a man who worked for a fur company transporting goods and men to and from remote areas.

Figure 8.5
Trading posts were places where people could gather and trade goods.

Figure 8.6

Dr. Beaumont at Fort Mackinac collecting fluids from Alexis St. Martin's stomach.

and blowing away significant pieces of skin and muscle. Parts of his lungs and stomach were burned and bleeding, and protruding out of the bullet hole. Internally, his diaphragm was ripped and his stomach wall had been blasted open, allowing partially digested food to flow out.

Dr. Beaumont patiently picked out some of the shot and pieces of clothing that had entered the wound; then he carefully poked the tissues back into place. The injury was so severe that Dr. Beaumont did not expect Alexis St. Martin to live more than a day or two. However, St. Martin was young and healthy, and under Dr. Beaumont's care, he survived and continued to improve. Dr. Beaumont moved his patient into his own home so he could watch him constantly. Day and night, Dr. Beaumont fed his patient, dressed his wounds, and charted his progress.

Unexpected Results

Alexis St. Martin recuperated, but very slowly. After a year, he was able to walk and feed himself, but the wound had not healed. Three years later, St. Martin still had a 2 cm (three-quarter inch) wide opening in his left side that led directly into his stomach. The edges of the tissues forming the wall of the stomach had grown to join the edges of the abdominal muscles and skin. Fortunately for St. Martin, this opening usually stayed closed. (Think of a buttonhole. It is usually closed, but you can open it wider.)

Sometimes a scientist can learn a lot from unexpected results. In this case, Dr. Beaumont realized that he had a wonderful opportunity to study the way the stomach functions. He began observing and experimenting on Alexis St. Martin's stomach in 1825, and continued to do so periodically for the next eight years. One of Dr. Beaumont's experiments on St. Martin is described in **Figure 8.7.**

Alexis St. Martin was not always a willing subject. He did not like being studied; the experiments were uncomfortable and inconvenient. After the first set of experiments in 1825, he returned to his home in Canada where he worked as a voyageur for the Hudson Bay Company, married, and fathered two children.

Dr. Beaumont begged him to come back, offering money, appealing to the fact that he would be contributing to medical science, and reminding St. Martin that he, Beaumont, had saved his life. Four years later, in 1829, St. Martin returned with his family and let Beaumont run his experiments for another four years. In 1833, St. Martin returned to Canada where he lived until his death at the age of 83.

Figure 8.7

This description is based on Dr. Beaumont's records as originally published in 1833. Therefore, it is written from the point of view of Dr. Beaumont.

Observations on the lining of the stomach...

May, 1825

The inner lining of the stomach is a light, pale pink color. It has a soft, velvety appearance and is covered with a very thin, transparent, sticky mucus. This mucus has appearance of new egg white.

I applied food to the lining of St. Martin's stomach and observed the effect through a magnifying glass. I could see many small, clear drops rising from tiny gland like bodies that lined the stomach. These drops of fluid came through the mucous layer. This fluid is acid.

stomach contains microscopic glands that release juice in the presence of food. Moreover, this juice is produced in the amount needed to dissolve the amount of food present.

William Beaumont, M.D.

Gut Reactions

Dr. Beaumont used St. Martin's stomach as a living laboratory to try to learn about digestion. In the 200 years since he did his famous experiments, we have learned much more about the digestive system. This lesson focuses on the stomach and the small intestine. As you read, think about the structure and function of both these structures.

Stomach Structure and Function

Down the esophagus, through the sphincter, plop! A wad of food lands in the stomach. The stomach is a sturdy muscular pouch located on the left side of the body, just below the diaphragm. It has two openings—one opening is the entrance and one is the exit.

When food reaches the stomach, it is a mixture of large chunks and tiny pieces. It contains protein, fat, and partially digested starch. Now what? The stomach has three important functions:

- It stores food.
- It churns and mixes the food.
- It begins to digest protein.

The structure of the stomach allows it to perform all of these functions.

Because this large organ can stretch to hold about 2 liters, we can eat quite a bit of food at one sitting. The walls of the stomach are made of smooth muscle cells, and the lining has folds. The muscle fibers contract and relax repeatedly, causing the stomach to churn. This action breaks up large pieces of food and mixes the food particles with stomach juices. The stomach makes 1.5 to 2 liters (about a half gallon) of this juice each day.

The stomach's juices are produced by the lining of the stomach. This juice is made up of water, enzymes, and acid. The *water* makes the food mixture more liquid. The *enzymes* in the juices break down proteins into amino acids. Carbohydrates and fats are not digested in the stomach.

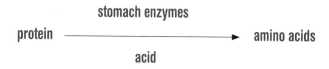

The *acid* in the stomach juices provides the environment the enzymes need to work. It also kills some of the bacteria that are swallowed along with the food. Stomach acid is so strong that it can burn your skin. It doesn't burn the walls of the stomach because the stomach lining is coated with a protective mucus.

Locate all the parts of the stomach in **Figure 8.8**.

Food usually stays in the stomach for three to five hours. It is ready to move on when it has become a thick, soupy mixture. The sphincter at the lower end of the stomach releases this mixture in spurts into the small intestine.

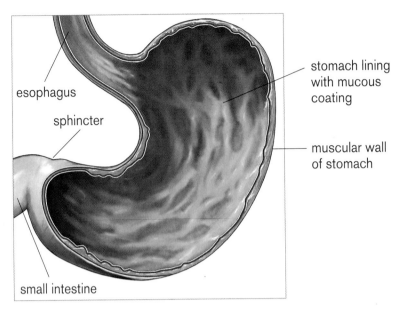

esophagus

sphincter

stomach lining
with mucous
coating

muscular wall
of stomach

small intestine

Figure 8.8

This shows a view of the stomach with a section cut away. Read each
of the labels and think about the function of each structure.

The Small Intestine's Function

It takes one to two days for food to travel from
one end of the small intestine to the other.
During this time, two important digestive
processes take place:

- The foods are completely broken down.

- Nutrients are absorbed.

Food could not be used by the body if it did
not go through the small intestine.

Very little of the food that arrives in the
small intestine has been digested. Although the
food particles are much smaller than they were
to start with, most are still too big to enter a cell.
Here in the small intestine, digestion is completed. The rest of the starch is broken down into
sugars. Proteins continue to be digested into
amino acids, and fats become individual molecules. The complete breakdown of these foods
involves many chemical reactions. Again, each
reaction requires a specific enzyme.

Some of these enzymes are produced in the
lining of the small intestine. Others are produced
in the pancreas and the liver. **Figure 8.9** shows the
location of the structures involved in digestion.

Pancreas, Liver, and Gallbladder

The **pancreas** releases pancreatic juice. This
juice neutralizes stomach acid; otherwise the
acid would destroy the lining of the intestine. It
also contains several different enzymes. Some
digest fats, some starch, and others protein.
Find the pancreas in **Figure 8.9**.

The **liver** is tucked up under the diaphragm
on the right side of the body. It performs many
important functions, one of which is to produce
bile. Bile does not contain any enzymes, but it
helps break fat into small droplets. It's easier
for enzymes to digest small droplets than large
ones. You can think of bile like a detergent that
"cuts grease."

Bile is stored in the **gallbladder** until it
is needed. The gallbladder is a small pouch
tucked up under the liver. After a meal, chemicals cause the gallbladder to contract, squeezing the bile into the small intestine. (Look at
Figure 8.9 and follow the path of bile from the
liver to the gallbladder to the small intestine.)

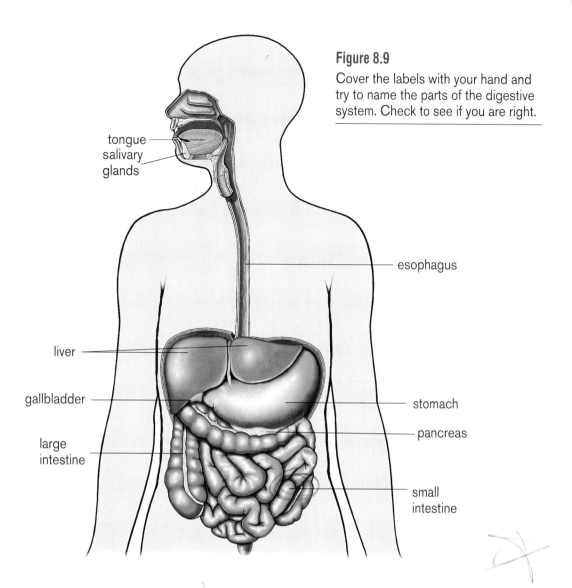

tongue
salivary glands

esophagus

liver

gallbladder

large intestine

stomach

pancreas

small intestine

The Small Intestine's Structure

The small intestine is a coiled "tube" tucked neatly into the lower abdomen. Smooth muscles in the walls of this tube contract and relax, squeezing and moving the food. As the watery mix of foods pass down this tube, they are digested and absorbed by the body. The structure of the small intestine allows it to perform these functions.

One structural characteristic of the small intestine is its length. The small intestine is the longest part of the digestive system—6 or 7 meters (18–24 feet) when uncoiled. It also has lots of surface area. If you looked at the inside of the small intestine, you would see that it is not smooth. Its lining has wavy folds and its surface looks velvety. That's because it is

covered with millions of fingerlike projections called **villi**. These villi provide lots of surface area for the absorption of nutrients. In fact, scientists have estimated that if an adult's small intestine, including every single villus, were laid out flat, it would cover an area about the size of a football field.

What happens to the nutrients once they have been absorbed into the wall of the small intestine? They don't just sit there; they have to circulate to all the cells in the body. Therefore, they need to enter a blood vessel. Each of the villi contains tiny blood vessels. Nutrients move into these tiny vessels and are carried away to other parts of the body. The cells finally get food! Study **Figure 8.10** to help you understand the structure of the small intestine.

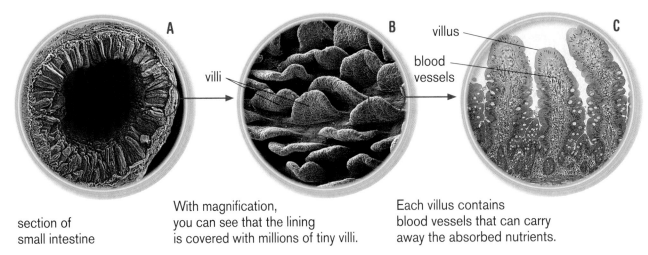

A — section of small intestine

B — With magnification, you can see that the lining is covered with millions of tiny villi.

C — villus / blood vessels — Each villus contains blood vessels that can carry away the absorbed nutrients.

Figure 8.10

Structure of the small intestine. The section of the small intestine circled in photo **A** is magnified in photo **B**. The section of the lining of the small intestine in photo **B** is magnified even more in photo **C** to show the structure of the villi. Photo credits: Dr. Gladden Willis/Visuals Unlimited, Inc.; Dr. Richard Kessel & Dr. Gene Shih/Visuals Unlimited, Inc.; Dr. Richard Kessel & Dr. Randy Kardon/Visuals Unlimited, Inc.

Moving On

The liquid mixture that reaches the end of the small intestine no longer contains usable nutrients. It will pass through another sphincter and enter the large intestine. Here it will go through the final steps of digestion.

Analysis Questions

1. What are the three main functions of the stomach?

2. Explain the purpose of each of these:
   a. muscles in the wall of the stomach
   b. mucus that covers the lining of the stomach
   c. acid in stomach juices
   d. enzymes in stomach juices
   e. sphincters

3. It is not good to take too much antacid for stomach "relief." Why do you think this is so?

4. What are two main functions of the small intestine?

5. How is bile important in digestion?

6. Explain why villi are important.

7. Here is a lunch menu:

white bread	apple
peanut butter	milk
grape jelly	chocolate chip cookie

   a. Which of these foods are digested, at least partly, in the stomach?
   b. Which of these foods are digested in the small intestine?
   c. Which foods from this lunch are still left in the liquid mixture when it moves on to the large intestine?

8. Which structure do you think is more important—your stomach or your small intestine? Explain.

9. What do you think will happen to the liquid mixture in the large intestine?

Extension Activities

1. **Models:** Villi increase the surface area of the small intestine. Think of a way to demonstrate this.

2. **Animal Kingdom:** Ruminants, such as cattle and sheep, have specialized digestive systems that allow them to digest their high fiber diets. Compare the digestive system of a ruminant with your own. How are they alike? How are they different?

Different Animals, Different Digestive Systems

All animals rely on the nutrients in their food for energy. Animals such as tapeworms "eat" by absorbing small molecules, so they don't have to digest food. Most animals, however, have some means of breaking down large molecules into small ones.

Animals such as worms, mollusks, arthropods, and vertebrates have digestive tubes running between two openings, a mouth and an anus. The tubes vary depending on the species' diet. In general, food is taken in through the mouth and goes through an esophagus that leads to either a crop, gizzard, or stomach, depending on the species. Crops and stomachs usually store the food; gizzards grind it. The mashed food flows through an intestine where nutrients are absorbed. Undigested waste leaves the body through the anus.

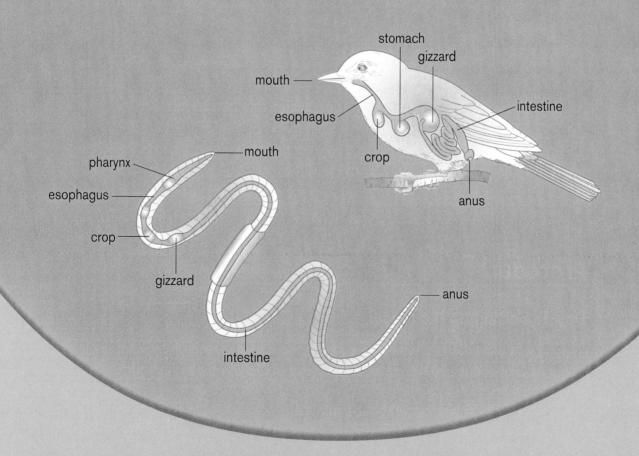

8-6

Crossing a Membrane

The food in the small intestine no longer looks like food. It is a soupy mix of molecules. These molecules must get into the cells that line the small intestine. Then they must cross more cells to enter the bloodstream. Molecules have to be pretty small to cross a membrane. How small is small enough?

Which nutrients are small enough to cross a cell membrane?

How can you show that molecules move when you cannot see them?

In this lesson, you will use a model to help you answer these questions. You will make a "cell" by putting water in a test tube and covering the opening with dialysis tubing. (As with cell membranes, only small molecules can cross dialysis tubing.) You will put the test tube upside down into a beaker of water that contains cornstarch and sugar. After a day or two, you will check to see if any cornstarch or sugar has gotten into your "cell."

Materials

For Day 1
- beaker
- rubber band
- test tube
- corn syrup solution
- cornstarch

- graduated cylinder, 10 mL
- stirring rod
- masking tape
- scissors
- 1 piece dialysis tubing, 6 cm

Procedure
(Work with a partner.)

Day 1: Setting up the Model

1. Soften the dialysis tubing so you can use it as a membrane.
   a. Put about 150 mL water and a piece of dialysis tubing in your beaker.
   b. Let the tubing soak about 1 minute.

c. Take it out and open it by rolling it between your thumb and finger.

d. Cut along the length of the tubing so you can open it flat.

2. Put your "cell" together.

   a. Get your test tube and rubber band.

   b. Use water from your beaker to add to your test tube—2 to 3 cm is plenty.

3. Cover the end of the test tube with the square of dialysis tubing. Attach it with a rubber band as shown in **Figure 8.11**.

4. Prepare the "nutrient soup" in the beaker.

   a. Use a graduated cylinder to measure 5 mL of the corn syrup solution, and then add it to the water in the beaker.

   b. Add about 0.5 teaspoon cornstarch to the beaker.

   c. Gently stir the solution with a stirring rod.

5. Put your cell upside down into the beaker. The level of liquid in the beaker should be slightly higher than the level in the cell. Add more water to the beaker if you need to. (See **Figure 8.12**.)

6. Use masking tape to put your names on your beaker. Put it where your teacher tells you. You will finish this experiment tomorrow.

7. Clean up your work area and answer analysis questions 1 through 5.

Figure 8.11

Figure 8.12

For Day 2
- model from Day 1
- two droppers
- two test tubes
- masking tape
- beaker
- iodine solution in dropper bottle

Sugar Test Station
- Benedict's solution in dropper bottle
- hot plate
- test tube holder
- safety goggles

Day 2: Looking for Evidence

Figure 8.13

1. Design a data table to record your results. You will need to read the rest of this procedure to find out what data to record. Have your teacher check your data table before you do the rest of this activity.

2. Prepare your materials.

   a. Label one test tube "inside cell" and the other "outside cell."

   b. Take your test tube out of the beaker and rinse the end with the dialysis tube under running water.

   c. Dry the outside of the tube, remove the dialysis tubing, and set your test tube back in the beaker as shown in **Figure 8.13**.

3. Get ready to test the liquids.

   a. Use one dropper to stir the solution in the beaker gently and to remove 15 drops. Put this liquid in the "outside cell" test tube. Put the dropper back in the beaker. (Do not mix up the droppers!)

   b. Use the other dropper to remove 15 drops of liquid from inside the cell. Put this "cell juice" into the test tube labeled "inside cell."

 Caution: You will work with hot liquids. Use the test tube holder and wear safety goggles.

4. Test both liquids for sugar at the sugar test station.

   a. **Put on your safety goggles.**

   b. Add ten drops of Benedict's solution to each test tube.

   c. Heat the test tubes in the water bath for two or three minutes.

   d. Record your results.

   e. Rinse out your test tubes.

5. Test both liquids for starch.

   a. Add 12 drops of liquid from inside the cell to one clean test tube and 12 drops from the beaker to the other clean test tube. Make sure the right liquid goes in the right test tube!

   b. Add ten drops of iodine solution to each test tube.

   c. Record your results.

6. Clean up your equipment and work space.

7. Answer the remaining analysis questions and write your conclusion.

Analysis Questions

1. What is the purpose of doing this activity?
2. What did you put:
   a. inside the cell?
   b. in the liquid outside the cell?
3. What results would you expect if you tested the *inside* of your cell on day 1:
   a. for sugar?
   b. for starch?
4. What results would you expect if you tested the *outside* of your cell on day 1:
   a. for sugar?
   b. for starch?
5. On day 2, you will test the liquids in the cell and in the beaker for both starch and sugar. What do you predict will be the results of these tests?
   a. inside the cell, test for starch
   b. inside the cell, test for sugar
   c. outside the cell, test for starch
   d. outside the cell, test for sugar
6. Look at your results.
   a. Where did you find sugar?
   b. Where did you find starch?
   c. How can you explain your results?
7. Compare your model cell with a cell in your intestine.
   a. List two ways they are alike.
    b. List three ways they are different.

Conclusion

Write a paragraph that answers the focusing questions.

Extension Activity

Design and Do: What other substances do you think might be able to cross the dialysis tubing? Plan a procedure to test your prediction. If your teacher approves, test your idea to see if you are correct.

Cell "Food"

Food in the small intestine does not look like food. It is a soupy mixture of sugars, amino acids, fats, vitamins, and minerals. These nutrients must get out of the small intestine and into the body.

Think about what happens to sugar molecules in the small intestine, for example. **Figure 8.14** shows four imaginary sugar molecules (green) in a section of the small intestine. (The green dots are much larger than an actual sugar molecule would be.) Molecule 1 is floating down the middle of the intestine. Molecule 2 is touching the villi that line the wall of the intestine. The third molecule has crossed a membrane and is inside a cell. The fourth molecule is flowing through a blood vessel and will soon move away from the intestine. (You cannot see the cells or small blood vessels in the illustration.)

Inside a Cell

Once a sugar molecule or any nutrient is in a cell, it may either be used by the cell or it may move on to another cell. If the cell "needs" the nutrient, enzymes go to work. They break it down and use it either for energy or to build new cell parts. For instance, it could become a part of a membrane, of the nucleus, or of any of the hundreds of other microscopic structures in a cell. (See **Figure 8.15**, arrow A.)

If the cell does not need the nutrient, the nutrient can continue its travels through the body. It will cross the cell and pass through the membrane into the next cell. In this way, nutrients eventually reach the bloodstream. Once they are in the bloodstream, they can be carried to all parts of the body. **Figure 8.15** shows these two possibilities.

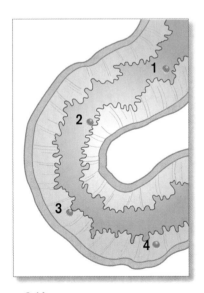

Figure 8.14

The green dots represent sugar molecules in the small intestine. Which ones look as if they have been absorbed?

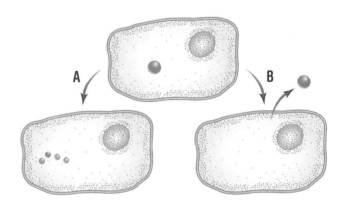

Figure 8.15

Once inside a cell, the sugar molecule may either be used up (arrow **A**) or it may move on to another cell (arrow **B**).

Diffusion

Why does a nutrient move into a cell? Because it is usually moving from a place of high concentration to a place of lower concentration. That is, it moves from where there are more of them to where there are fewer. **Figure 8.16** illustrates what is meant by the term **concentration**. It shows two beakers of water. One teaspoon of sugar is being added to beaker *A*; 3 teaspoons are being added to beaker *B*. The sugar water in beaker *B* will be more concentrated than the sugar water in beaker *A*.

Molecules move from a place of high concentration to a place of low concentration as the result of a process called **diffusion**. Molecules can diffuse into a cell or out of a cell. Diffusion stops when there are equal concentrations of molecules on both sides of the membrane. This is illustrated in **Figure 8.17**.

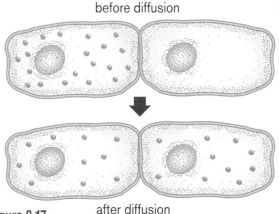

before diffusion

after diffusion

Figure 8.17

Diffusion results in equal concentrations of molecules on both sides of a membrane. Besides sugar, what kinds of molecules could the green dots in this drawing represent?

Membranes

Molecules can diffuse across membranes because membranes are not solid; they contain extremely tiny openings. Small molecules can get through these openings, but large ones cannot. That's why sugar molecules can pass through a membrane but starch molecules cannot. (See **Figure 8.18**.)

Diffusion is an extremely important process. Many substances move in and out of cells by diffusion. You will learn more about diffusion in future lessons.

Figure 8.16

Both beakers contain 100 mL of water. Which beaker will contain the most concentrated sugar water?

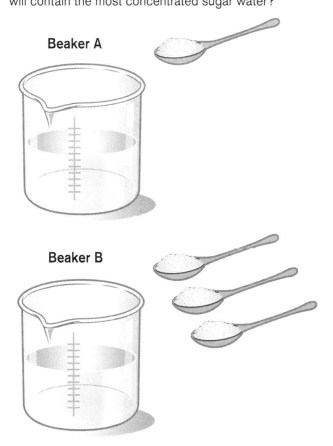

Beaker A

Beaker B

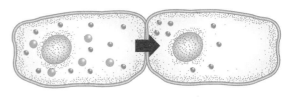

● = sugar ◉ = starch

Figure 8.18

Cell membranes are not solid. They have very small openings that permit small molecules to enter and leave a cell.

Includes sci**LINKS**
NSTA

Topic: Diffusion
Go to: www.scilinks.org
Code: MSLS3e205

Analysis Questions

1. What are two ways that a cell might use a nutrient?

2. Define the word diffusion.

3. Choose one of these models and use it to explain diffusion. You can either write your explanation or illustrate it.

    a. a bottle of perfume
    b. helium in a typical rubber balloon
    c. tea bag in water

4. **Figure 8.19** shows three cells.

    a. Which cell has the highest concentration of sugar molecules?
    b. Which cell has the lowest concentration?
    c. Explain which way the sugar molecules will diffuse.

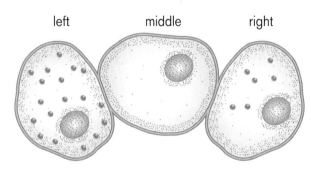

left   middle   right

Figure 8.19

5. Two students did another experiment with dialysis tubing. (See **Figure 8.20** for the set-up.) They tested the solutions on the second day. Predict the results of these four tests.

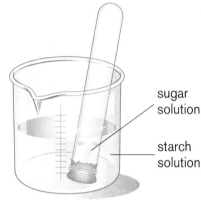

sugar solution

starch solution

Figure 8.20

    a. Sugar Test on the inside solution
    b. Sugar Test on the outside solution
    c. Starch Test on the inside solution
    d. Starch Test on the outside solution

6. Why does food have to be digested before it can be used by the body?

7. When you eat, who are you feeding, you or your cells? Explain.

Extension Activity

Going Further: Set up the model cell with another set of conditions. Choose one of the conditions in **Figure 8.21** to try. Predict what you think will happen. Explain your results.

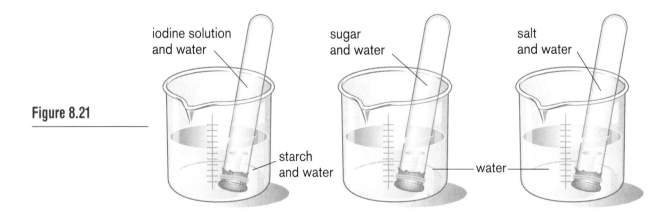

Figure 8.21

iodine solution and water

starch and water

sugar and water

water

salt and water

8-8

The Final Step: The Large Intestine

The "food" that leaves the small intestine has been digested and the nutrients absorbed. Only the "leftovers" remain. These leftovers will travel through the large intestine and, eventually, leave the body.

Structure

The large intestine is much shorter than the small intestine. It is only about 1.5 meters long (4 or 5 feet). It travels up the right side of the body, crosses the abdomen just below the stomach, and makes an abrupt turn down the left side. **Figure 8.22** shows the location of the large intestine.

Unlike the small intestine, the large intestine does not have villi; its lining is smooth. Like the other parts of the digestive system, it is coated with mucus and its walls contain smooth muscle fibers. When these fibers contract, they move the undigested food along. Sometimes

you can feel a "wave" of movement when this happens. It takes about 12 to 24 hours for food to travel through this organ.

The last part of the intestine is called the **rectum**. This is where wastes are stored until they leave the body through the anus. There are two sphincter muscles in the **anus**. One is made of smooth muscle fibers and the other of skeletal muscle. This makes it possible to control the opening and closing of the anus.

The **appendix** is a 9-cm (3.5-inch) pouch near the beginning of the large intestine. It is not an important part of the digestive system. However, it may swell and fill with pus, causing pain in the lower right abdomen. The pain, along with fever and nausea, are symptoms of **appendicitis**. If the swollen appendix is not removed, it will burst, releasing bacteria into the abdomen and causing serious infection. If you ever think you have appendicitis, see a doctor immediately!

Figure 8.22

The large intestine is the end of the digestive system.

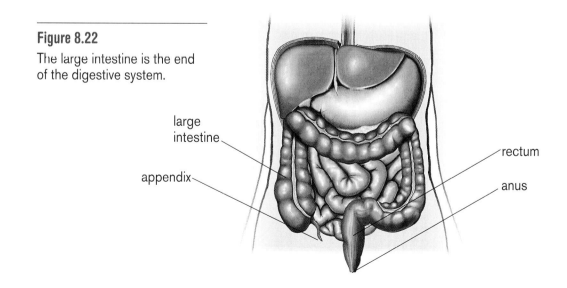

large intestine

appendix

rectum

anus

Functions

Two important functions of the large intestine are absorption of water and storage of "left-overs." The water comes from the liquids we drink, the foods we eat, and digestive juices (saliva, stomach juice, and pancreatic juice). The large intestine reclaims the water so it is not lost along with the waste materials.

The large intestine stores the leftovers, allowing us to choose the time when we get rid of the waste. The large intestine does not produce enzymes. No digestion takes place here. These two functions are illustrated in **Figure 8.23**.

Feces

The scientific term for the "leftovers" is **feces**. If you analyzed the content of feces, you would find that it is 75% water. The other 25% is bacteria, fiber, undigested food, mucus, and dead cells. Feces are brown because of the bile from the liver that was mixed into the food.

The odor of feces is due to the bacteria. Hundreds of kinds of bacteria live in the intestine and feed on the leftovers, producing gas. Some foods, such as beans and cabbage, contain molecules we cannot digest. These foods leave lots of leftovers for bacteria so lots of gas is produced. In large amounts, these

Figure 8.23

As the wastes travel through the large intestine, water is absorbed.

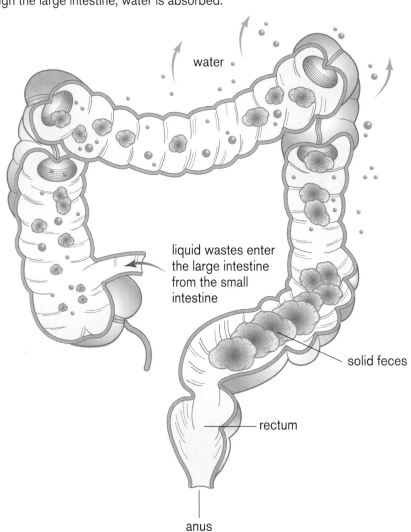

water

liquid wastes enter the large intestine from the small intestine

solid feces

rectum

anus

gases feel uncomfortable and may be mistaken for a stomachache.

Bowel movements remove the feces from the large intestine. Many people think that a "once-a-day" bowel movement is essential. Actually, there is no rule for the right number of bowel movements. Some people are comfortable with a bowel movement once every two or three days.

Think for a minute about what the digestive system does. Starting with foods of all types, it removes the nutrients and eliminates the waste. Without this system, your cells (and therefore you) would starve.

Analysis Questions

1. What are the two main functions of the large intestine?

2. How would your life be different if you did not have a large intestine?

3. Compare and contrast the large intestine and the small intestine. List as many ways as you can that they are alike and different.

4. Why do we say that the digestive system is a system?

5. HB Connections:

    a. Why do other body systems need the digestive system?

    b. Why does the digestive system need other body systems?

8-9

Pains, Pangs, and Problems

Systems don't always work smoothly. Any time one part of a system doesn't work properly, it can affect the functioning of the system. Think about an automatic sprinkler system, for example. If a sprinkler head breaks, the water may shoot up like a fountain, flooding some areas while other areas remain dry. If there's a sensor in the system and it doesn't work, the sprinklers may go off in the middle of a rainstorm. What would happen if there's a break in the underground pipeline? If the water pump breaks? If there's an electrical outage? As you can see, depending on the part of the system that malfunctions, it can either affect the functioning of the entire system or one part of the system.

When something interferes with the functioning of a body system, health problems result. Whenever there's a problem, it's important to find its cause. If you are aware of what causes digestive problems, you may be able to prevent them. Here is some information about a few problems that can affect the stomach and intestines.

Stomach

We all swallow some air when we eat. If a person swallows too much air, gas builds up in the stomach. Lots of air can cause stomach pain. One way to get rid of this extra gas is by *burping*. Even though burping is a "natural action," it is not considered polite in our culture.

About every 20 seconds, the smooth muscles in the walls of the stomach contract, churning the contents of the stomach. When there's air in the stomach, the churning action can cause the stomach to rumble loud enough for someone sitting next to you to hear. When an empty stomach churns, *hunger pangs* are felt.

If you eat too much, the muscle fibers in the walls of the stomach are overstretched, causing pain. Overeating, therefore, is one cause of *stomachaches*. *Ulcers* are another cause of stomachaches. The acid and churning is hard on the lining of the stomach. New cells are constantly being produced to replace the damaged ones. If the lining is not repaired promptly, a sore spot called an ulcer develops.

When the stomach muscles contract sharply, they may force the food up, causing a person to *vomit*. This reaction can be triggered by drugs, viruses, bacteria, or even some motions. The burning sensation in the mouth and throat is from stomach acid.

Intestines

If feces move through the large intestine too quickly, there is not enough time for the water to be absorbed, and the feces will be too liquid, resulting in *diarrhea*. If a person, especially a young child, has diarrhea for too long, he or she will become dehydrated. Diarrhea can result from many things, including an infection, being nervous, or eating spoiled food.

If feces movement is too slow, too much water is taken up and *constipation* is the uncomfortable result. A high fiber diet can help prevent constipation. Fiber retains water to help keep the feces moving.

These are some of the most common problems that affect the digestive system. There are many others. Whenever you hear about a problem, ask yourself, "What causes this problem? How does it affect the system? Could it be prevented?"

Summary Chart

Make a chart summarizing the information just presented. List the problems that can affect the digestive system, the causes of the problems, and how the problems can be prevented. Indicate whether or not the problem is contagious.

Extension Activity

Health Application: Look up other digestive system problems to include on your chart. Examples of problems include tooth decay (cavities), gum disease, salmonellosis, cholera, indigestion, hemorrhoids, and gallstones. You may know of others you want to research.

Different Structures, Different Functions

Stomach, small intestine, and large intestine—these three body organs are important parts of the digestive system. Each organ has a different role in the digestive process. Each one also has its own specialized structure.

What clues are helpful for identifying sections of the small intestine, large intestine, and stomach?

You will observe sections of the small and large intestines with a microscope, along with three specimens. The specimens are from the stomach, small intestine, and large intestine, but you will not know which is which. They are labeled A, B, and C. Use what you have learned about the digestive system to identify the specimens. In the process, you will be answering the focusing question.

Materials

- specimens A, B, and C
- magnifying lens
- dissecting tray
- metric ruler
- soap
- Handout, *Different Structures, Different Functions*

- paper towels
- microscope

Prepared microscope slides
- small intestine
- large intestine

Procedure   *(Work in a group with two other students.)*

1. As a group, fill in the top part of the handout. If you don't agree on an answer, review the readings to check your facts.
2. Share responsibilities. Decide who will do what.

   recorder: reads directions and fills in the handout while group works (R)
   handler: carefully handles specimens and holds them for others to see (H)
   trotter: gets materials and returns them neatly to where they belong (T)

3. *Trotter:* Get a dissecting tray, a metric ruler, a magnifying lens, and *one* specimen. It does not matter which specimen you inspect first.

4. *Recorder:* Read the directions on the handout and fill in the table while the group works.

5. *Trotter:* When the group finishes with a specimen, return it and get a different one.

6. When you have looked at all three specimens, clean up your work space. The handler should wash any dirty equipment when washing his or her hands.

7. Two microscope slides have been set up for you to view. One shows the lining of the small intestine; the other shows the lining of the large intestine. For each microscope slide, do the following:

   a. Sketch what you see.

   b. Record the magnification.

   c. Use arrows to point to examples of cells.

   d. Label the inside and outside surfaces.

   e. Title your sketches.

8. Answer the analysis questions and write your conclusion.

Analysis Questions _ _ _ _ _ _ _ _ _

1. What is specimen A? (Provide evidence for your decision.)

2. What is specimen B? (Provide evidence for your decision.)

3. What is specimen C? (Provide evidence for your decision.)

4. Describe the main difference between the small and large intestines when viewed through the microscope.

5. What do you think you will remember the most about doing this activity?

Conclusion

Write a paragraph that answers the focusing question.

Extension Activity

Going Further: The tongue is another part of the digestive tract that you can inspect. Buy a beef tongue—whole, not sliced—at a food store. Decide how it fits into the cow's mouth. Make observations of the outer surface and the muscle tissue inside the tongue. Think about function while you look at the tongue's structure.

Food Tube

You started this chapter by making a sketch that illustrated what you knew about the digestive system. Then you studied the major organs that make up the digestive system and learned about the functions of each of the organs. You studied enzyme action and diffusion, and you read about a famous experiment that helped us understand more about the stomach. Now you are ready to reconsider the questions you answered at the beginning of the chapter.

Why is the human digestive system made up of so many different parts?

Which parts of the digestive system do you think a person could live without? Explain.

You will make an illustration of the digestive system again. It does not have to be realistic. Instead, make a "food tube." Imagine the digestive system as it would look if it were straightened out to make one long tube. Think about the answers to the focusing questions while you work on your food tube.

- Read the procedure and decide what materials you need.

Procedure

1. Make a poster of the food tube. It should include the following things:
   a. All the structures of the tube drawn in the correct order and to scale. (See the chart on the next page.)
   b. The functions of each structure.
   c. Arrows showing substances that move into the "tube" from "helper" organs. (And show the locations of these other organs.)
   d. Arrows showing substances that leave the "tube" to enter the bloodstream.

2. Make sure your food tube is neat so someone else can understand it. Color it for clarity.

The food tube in an adult averages about 890 cm long. But instead of making your food tube 890 cm long, make it 89 cm long. Here are the measurements you will need to complete this task.	
25 cm	esophagus
150 cm	large intestine
15 cm	mouth (lips to esophagus)
680 cm	small intestine
20 cm	stomach

3. Your food tube should summarize all the important information about the digestive system. Check it to make sure it is complete and accurate.

Conclusion

Write a paragraph in answer to each of the focusing questions.

> Although the focus of this chapter has been on the digestive system, the muscular system and circulatory systems were also mentioned.
> No system acts alone. In chapter 10, you will learn about the role of the circulatory system.

Tract Records

It takes one to three days for food to travel the length of the human digestive system. Birds process their foods much more rapidly. A shrike can digest a mouse in three hours. Berries pass through the digestive tract of a thrush in just 30 minutes. Why do you think it is important for a bird to be able to digest its food quickly?

3 days

3 hours

Chapter 9

Respiratory System

What happens to the air inside your lungs?

Using X-Rays as Evidence

9-1

Take a deep breath. Hold it. Let it out. When you inhale, your lungs fill with air. When you exhale, you empty your lungs. Have you ever wondered what your lungs look like? What about the other parts of the respiratory system? What do they look like? You cannot inspect your respiratory system, but there are other ways to learn about its structure. For example, you can look at x-ray films of the chest. Based on what you observe, you can infer something about the structure of the respiratory system.

What can you learn about the respiratory system by looking at chest x-rays?

There are several parts to this activity. First, consider what you already know about the structure of the respiratory system by sketching the respiratory system and labeling the structures. Then inspect a chest x-ray and see what evidence it provides about lungs and other parts of the respiratory system. Finally, consider how a health problem such as pneumonia can help you understand how the body works. Keep the focusing question in mind while you work.

Materials

- one sheet of unlined paper
- microscope
- red colored pencil
- microscope slide: *Streptococcus pneumoniae*
- Reading, *Pneumonia: A Lung Disease*, page 220

X-Rays
- normal lungs
- lungs with pneumonia

Procedure

Part A: Structure of the Respiratory System

1. Talk about the respiratory system with your partner. Tell each other where you have learned about the respiratory system. Did you study it in school? If so, what activities did you do to help you understand how the respiratory system works?
   a. Along the left side of your sheet of unlined paper, list all the parts of the respiratory system that you and your partner can recall.
   b. Sketch the respiratory system and label the parts.
   c. In two or three sentences, explain how the respiratory system works. (Write your explanation on the bottom or on the back of the page with your sketch.)
   d. Have your teacher check your work before you continue.

2. Get an x-ray film of normal lungs. Only touch the edges of an x-ray—the oils and dirt on your fingers will damage the x-ray.
   a. Observe it by holding it up to the light. Imagine that you are using x-ray vision to look through a person's back.
   b. Answer analysis questions 1 and 2.

3. Look at your sketch of the respiratory system. Are there any changes you could make so it would be more accurate? If so, make the corrections with the red colored pencil.

Part B: A Respiratory System Problem

1. In the United States, one or two people out of 100 develop pneumonia each year. Read about this common lung problem in the reading: *Pneumonia: A Lung Disease.*
2. Answer analysis questions 3 through 5.
3. Observe the chest x-ray of someone who has pneumonia while you answer analysis question 6.
4. Look at the microscope slide of *Streptococcus pneumoniae*. This kind of bacteria is often the cause of pneumonia.
   a. Remember, bacteria are very small. Be patient as you search for them on the microscope slide.
   b. Answer analysis questions 7 and 8.
5. Answer analysis question 9 and write your conclusion.

Analysis Questions _ _ _ _ _ _ _ _ _

1. Look for a pattern to the colors of the tissues as they appear on an x-ray.
   a. What tissues look white?
   b. What tissues look black?
2. Look at the x-ray of normal lungs.
   a. Sketch the shape of a lung.
   b. Is a lung hollow? Use evidence from the x-ray to explain your answer.
   c. How do you think the size of these lungs compares with the size of your own?
   d. Name at least four body structures, other than lungs, that you can see in this x-ray.
3. What happens to the lungs when a person has pneumonia?
4. What can cause pneumonia?
5. Should antibiotics be used to treat pneumonia? Explain.
6. Look at the x-ray of the lungs with pneumonia.
   a. What is the main difference between the lungs in this x-ray and the normal lungs?
   b. Do you think this person is a baby, child, teen, or adult? Explain.
7. Sketch what these bacteria look like through the microscope. Remember to record the magnification.
8. How do you think *Streptococcus pneumoniae* gets into a person's lungs?
9. What other things could you do to learn about your own lungs?

Conclusion

Think about what you observed on the x-rays, and then write an answer to the focusing question.

Extension Activity

Careers: Inhalation therapists work with people who have pneumonia and are admitted to a hospital. Find out what inhalation therapists do. Maybe you can visit one and see the types of equipment used to treat respiratory problems.

Pneumonia: A Lung Disease

Infections can develop anywhere in the body, including in the lungs. If an infection develops on the skin or in the mouth, it is easy to diagnose because it can be observed. Infections inside the body can be harder to detect. One of the most common lung infections is pneumonia. You may even have heard a conversation that sounded similar to this one:

"You missed more than a week of school. You must have been really sick."

"I was; I had pneumonia. I felt awful."

"Didn't the doctor give you antibiotics? I had pneumonia last year. As soon as I started on antibiotics, I felt much better."

"The doctor said antibiotics wouldn't do any good because pneumonia is caused by a virus."

"What's that got to do with anything? Antibiotics sure helped me."

Who's right? Is pneumonia caused by a virus? Will antibiotics help someone who has pneumonia?

Symptoms

A cough is a common symptom of pneumonia. In addition, it may hurt to breathe, and a fever with chills is likely to develop. Sometimes the symptoms are so mild that the person may think he or she has a cold. At other times, the symptoms can be so severe that death results. The exact symptoms depend on how much of the lung is affected and on the cause of the disease.

Causes

Pneumonia is not one single disease. If someone has pneumonia, that simply means that he or she has inflamed lung tissue. It can be caused by bacteria, or by viruses, or from inhaling poisonous gas. Most of the time pneumonia is caused by bacteria.

Imagine that you have just inhaled a big breath of air and, along with the gases, many bacteria landed in your lungs. The lungs are warm and moist, making them an ideal environment for bacteria. Luckily, white blood cells usually destroy the dangerous microbes before they "settle in." However, when people have been sick or if their bodies are under stress, their immune systems do not work as well as they should. This is when the microbes can take over.

One bacterium soon divides into two, and the two divide again and again until there's a whole colony of bacteria living in the lung. (See **Figure 9.1.**) As the numbers of bacteria grow, they produce more and more waste products. These wastes irritate the lung cells. As a result, the lung produces fluids to try to rinse away the wastes. These fluids, however, are also good for bacterial growth so the infection spreads.

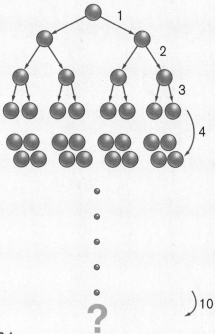

Figure 9.1

If you start with one bacterium, how many bacteria would you have after ten rounds of cell division?

Diagnosis and Treatment

When pneumonia is suspected, the doctor listens for fluid in the lungs with a stethoscope. If there is fluid, chest x-rays can show which parts of the lung are affected. "Double" pneumonia means that there is fluid in both lungs.

Next, it is important to decide what is causing the pneumonia. Laboratory tests on mucus or blood can determine if the infection is bacterial or viral. If the irritation is due to bacteria, antibiotics may be prescribed. Viral pneumonia is more difficult to treat because viruses are not affected by antibiotics.

No matter which type of pneumonia a person has, it is important to see a doctor. If the condition is not treated, lung damage may result. Before antibiotics were available, pneumonia was a major cause of death. It is still a serious illness for the elderly and for people who have other illnesses.

Breathing Parts

The human respiratory system is designed to take in oxygen and get rid of carbon dioxide. As with other systems, the respiratory system is made up of several parts and each part has a specific "job." When all the parts are working, the exchange of gases goes smoothly.

Getting In:
The Nose and Mouth

The **nose** is the main entrance to the respiratory system. Air passes through this structure on the way to the lungs. Air can be so dry, dirty, and hot (or cold) that it can damage the lungs. It is the job of the nose to filter, moisten, and warm or cool the air before it reaches the lungs.

Short hairs at the front of the nose act as a filter, trapping pollen, dust, and other small particles that may be in the air. A *mucous membrane* lines the nose. It moistens the air to prevent the fragile lung tissue from drying out. Small particles may also be trapped by this moist surface. The mucous membrane contains many capillaries. As blood circulates through these microscopic vessels, it warms the air.

Think about how "scratchy" your throat feels when your nose is plugged and you are forced to breathe through your mouth. The mouth can serve as an emergency passage for air but it is not designed for this purpose.

Moving On:
The Trachea and Bronchi

After air passes through the nose or mouth, it enters the **larynx**. Sometimes the larynx is called the voice box or Adam's apple. It is the firm structure at the top of your throat. (Look at **Figure 9.3** and locate the larynx.) Air then flows into the **trachea**. The trachea, or windpipe, is in

Figure 9.2

How would you describe the air you breathe?

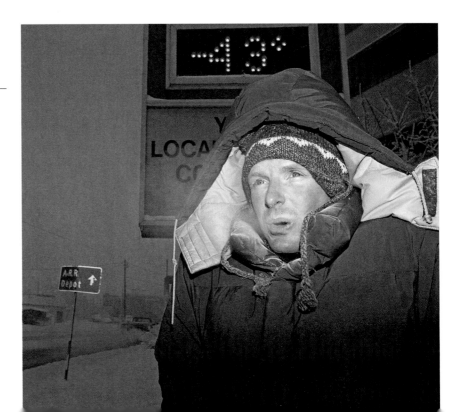

front of the esophagus. Look at **Figure 9.3**; you can see that the openings to these two tubes are close together.

It is important that air goes down the trachea and food goes down the esophagus. In order to prevent a mix-up, we have a built-in safety device, the **epiglottis**. It is a flap of tissue that folds down to cover the opening to the trachea when you swallow. It lifts when you breathe. Look at the position of the epiglottis during swallowing and breathing as shown in **Figure 9.3**.

If food gets into the trachea by mistake, it can block the passage of air and a person can suffocate. That's why food in the trachea causes a reflex reaction—choking. This forces the food back up and out of the trachea.

You can feel the trachea where it continues down along the front of the throat. The larynx and trachea feel firm because they are made of cartilage. (See **Figure 9.3**.) Most other structures are made of smooth muscle and connective tissue.

Once in your chest, the trachea branches to form two **bronchi**. One bronchus goes into each lung. The bronchi divide again and again into smaller and smaller bronchial tubes.

The trachea and bronchi are lined with cilia. **Cilia** are fine hair-like structures on the outside of some cells. They help keep the lungs clean by waving back and forth, moving mucus, along with dust and pollutants, up the bronchi and out of the respiratory system. (See **Figure 9.4**.) When the mucus reaches the top of the larynx, it goes down the esophagus. Each day, we swallow about one liter of this mucus.

Figure 9.3

Notice that the trachea is in front of the esophagus. Look at the position of the epiglottis in each of these drawings.

Swallowing

Breathing

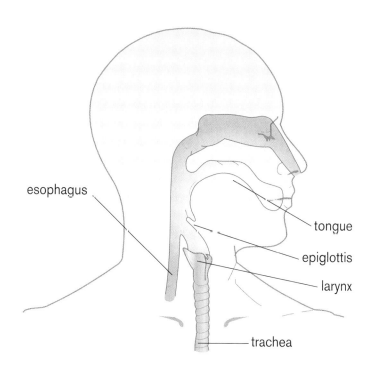

esophagus

tongue

epiglottis

larynx

trachea

food

food particle

air

Sometimes microorganisms such as a cold virus may get into the bronchi. When this happens, an infection may develop, causing a condition called *bronchitis*. (Bronchitis means inflammation of the bronchi.)

The Destination: Lungs

At the ends of the smallest tubes are microscopic **air sacs**. In some ways, they are like tiny balloons. Each lung is made up of millions of these little air sacs and tubules (very tiny tubes that are sometimes called bronchioles). (See **Figure 9.5**.) All the air in lung tissue makes it so light that a piece of lung tissue will float in water.

Because the lungs are made up of air sacs, there is more surface area for the exchange of gases. If it were possible to open all of the air sacs in a pair of lungs and spread them out, they would cover a space at least 60 square meters in size. That's a lot of space! (How many square meters of floor space are in your classroom?)

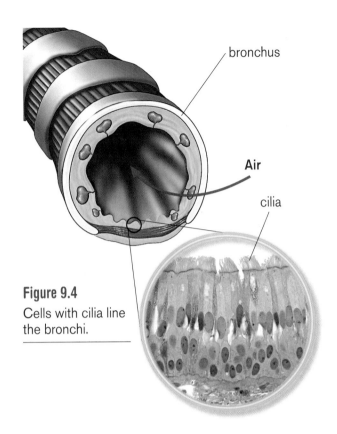

Figure 9.4
Cells with cilia line the bronchi.

bronchus

Air

cilia

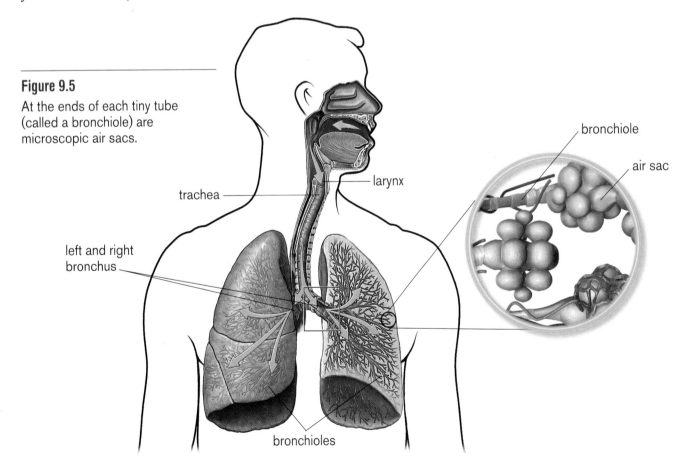

Figure 9.5
At the ends of each tiny tube (called a bronchiole) are microscopic air sacs.

bronchiole

air sac

larynx

trachea

left and right bronchus

bronchioles

Moving the Air: Ribs and Diaphragm

Lung tissue does not contain muscle, so lungs cannot move by themselves. You need to use your diaphragm and ribs to expand and contract your lungs. These actions are illustrated in **Figure 9.6**.

The **diaphragm** is a sheet of muscle just below the lungs. When it contracts, it flattens and air enters the lungs. When it relaxes, it pushes up on the lungs and forces air out. If your diaphragm contracts more vigorously, you get the hiccups. When you hiccup, you inhale in short gasps that sound very different from normal breaths.

Ribs form the rib cage. Muscles extend from rib to rib. When these muscles contract and relax, the rib cage moves in and out. This action also helps move the air through the lungs.

One Final Word

Two bands of tissue cross the inside of the larynx. These bands are the vocal cords. When air passes between the vocal cords, you can make sounds. A strong blast of air will produce a shout, but very little air is needed to whisper. You can turn the sound of moving air into words by using your jaw, lips, and tongue. (Try to say "How are you?" without moving your jaw, lips, or tongue. You will find that you need more than just your vocal cords in order to talk.)

You have just read about the parts of the **respiratory system**. When all the parts are working, pumping air in and out of the lungs, we can breathe. But that still doesn't explain how the respiratory system works. In the next lesson, you will learn what happens to the air inside the lungs.

Figure 9.6

How would you describe the position of the diaphragm when a person inhales? Exhales?

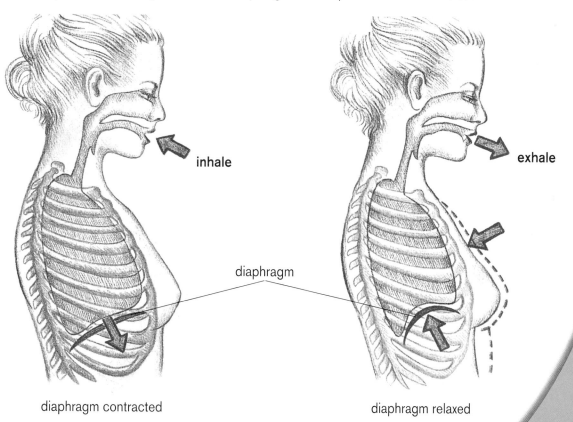

inhale

exhale

diaphragm

diaphragm contracted

diaphragm relaxed

Analysis Questions

1. Make a chart entitled, "Parts of the Respiratory System and Their Functions."

   a. In the first column, list the parts of the respiratory system.

   b. In the second column, describe the function of each of the parts.

   c. In the third column, describe something about the structure of each part that makes it possible for it to perform its function.

2. Think about the trachea and the esophagus.

   a. List at least two ways they are alike.

   b. List at least three ways they are different.

3. Why is it better for a lung to be made of millions of air sacs instead of being one big sac like a balloon?

4. Make a list of problems which can affect the respiratory system that are mentioned in the reading. Circle the problems that are contagious.

5. HB Connections:

   a. What does the respiratory system do for other body systems?

   b. Why does the respiratory system need other body systems? (Use at least three examples of other systems to explain your answer.)

6. When you talk, is air entering or leaving your lungs? Explain.

Extension Activities

1. **Going Further:** Call a meat packing plant (slaughterhouse) and ask for an animal's lung or for a piece of one. Identify as many structures as you can. Start with the largest tube and follow it down. How many times does it branch? What size is the smallest tube you can find? Look at a piece of the tissue under a dissecting microscope.

2. **Health Application:** Look up health problems that can affect the respiratory system. Make a chart that summarizes the causes of the problems, how the problems can be prevented, and whether or not each problem is contagious. Examples of problems include emphysema, asthma, laryngitis, and nosebleeds. You may know of other problems you want to research.

Try to breath without moving your ribs.

Gas Exchange

A person inhales about 2,000 liters of air each day—that's a lot of air! How much air do you inhale during one class period? How much air do you think your classroom holds? What is air anyway? What happens to air when we inhale?

The Air We Inhale

Some people think air is all oxygen. Actually, air contains a variety of gases as well as many small particles. Oxygen isn't even the most abundant gas in air. Air contains more nitrogen than anything else. Air is about 79% nitrogen, 20% oxygen, and 1% other gases. Water vapor is also in air. The amount of water varies from place to place and from day to day. For example, there is more moisture in the air near an ocean than in the air over a desert. The jar labeled "Inhaled Air" in **Figure 9.7** illustrates the normal content of air. Notice that carbon dioxide is written as CO_2.

That's because carbon dioxide is made up of one carbon and two oxygen atoms. The "C" stands for carbon. The number 2 means there are two oxygen atoms, so oxygen is written as O_2. That's because two oxygen atoms are bonded together when they are floating in air.

When the normal balance of gases changes, it may be because of pollution. These gaseous pollutants come from industries, automobile emissions, and chemicals we use at home and at work. Dust, tiny particles of sand, and salt are examples of solid pollutants. Some pollutants are dangerous; they can harm plants and cause health problems for humans and other animals. When we breathe, we inhale it all—the normal gases along with the pollutants.

The Air We Exhale

The air you breathe out is not the same as the air you breathe in. Breathe on your hand for three or four breaths. The air feels warm, doesn't it? It also contains more moisture and a

Figure 9.7

How does the gas content of the air you inhale differ from the air you exhale?

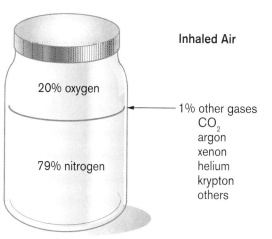

Inhaled Air

20% oxygen

1% other gases
CO_2
argon
xenon
helium
krypton
others

79% nitrogen

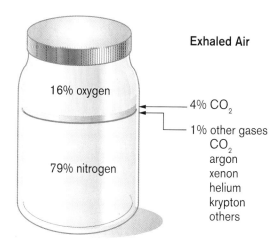

Exhaled Air

16% oxygen

4% CO_2

1% other gases
CO_2
argon
xenon
helium
krypton
others

79% nitrogen

different concentration of gases than the air you inhaled. Exhaled air is still 79% nitrogen and 1% other gases; but it contains more carbon dioxide (4%) and less oxygen (16% instead of 20%). **Figure 9.7** summarizes information about the gases in the air we inhale and exhale.

Something happens to the air while it is inside your lungs; gases are exchanged. Your body keeps some of the oxygen and gets rid of carbon dioxide. Do you know how this exchange takes place?

In the Lungs: A Gas Exchange

When air enters the lungs, it goes through the bronchial tubes and into the air sacs. Each air sac is surrounded by a network of blood vessels called capillaries. (See **Figure 9.8**.) As blood flows through these capillaries, it picks up oxygen from the air sacs.

At the same time, carbon dioxide diffuses out of the blood and into the air sacs. The exchange has taken place! This process is summarized in **Figure 9.9**. Study the drawings; make sure you understand the connection between the circulatory and respiratory systems.

Stop and ask yourself—why? Why does this gas exchange take place? Why does the blood that enters the lungs have extra carbon dioxide (CO_2)? Where does it come from? And what happens to the oxygen you inhale?

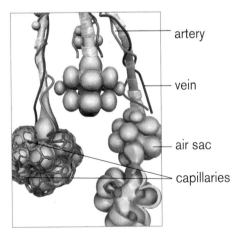

Figure 9.8
The air sacs are enclosed by a network of capillaries.

artery

vein

air sac

capillaries

The Journey of an Oxygen Molecule

It isn't easy to understand what happens to oxygen once it gets into the body. So, to help you picture what is going on, imagine that you are an oxygen molecule ... you have just been inhaled—whoosh! You zoom down the trachea and into a bronchial tube. Not even a second has passed, but you are already moving through a series of smaller and smaller bronchial tubes. You can't see where you're going because it's dark in a lung. Whop! You hit a dead end; that's because you're in an air sac. You feel yourself being drawn to the wall of the air sac. Whoops! You just crossed a membrane, and another one. You realize that you went through the wall of the air sac. Well, that shouldn't be surprising; it's not very thick. But there's no time to think about what's happening because you've just been swept up by a red blood cell. (Look at **Figure 9.10**. It's like a road map of this trip. Can you find where you are now?)

Now, you cruise along in the bloodstream. You make a quick trip through the heart and move out through the aorta. Everything is happening so fast! You feel like you're in a river, flowing with the current. You zip through

Figure 9.9
The arrows in these drawings show the directions that gas molecules move in air sacs and capillaries. How many differences can you find between the two drawings?

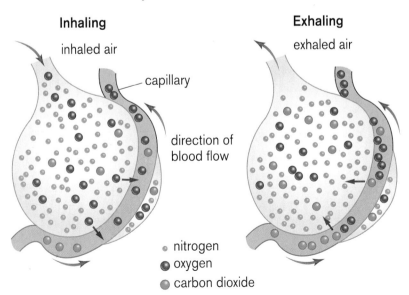

Inhaling

inhaled air

capillary

direction of blood flow

Exhaling

exhaled air

. nitrogen
● oxygen
● carbon dioxide

a series of smaller and smaller arteries when suddenly the cell you're in bends and twists. You realize that you're in a very small vessel—why, it's a capillary! Again, you feel yourself being drawn away. The hemoglobin molecule lets go of you and ... whoosh! ... you diffuse out of the blood cell, through the capillary wall, and into a body cell. (Check **Figure 9.10** again. Can you find where you are now?)

This is interesting. You've never been inside a cell before, but there's no time to look around. An enzyme grabs on to you. Here comes a sugar molecule. The enzyme grabs the sugar molecule too, and it comes apart. Wham! Energy is released. A chemical reaction just took place. You were part of the reaction, so you are no longer an oxygen molecule. You have become either a part of a carbon dioxide molecule (CO_2) or part of a water molecule (H_2O). That's the end of your journey as an oxygen molecule.

Inside the Cells

Well? Do you understand why you have to breathe? You breathe to take in oxygen, and your cells use the oxygen to produce energy. Of course, oxygen alone isn't enough. Cells also need sugar. The chemical reaction that takes place inside a cell when energy is produced can be summarized like this:

sugar + **oxygen** ⟼ **carbon dioxide** + **water** + **energy**
($C_6H_{12}O_6$)   **(O_2)**   **(CO_2)**   **(H_2O)**

This reaction is extremely important. Cells die when they don't have energy and in order to produce energy, they need sugar and oxygen. If too many of your cells die, you will die, too. Combining sugar and oxygen to produce energy is called **cellular respiration**. This reaction is illustrated in **Figure 9.11**.

Oxygen is used in this reaction and carbon dioxide is produced. The carbon dioxide is a waste product. It has to be carried away from the cells and removed from the body.

Includes *sci*LINKS®
NSTA

Topic: Cellular Respiration
Go to: www.scilinks.org
Code: MSLS3e229

Figure 9.10

An oxygen molecule on its journey to a body cell.

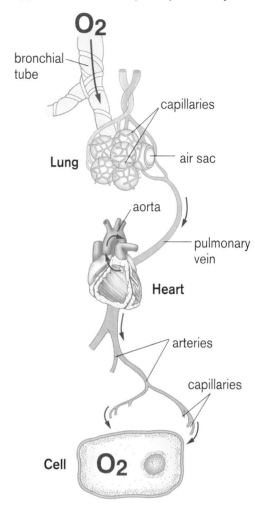

O₂

bronchial tube

capillaries

Lung

air sac

aorta

pulmonary vein

Heart

arteries

capillaries

Cell   O₂

Figure 9.11

With the help of enzymes, cells use sugar and oxygen to produce energy.

ENERGY

Sugar

+ O₂

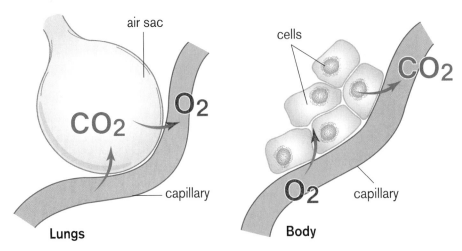

Figure 9.12
Gases are exchanged in the lungs and in body cells. Remember, CO_2 stands for carbon dioxide and O_2 stands for oxygen.

Gas Exchange—In Summary

By now you should realize that gases are exchanged in two places: the lungs and the cells. In both of these places, gases diffuse across membranes. If it were not for these gas exchanges, our cells would not get the oxygen they need, and we would not be able to get rid of carbon dioxide. (See **Figure 9.12**.)

Analysis Questions

1. What are the main differences between exhaled and inhaled air?
2. Explain what is meant by the term cellular respiration.
3. Think about the carbon dioxide you exhale with every breath.
   a. Where does it come from?
   b. Sort the following structures into two lists, those that produce carbon dioxide and those that do not.

      fiber molecule
      bone cell
      smooth muscle cell
      plasma
      white blood cell
      lung cell
      cell in the stomach lining
      saliva

4. Why does the body need oxygen?
5. A *carbon dioxide* molecule is made up of *carbon* and *oxygen*. What makes up each of these substances? (You may need to refer back to Chapter 7.)

   a. sugar   e. protein
   b. fat   f. cholesterol
   c. water   g. fiber
   d. starch   h. amino acid

Homework: It is your turn.

Imagine that you are a carbon dioxide molecule. You are in a muscle cell in someone's gastrocnemius. Describe your journey from the cell to the outside air.

Extension Activity

Animal Kingdom: Get a fish head from a fisherman, a restaurant, or a fish market. Find the gills and inspect them with a magnifying glass. Submerge the head under water and look at the gills. Then do some research to find out how gills work.

Do Fish Have Noses?

Different organisms have evolved different ways of taking in the gases they need and getting rid of waste gases. A single-celled organism doesn't need a specialized system for gas exchange. Oxygen can diffuse across the membrane into the organism and carbon dioxide can diffuse out.

Gas exchange is more of a challenge for larger animals. Some animals, such as fish, absorb oxygen from the water through gills. (No, they do not have noses.) Animals that take in oxygen through their skin, such as frogs and earthworms, have to live in moist environments. Mammals, birds, and reptiles have lungs.

Cigarette Smoke

We inhale because we need oxygen, but there are other kinds of molecules in each breath of air. Sometimes we inhale air that contains lots of pollutants. Think about the air a smoker inhales. Obviously, smoke can't be good for the lungs, and if it isn't good for lungs, it isn't good for the smoker. In fact, medical researchers estimate that a regular smoker probably shortens his or her life by about 5 ½ minutes for each cigarette smoked. The average smoker smokes 15 to 20 cigarettes a day. What does all this smoke do to lungs?

How can a model help us understand the effect of cigarette smoke on a person's lungs?

Use a lung model to "smoke" a cigarette. Observe the effect of one cigarette on the model. Keep the focusing question in mind while you work.

Lung Model
- 7–9 cotton balls
- rubber tubing, 8 cm long
- modeling clay
- clear plastic 10- to 16-oz bottle with small hole punched near the top

- cigarette
- matches
- can with sand for cigarette stubs

Procedure   *(Work in a group with one or two other students.)*

Part A: Making a Lung Model

1. Get the materials you need to make a lung model.
2. Gently pull the cotton balls apart to make them fluffy. (See **Figure 9.13**.)
3. Carefully push the cotton balls into the bottle. Do not pack them in.
4. Use clay to hold the tubing in the neck of the bottle. About 3 cm of tubing should stick above the top of the bottle. (See **Figure 9.13**.)

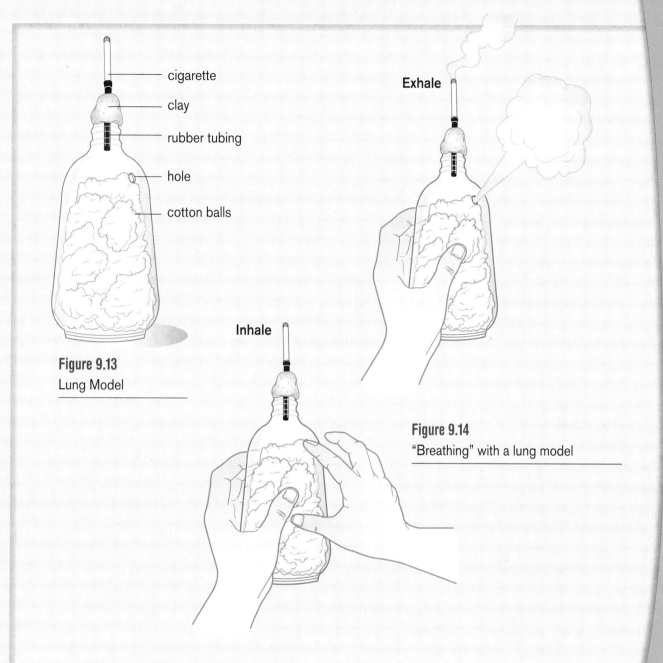

Figure 9.13
Lung Model

Figure 9.14
"Breathing" with a lung model

cigarette

clay

rubber tubing

hole

cotton balls

Exhale

Inhale

Part B: Practicing Breathing

Each person in your group should "practice breathing" as described next before the group continues. (See also **Figure 9.14**.)

1. *Exhale:* Squeeze the air out of the bottle. Leave the small hole uncovered. Watch how the air gets out of the bottle.
2. *Inhale:* Cover the small hole with your finger. Slowly release the bottle. Watch how air enters the bottle.
3. Repeat the process. Exhale. Inhale. Exhale. Inhale. Breathing should be slow and steady. (Let the bottle breathe 8–9 breaths a minute.)

Part C: "Smoking" the Cigarette

1. Have one person take the lung model to your teacher. Your teacher will position a cigarette in the rubber tubing.
2. Have the teacher light the cigarette while another person holds the lung model. Do so by following these steps:
   a. Squeeze bottle (exhale).
   b. Light cigarette.
   c. Cover hole and release the bottle; the air will enter through the cigarette (inhale).
3. "Inhale" and "exhale" with the model at the rate of 8 or 9 breaths a minute. Continue until the cigarette is burned. Remember to "breathe" slowly.
4. Put the burnt cigarette stub in the disposal can.

Part D: Finishing Up

1. Closely observe the cotton. Record any changes.
2. Compare your lung model with others. Make sure you look at the model that has not "smoked" any cigarettes.
3. Clean up and put away all of your equipment.

Analysis Questions ＿ ＿ ＿ ＿ ＿ ＿ ＿ ＿

1. Analyze the model.
   a. List at least three ways this is like a real smoker's lung.
   b. List at least three ways this differs from a real lung.
   c. Why were you supposed to cover the hole when you released your grip on the squeeze bottle?
2. What was the purpose of using one lung model without a cigarette?
3. List three differences between a lung model that smoked a cigarette and one that did not.
4. Imagine you are looking at the cotton in two lung models that smoked. The cotton balls in one bottle are very brown; the cotton balls in the other are a light yellow. Give at least two reasons why they could look so different.

Read the introductory paragraph for this chapter again before you answer analysis questions 5 and 6.

5. About how many cigarettes does the average smoker smoke:

   a. each day?

   b. every week? (Use the larger number and show your work.)

   c. every year?

6. If the figures are correct, how much is a smoker's life shortened by smoking 20 cigarettes a day:

   a. for one day?

   b. for one year?

   c. for 20 years?

7. Your friend reads you a sentence from an article. It says, "There is no definite proof that cigarette smoking is bad for a person's health."

   a. List at least three questions you could ask your friend about this article.

   b. Explain to your friend why you agree or disagree with that sentence.

8. Many states have laws that make it illegal to sell cigarettes to anyone under 18 years of age. Do you think these laws are necessary? Explain your answer.

Conclusion

Write a complete answer to the focusing question.

Extension Activities

1. **Going Further:** Make a lung model at home and use it to "smoke" an unfiltered cigarette. How do the cotton balls look compared with the ones that were exposed to filtered smoke? (Ask a parent to help you if you decide to try this experiment.)

2. **Going Further:** Look at cotton fibers from clean cotton under the low power of the microscope. Compare these fibers with fibers from cotton that was used in a lung model. Describe the differences.

Interviewing a Smoker

You can read about cigarettes, you can listen to news documentaries about smoking, and you can study the research findings about the effects of smoking, but there will still be something missing. That "something" is the smoker's point of view. Some people prefer not to talk about their smoking habits, but many people are willing to share their thoughts on the subject.

What do people who smoke say about the hazards of smoking?

Interview someone who smokes cigarettes and then share what you learn with your classmates. Use the class findings to answer the focusing question.

Materials

- Handout, *Interviewing a Smoker*
- Handout, *Results of Smoker Interviews*
- scissors

Procedure

Part A: Getting Ready

1. Read the questions on the handout, *Interviewing a Smoker*.
   a. Note any questions that you think someone might be uncomfortable answering.
   b. Think about any questions you might want to add to this form.
   c. Answer analysis questions 1 through 3.

2. Decide on an adult you would like to interview. Choose someone who you think will feel comfortable talking about smoking.

3. Think about what you are going to say when you contact the person you want to interview. Remember to explain that this is a class assignment. Tell them that it is all right if they choose not to answer a question.

4. As you interview, record the answers on the handout. Remember to thank the person when you are done.

Part B. Summarizing the Interviews

1. In the margin by each question, jot down the approximate age of the person you interviewed.
2. Cut apart each interview question with its response. Keep the questions in order when you do so. Note: Each part of question 7 will be on a separate piece of paper. You will end up with 13 strips of paper.
3. In class, place your questions and responses at the appropriate station. When all students have done this, everybody's responses to question 1 will be at one station, those for question 2 at another, and so on.
4. With a partner, summarize the responses for the question that your teacher assigns to you.

   Look for patterns. Does the age of the smoker affect the responses? Be prepared to present your summary to the class.
5. Listen to other students summarize the responses. While you listen, takes notes on the handout, *Results of Smoker Interviews*. Ask questions if you need further information to write a good summary.
6. After the class discussion, answer analysis questions 4 through 6, and write your conclusion.

Analysis Questions _ _ _ _ _ _ _ _

1. How do you predict adults will answer:
   a. interview question 3?
   b. interview question 9?
2. Which question do you think looks the most interesting? Why?
3. Imagine you just asked Mrs. Harris if you could interview her. She seems angry and says, "No, I'd rather you didn't."
   a. Why do you think she might be angry?
   b. What will you say to her now?
4. What are the five main reasons people said they started to smoke?
5. What are the three main reasons smokers try to quit?
6. What was the most interesting thing you learned by doing this activity?

Conclusion

Think about the interview you did and the class discussion of the interview responses. Then write a summarizing paragraph to answer the focusing question.

Smoking and the Human Body

In addition to doing this reading, look at two microscope slides. It does not matter which part you do first. Analysis questions 1 through 3 relate to the microscope slides; the other questions relate to the reading.

Everybody knows that smoking cigarettes is a health hazard. It has been reported over and over again that smoking increases a person's chances of getting lung cancer and emphysema. Actually, respiratory problems are not the only health hazards associated with cigarettes. Smoking has widespread effects on the entire body.

One Cigarette

The damage to a smoker's body starts with just one cigarette. That's right. Researchers have monitored people before and after they smoke a cigarette and found that even one cigarette speeds up the heart rate, increases blood pressure, drops the temperature of fingers and toes, and interferes with gas exchange in the lungs.

The more cigarettes a person smokes, the more health problems they are likely to develop. After years of smoking, the average smoker is 20 times more likely to die from cancer than a nonsmoker. Smokers also double their chances of dying from a heart attack.

Smoke

The smoke from cigarettes causes these health problems. This smoke contains hundreds of chemicals. Three of the most dangerous ones are tars, nicotine, and carbon monoxide.

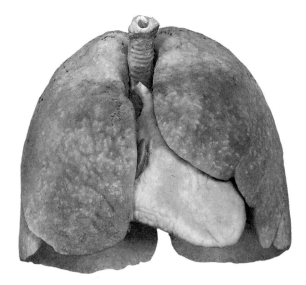
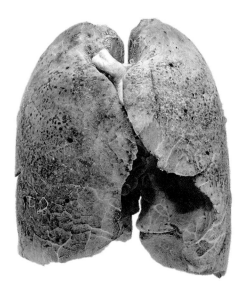

Figure 9.15
The image on the left shows healthy lungs. The image on the right shows the lungs of a smoker.

If you used the lung model to smoke a cigarette, you saw the tars build up on the cotton. Tars also build up in human lungs. Tars are brown, sticky substances that irritate the cells in the respiratory system. They also interfere with the exchange of gases that takes place in the air sacs.

Nicotine is the addictive chemical in cigarettes. It acts on the body's nervous system. Once they begin smoking, most people crave the feeling they get from nicotine. This craving can make it very hard for people to stop smoking once they start. They become addicted to nicotine.

Carbon monoxide (CO) is one of the gases in the smoke. When it gets in the lungs, hemoglobin in the red blood cells picks it up instead of oxygen. If too much carbon monoxide is present, the body does not get the oxygen it needs.

Effects of Smoking

Doctors say that one of the best things you can do for your health is not smoke. Everyday they see people suffering from health problems that are the result of cigarettes. **Figure 9.16** illustrates some of the most common health problems as well as other hazards associated with smoking.

Special Situations

Studies have shown that *nonsmokers* can get sick from other people's smoke. Sometimes the level of carbon monoxide in a smoke-filled room gets so high that it violates air pollution standards. How many times have you found yourself breathing someone else's cigarette smoke?

This secondhand smoke affects your body whether or not you realize it. Your heart rate increases and your blood pressure goes up. Many people also complain of sore eyes, headaches, itchy throats, and stuffy noses.

Pregnant women need to be especially aware of the problems associated with cigarette smoke. Nicotine and carbon monoxide can slow a baby's growth. Children born to women who smoke are often smaller than average at birth. This is a concern because small babies are more likely to have health problems during infancy.

Quitting

It does pay to quit. Almost immediately, the ex-smoker will notice that food tastes better, clothes smell fresher, and coughing stops. People who quit smoking, regardless of their age, are less likely to die from smoking-related illness than those who continue to smoke. Quitting is not easy. Yet this is one time when being a quitter is something to boast about!

Chewing Tobacco

You may have wondered if it is any safer to chew tobacco than it is to smoke it. It may be easier on the lungs, but it is not "safe." Tobacco chewers suffer a variety of other problems, some of which include loss of teeth, bad breath, nicotine addiction, mouth sores, gum disease, and mouth cancer.

Chewers can also be social rejects. Many people find chewing and spitting to be disgusting habits. So, is chewing tobacco "safe"? Definitely not!

Marijuana

During any discussion of the dangers of cigarette smoke, someone will probably ask about marijuana. The smoke from marijuana contains even more cancer-causing substances than cigarette smoke; it also contains more tars. Thus, people who smoke marijuana may have an even higher risk of developing cancer than people who smoke tobacco. They also may have a higher risk of getting other respiratory problems such as bronchitis and asthma.

Includes sciLINKS NSTA

Topic: Respiratory Diseases
Go to: www.scilinks.org
Code: MSLS3e239

Effects of Smoking Cigarettes

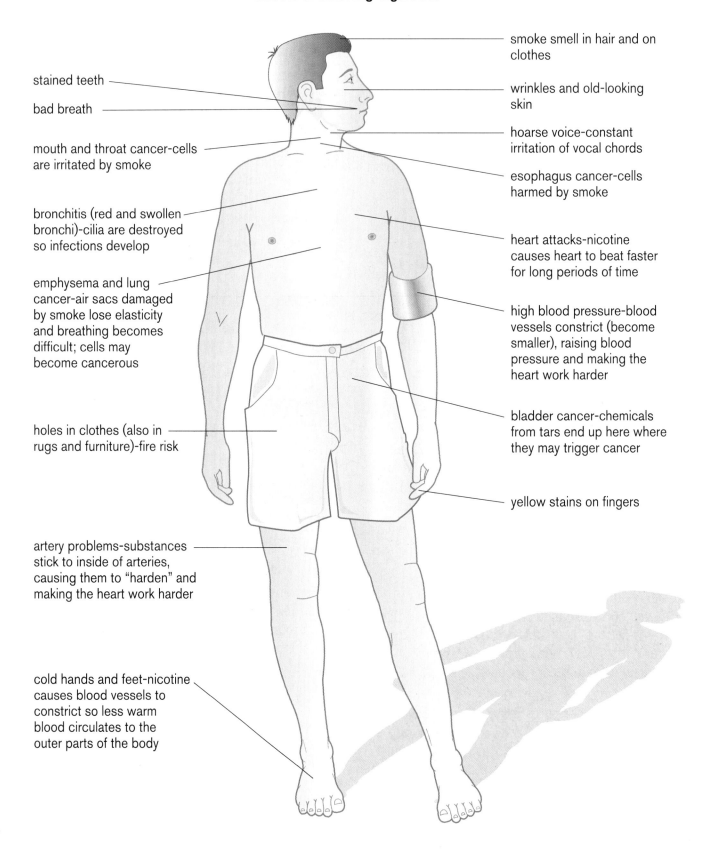

stained teeth

bad breath

mouth and throat cancer-cells are irritated by smoke

bronchitis (red and swollen bronchi)-cilia are destroyed so infections develop

emphysema and lung cancer-air sacs damaged by smoke lose elasticity and breathing becomes difficult; cells may become cancerous

holes in clothes (also in rugs and furniture)-fire risk

artery problems-substances stick to inside of arteries, causing them to "harden" and making the heart work harder

cold hands and feet-nicotine causes blood vessels to constrict so less warm blood circulates to the outer parts of the body

smoke smell in hair and on clothes

wrinkles and old-looking skin

hoarse voice-constant irritation of vocal chords

esophagus cancer-cells harmed by smoke

heart attacks-nicotine causes heart to beat faster for long periods of time

high blood pressure-blood vessels constrict (become smaller), raising blood pressure and making the heart work harder

bladder cancer-chemicals from tars end up here where they may trigger cancer

yellow stains on fingers

Figure 9.16

Of course, some of the greatest dangers of smoking marijuana involve the nervous system. Any substance that affects the mind must be handled cautiously or not used at all.

A Puzzle

It's a puzzle. Why do people even start to smoke when they've heard about all the health problems that can result? In fact, many physicians say one of the best thing you can do for your health is to not smoke. And yet people continue to smoke. Few people take up smoking as adults. Most new smokers are teens. Why do you think this is so? Why don't young people use all the evidence to make smart decisions?

Analysis Questions

1. Look at the microscope slide made from a nonsmoker's lung.
   a. Sketch a few typical cells.
   b. Describe what you see.

2. Look at the microscope slide made from a smoker's lung.
   a. Sketch a few typical cells.
   b. Describe what you see.

3. Compare the lung tissue you observed on the two microscope slides.
   a. List several ways they are alike.
   b. List several ways they are different.

4. Decide if these four statements are true or false. If a statement is false, explain why it is false.
   a. Smoking one cigarette does not affect a person.
   b. A person who smokes heavily will die from lung cancer.
   c. Carbon dioxide is a dangerous gas in cigarette smoke.
   d. Smoking cigarettes can make a person's feet feel cold.

5. Name three of the most harmful chemicals found in cigarette smoke.

   Refer to **Figure 9.16**, "Effects of smoking cigarettes," to answer questions 6 and 7.

6. If someone has just started to smoke, which of the problems shown on the diagram might he or she notice?

7. Every body system is affected by cigarette smoke.
   a. List at least three health problems that involve the respiratory system.
   b. List at least three other systems that are affected.

8. What do you think is the biggest health problem that can result from:
   a. smoking cigarettes?
   b. chewing tobacco?
   c. smoking marijuana?

9. Your friend just started to smoke. You try to talk her out of it, but she tells you she is not worried about health problems because her uncle has smoked for 25 years and he's fine. What would you say to her next?

Extension Activity

Health Application: You know that smoking can cause cancer when you get older, but did you know that it also has bad effects on your body right now? Research how smoking can affect a teenager's health and lifestyle.

Clean Air, Dirty Air

Even when the air smells absolutely terrible, you breathe. You cannot help it. Oxygen is so important to the body that you breathe even when you know it is dangerous to do so. The healthiest thing you can do for your lungs is to avoid dirty air. So, what is dirty air?

Dirty Air

"Dirty air" is another way to say air pollution. It can be caused by thousands of different chemicals and gases in the air. Even the gases we normally breathe can be dangerous in high concentrations. **Table 1** lists some of the gases found in air. You can see that gases such as carbon monoxide and butane are normally found in air. However, at high concentrations, they can cause health problems or even death.

Air pollutants come from many sources. Cigarette smoke is one example of a pollutant. Dust, sand, and salt from ocean spray are pollutants carried by the wind. The burning of fuels and wastes are major sources of air pollution. Even woodstoves and fireplaces contribute to dirty air.

Table 1

Gases in the Air*	
nitrogen	hydrogen
butane	nitrogen dioxide
argon	acetylene
neon	carbon dioxide
ethane	ethylene
methane	carbon monoxide
propane	krypton
oxygen	formaldehyde
helium	isobutane
ozone	nitrous oxide
xenon	sulfur dioxide
ammonia	

*This is not a complete list.

The Air You Breathe

Many cities have serious air pollution problems. For days on end, levels of pollutants may be so high that they endanger people's health. You can judge the quality of the air in a location by looking at its air quality rating. In order to arrive at the rating, the levels of five substances are measured. These five substances are sulfur dioxide, nitrogen dioxide, carbon monoxide, ozone, and solid particulates. **Table 2** explains the air quality rating system. **Table 3** gives information about the five pollutants that are monitored.

Air pollution is a serious problem that is not going to go away. Be alert to the condition of the air in your own community.

Table 2.

Air Quality Rating System		
Quality	**Rating**	**Health Effects**
Good	0-50	This is the cleanest air.
Acceptable	51-100	Okay, but could be better.
Poor	101-199	Unhealthful. People may get sore eyes and headaches.
Extremely Poor	200-299	Very unhealthful. Healthy people complain of sore eyes and headaches. People with heart or lung problems should stay inside.
Dangerous	300-399	Hazardous. Triggers asthma, bronchitis, and heart attacks. Everyone should stay inside.
Extremely Dangerous	400-500	Hazardous. Causes death in the ill and elderly. Healthy people feel sick.

Air pollution in Los Angeles, California

How many polluters do you see pictured here?

Table 3.

Air Pollutants	
Pollutant	**Description**
carbon monoxide (CO)	Carbon monoxide is an invisible, odorless gas. It combines with hemoglobin in the blood. When this happens, blood cannot carry enough oxygen. Large amounts of carbon monoxide cause death. Small amounts cause dizziness, headaches, and fatigue.
sulfur dioxide (SO_2)	Sulfur dioxide is a gas that irritates the eyes, nose, and throat. It causes breathing problems and can damage the lungs. It kills plants.
nitrogen dioxide (NO_2)	Nitrogen dioxide irritates the eyes and nose. It can cause respiratory problems such as bronchitis and pneumonia.
ozone (O_3)	Ozone is formed when nitrogen oxides are exposed to the energy from sunlight. Because sunlight is needed to produce ozone, ozone levels increase during the day and decrease at night. This gas irritates the lungs and causes eyes to "burn." It damages plants and materials such as rubber.
solid suspended particles	These include dust, smoke, soot, ash, and pollen. Some dust particles can be radioactive, poisonous, and/or cancer causing. These particles may stay suspended in the air or settle to the ground. They darken the air and cause objects to become dirty. They also interfere with normal breathing and may injure the lungs.

Analysis Questions

1. Look at the gases listed in **Table 1**.
   a. List at least three of these gases you have heard of before.
   b. List three of these gases you have never heard of.

2. List at least five places or things in your community that add pollutants to the air you breathe.

3. Read the information in **Table 2**.
   a. What is the air quality rating on an unhealthful day?
   b. What symptoms might you experience if the rating were 230?

4. Read **Table 3**. Which of the pollutants can cause respiratory problems?

5. We exhale *carbon dioxide*, which is made up of *carbon* and *oxygen*. What makes up each of these pollutants?
   a. carbon monoxide
   b. sulfur dioxide
   c. nitrogen dioxide
   d. ozone

Extension Activities

1. **Health Application:** Cut out newspaper articles about air pollution for one week. Make a list of the pollutants that are mentioned. According to your articles, why are these substances a concern? What is causing the pollution?

2. **Community Connections:** See if your local newspaper reports air quality readings for your area. (These are often on the page with the weather.) If you find air quality readings, record the values for at least one week. Make this a bigger project by graphing the air quality for one month or more. Look for patterns.

Chapter **10**

Circulatory System

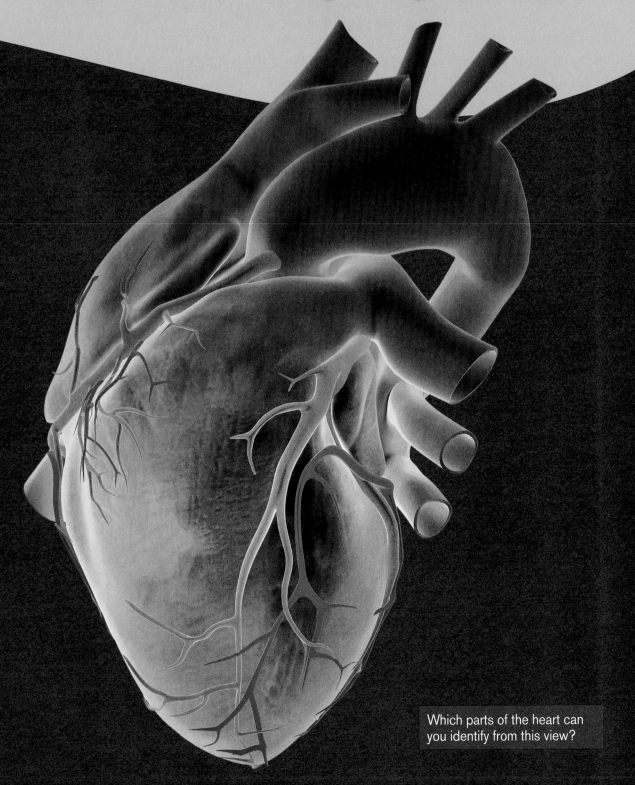

Which parts of the heart can
you identify from this view?

Heartbeats and Stethoscopes

One way to learn about the heart is to feel it beating; another way is to listen to it. Doctors and nurses use both of these methods. The stethoscope is the instrument they use to listen to a heart. Have you ever listened to your own heart? If so, did you know what you were hearing?

What causes heart sounds?

How many different heart sounds can you hear?

During this activity, you will listen to your own heart with a stethoscope. Think about what you are hearing, read *Lub Dub* and then answer the focusing questions.

- rubbing alcohol
- cotton ball
- stethoscope
- Handout, *Heartbeats and Stethoscopes*

Procedure

1. Work quietly so people will be able to hear their heart sounds.
2. Moisten a cotton ball with rubbing alcohol and use it to clean the earpieces on the stethoscope.
3. Listen to your own heart with the stethoscope. If you can't hear anything, tell your teacher. There may be a problem with the stethoscope.

Figure 10.1

The earpieces of some stethoscopes can be tilted to make them more comfortable.

4. Read *Lub-Dub*.

5. Complete the handout and then write your conclusion.

Conclusion

Answer the focusing questions. When you are done, review your answers. Do they say what you want them to?

Extension Activity

Health Application: Talk to a health professional about heart murmurs. Find out what causes them, how they are detected, and how they are treated.

Lub-Dub

When blood moves from one part of the heart to the next, it passes through a valve. The valve snaps shut to keep blood from flowing backward. There are several valves in the heart. Each valve makes a distinctive sound. These are the "lub-dub" sounds we call the heartbeat.

Heart murmurs are often caused by valves that do not close completely. The "murmur" is the sound of blood squeezing through the opening. Many people have murmurs. Some murmurs are normal and do not cause any problems; others are more serious. They may indicate that the heart is not functioning properly. People with serious valve problems may have surgery to replace their faulty valves with artificial ones.

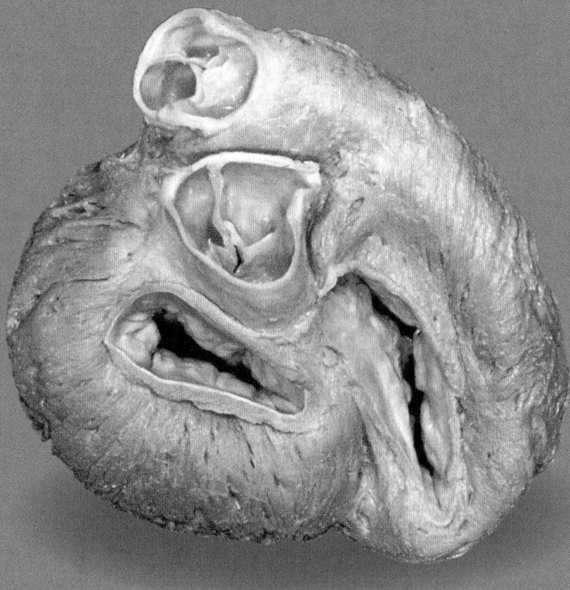

Figure 10.2

In this photograph, you can see the valves in the vessels of a human heart.

Using Pulse Rate as Evidence

Blood. It flows throughout the body, bringing nutrients and oxygen to all the cells. In general, we are unaware of this continuous flow. However, there are several places on the body where you can actually feel a "wave" of blood as it passes. These places are called **pulse points**.

You have probably taken your pulse at some time in your life. Do you remember why? Did you do anything with the information once you gathered it? **Pulse rate** (the number of pulses per minute) provides valuable information about the fitness of your heart.

What factors affect your pulse rate?

To answer this question, you will find your resting pulse rate, and then you will design an experiment to find out what affects pulse rates.

Materials

- watch or clock with second hand
- blank paper
- Reading, *Exercise for Your Heart*, page 253
- other materials you decide you need
- graph paper

Procedure

Part A: Resting Pulse Rate

1. Find the pulse point on your wrist. (See **Figure 10.3**.)
2. Count your pulse for 15 seconds. Record the count on your paper.
3. Do this twice more.

4. Multiply each 15-second pulse count by 4 to find the number of pulses in 60 seconds. The number of pulses per minute is called *pulse rate*.

5. Calculate your average resting pulse rate.

Part B: Exercise and Pulse Rate   *(Work with a partner.)*

1. Think of factors that might affect a person's heart rate. Work with your partner to create a list of questions about those factors.

   a. Which of your questions can be tested by doing a classroom experiment? Put a check mark next to those questions.

   b. Select one of your questions to test in a classroom experiment. Remember that a good testable question can be answered by a fair test where you test only one variable and keep all the others the same.

   c. Get approval for your testable question from your teacher

   d. On your paper, under the heading "**Testable Question,**" write the question you selected for your classroom experiment.

2. What do you already know about your question? Where can you find out what other scientists have already discovered about your question?

   a. Read *Exercise for Your Heart*.

   b. Optional: Conduct research at the library and on the Internet to learn more about factors that affect a person's heart rate.

3. Write a **hypothesis** for your testable question, using the *"If, then"* format.

   "*If* ____ *[what is manipulated in the experiment]* ____, **then** ____ *[How you think it will respond]* ____.*"

4. Design your experiment. Make sure it is a fair test, meaning you only test the variable from your testable question. Think about how you are going to control all the variables so you only test one.

   a. What are all the variables you can think of in this experiment? Write them down on your paper.

   b. What is the one variable that you will be testing in your experiment? This is called the *manipulated* or **independent variable**. In your experiment, the independent variable is the one you will be deliberately changing or *manipulating*.

   c. What will you be measuring in your experiment? This is called the *responding* or **dependent variable**. In your experiment, the dependent variable is the variable that reacts or *responds* when the independent (manipulated) variable is changed.

Figure 10.4

How does exercising with additional weight affect your pulse rate?

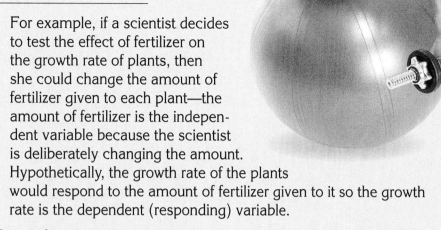

For example, if a scientist decides to test the effect of fertilizer on the growth rate of plants, then she could change the amount of fertilizer given to each plant—the amount of fertilizer is the independent variable because the scientist is deliberately changing the amount. Hypothetically, the growth rate of the plants would respond to the amount of fertilizer given to it so the growth rate is the dependent (responding) variable.

   d. Except for your independent variable, how would you recommend controlling the rest of the variables? These are called **controlled variables**.

5. Write down your procedure, step by step.

   a. Be sure to include repeating your experiment at least three times. This is an important step to make sure your results are consistent and not just an accident.

   b. Ask another team to check your procedure. If they don't understand exactly what you plan to do, you may have to rewrite some of your steps.

   c. Have your teacher check to make sure that you have designed a safe experiment.

6. Design a data table to record your data.

7. Collect your data. Be sure to keep all your variables controlled, except the independent variable (the one you are deliberating changing) and the dependent variable (the one that responds to the independent variable.)

    If appropriate, calculate averages from your repeated experiments (trials).

8. Construct an appropriate graph (histogram, bar graph, line graph) to summarize and present your data. Be sure you know which is the **independent variable** (plotted on the x-axis) and which is the **dependent variable** (plotted on the y-axis).

9. On your paper, write a conclusion by answering the following questions:

Conclusion

   a. What does your data tell you? Does it support your hypothesis? Why or why not?

   b. What other explanations could explain your results? What errors could affect your results?

   c. What other questions do you have about pulse rates now that you've completed your experiment?

10. Answer the analysis questions.

Analysis Questions _ _ _ _ _ _ _ _

1. Look at your resting pulse rate data.

   a. What is your average resting pulse rate?

   b. How many times does your heart beat in one minute?

   c. If you took your pulse on the inside of your ankle, how would that compare with the pulse rate you got from your wrist?

2. About how many times does your heart beat each day? Show your calculations.

3. If a person gains weight, how can that affect his or her heart?

4. Your pulse rate goes up during exercise. Other things also affect pulse rate.

   a. List at least three things you think could cause your pulse rate to be slower.

   b. List at least three things you think could cause your pulse rate to be faster.

   c. You could design an investigation to find out if some things you listed above (in a and b) are true. Circle the ones you think you could check by designing an investigation.

5. What aerobic activity do you do the most often?

6. Why do you think your pulse rate increases when you exercise?

Conclusion

Write complete answers to the focusing question. Use evidence from your data to support your answer.

Extension Activity

Animal Kingdom: Try to measure the heart rates for other organisms. Take the pulses of cats, gerbils, birds, and other cooperative animals. How will you measure their heart rates? Be gentle!

Exercise for Your Heart

Did you know that your heart benefits from exercise? Like skeletal muscles, the **cardiac** muscle also can become stronger when it is given a "workout." Which of these exercises do you think are best for your heart?

swimming	lifting weights	softball	biking
bowling	sprinting	jogging	volleyball

> The word **cardiac** always refers to the heart.

On a piece of scratch paper, write down your choices. You may want to change your mind as you read the next few sections. At the end of this reading, you will learn which choices are best for your heart.

Choosing "Hearty" Exercises

Almost any exercise is better for improving heart fitness than doing nothing, but some exercises are better than others. Good heart exercises are said to be **aerobic,** that is, the muscles get plenty of oxygen while they are working. Here are three clues to help you decide whether or not an activity is aerobic.

1. It raises your pulse rate to about 150 beats per minute, but not higher. You do not want your heart to "race."
2. It keeps the heart rate at about 150 beats per minute for 10 to 20 minutes.
3. It is continuous; it does not involve a lot of starting and stopping.

In order to benefit the heart, people should do aerobic exercises at least three times a week.

A Healthy Heart

Aerobic exercise helps your heart and blood circulation in several ways.

1. It strengthens your cardiac muscle.
2. It causes your resting heart rate to slow down. (A strong heart does not have to beat as often because it pumps more blood with each beat.)
3. It develops arteries that go to the heart so if one of them becomes blocked there will not be as much damage to the heart muscle.
4. It improves circulation throughout the body.

Figure 10.5

Activities like bicycling are not only fun, they can also be good for the heart.

Do you think this activity is aerobic?

One way to check the fitness of your heart is to check your pulse rate before and after you exercise. That's because you can feel a pulse every time your heart beats. During exercise, your heart beats faster; as soon as you stop, your heart gradually slows down. The time it takes for your pulse rate to return to normal indicates your heart's fitness. If your heart is in good shape, it will return to its normal pace fairly quickly.

Improving Heart Fitness

You can improve the health of your heart by doing aerobic exercises. Look at this list of activities again. Which ones are aerobic?

swimming	lifting weights	softball	biking
bowling	sprinting	jogging	volleyball

If you said that swimming, jogging, and biking are aerobic, you're right. Look at the clues for determining whether or not an exercise is aerobic, and decide why the others are not as good for your heart.

Are you exercising for your heart? If you do, you will feel better, and possibly live longer.

Heart Rate

Every animal has a characteristic heart rate. Here are some examples of heart rates.

A masked shrew

30	codfish
600	hummingbird
200	rabbit
70	human
400	mouse
25	elephant
800	shrew

Look at the data for the mammals. Can you see an association between animal size and heart rate? Codfish and rabbits are about the same size, but notice how much difference there is in their heart rates. That's because their body systems work so differently. Rabbits, like all mammals, maintain a constant body temperature. A fish's body temperature changes with the temperature of the water.

Heart Parts

While you can learn some things about the heart by feeling it beat and by listening to the "lub-dub" sounds, you can learn much more about the heart by examining one. What does a heart look like? What does it feel like? And, most importantly, what clues can you find about the way the heart works from the way it is constructed?

Which parts of a heart can you locate and name?

How would you describe a preserved heart to a person who hasn't seen one?

In this activity, you will inspect and label a preserved heart to answer these questions.

- 21 pins
- paper towels
- dissecting pan
- preserved heart (pig, sheep, or cow)
- red and blue markers or colored pencils
- Handout, *Observation Guide* (one per group)
- Handout, *Heart Parts* (one per person)

- scissors
- soap

Procedure
(Work in a group of four students.)

1. Decide who will assume each of these roles.

   reader: reads checklist and keeps group on task

   cutter: cuts apart list of heart parts as needed

   pinner: pins name to correct spot on preserved heart

   handler: carefully handles and positions heart so everyone can see it

2. Have the *cutter* prepare the group's *Observation Guide* by circling the heading "Part A. Outside of Heart" in red and by drawing a red line through the bars by the listing of heart parts. Using blue, do the same thing for the heading and list that goes with Part B.

3. Follow the directions while the *reader* reads from the *Observation Guide*.

   a. When you are supposed to label a heart part, have the *cutter* cut that word off the list.

   b. Once everyone agrees on the location of the part, have the *pinner* pin the word in place.

4. When your group has finished labeling the preserved heart, label your *Heart Parts* diagram. Compare the real heart with the drawing as you label the structures.

5. Remove the pins. Return the heart to the plastic bag and clean up your work space.

Conclusion

Write a sentence in answer to each of the focusing questions.

Extension Activity

Going Further: Ask a butcher or a meat packing plant manager for a fresh heart. Try to find the same structures on it that you found on the preserved heart. How are the two hearts alike? How are they different?

The Heart: A Double Pump

The heart, blood vessels, and blood make up the **circulatory system**. This system is like a delivery system; it delivers nutrients and oxygen to every cell in the body, and carries away the waste products. The heart keeps the whole system moving. In this reading, you will learn about the structure and function of this powerful organ.

Location of the Heart

The fist-sized muscular organ known as the heart is located inside the chest behind the sternum. The sternum, together with the ribs, form a bony cage that protects the heart. The heart is not in the exact center of the chest; it lies slightly more to the left of the sternum than to the right. (See **Figure 10.6**.) About once a second (sometimes more often, sometimes less) the muscles in the wall of the heart contract. This repeated contraction and the relaxation that follows is what we call the heartbeat. If you could watch the heart beat, it would look like a pulsing, writhing mass of muscle.

Try to imagine what is going on inside the heart when it beats. With each beat, the chambers of the heart squeeze closed and then reopen. When the chambers open, they fill with blood; when they close, they squeeze the blood out. Open and fill, close and empty; open and fill, close and empty. This pumplike action is repeated over, and over, and over again. On the average, the heart pumps blood about 65 times a minute. And, with each pump, 2 or 3 ounces of blood spurts from the heart. After the blood circulates through the body, it returns to the heart, only to be pumped back out again. (See **Figure 10.7**.)

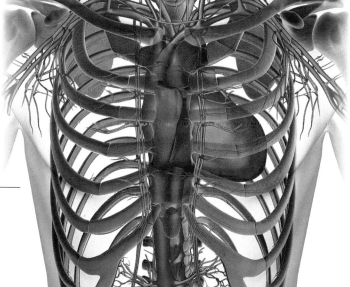

Figure 10.6
The heart is protected by the rib cage.

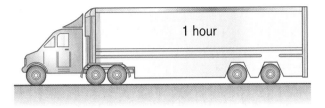

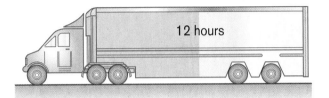

Figure 10.7

At the rate of about 65 beats a minute and with 2 or 3 ounces of blood being pumped with each beat, the heart moves about 5 quarts of blood in a minute. In one day, the heart could pump enough liquid to fill a 2,000-gallon tanker.

A Double Pump

If you think about it, you may wonder why the heart needs *four* chambers. It might seem that *one* chamber would be enough. Blood could flow into this single chamber and then be squeezed out. While this design might work as a pump, it would not be an efficient heart. That is because a human heart actually works like two pumps; there is a pump on the left and a pump on the right.

The pump on the left sends oxygen-rich blood to the entire body. The pump on the right sends oxygen-poor blood to the lungs. This explains why the left ventricle is so much more muscular than the right. When it contracts, it has to send blood to cells as far away as the toes. The right ventricle only has to get blood to the lungs, and that's a much shorter

distance for blood to go. Study **Figure 10.8** and try to understand what it is illustrating.

When it returns to the heart from the body, blood enters the atria—but which one? Think about it for a minute. The blood that has circulated through the body is low in oxygen. It returns to the right atrium so it can be pumped to the lungs. The blood that returns from the lungs to the heart is rich in oxygen. It is ready to

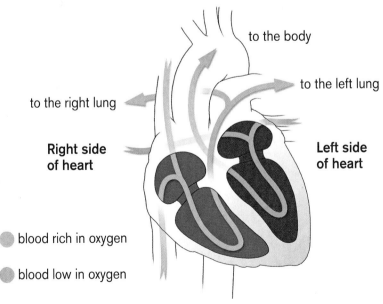

Figure 10.8

Blood that leaves the right ventricle goes to the lungs. Blood from the left ventricle goes to the rest of the body.

go to the rest of the body, so it returns to the left atrium. The return of blood to the heart is illustrated in **Figure 10.10a**. Does it make sense to you?

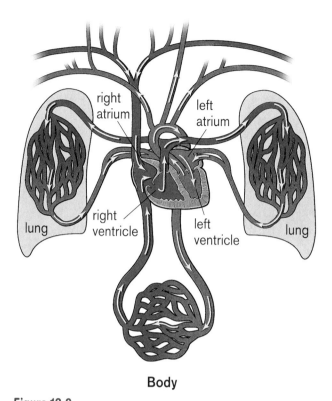

Figure 10.9

This drawing summarizes what happens when blood is pumped out of each side of the heart.

Check your understanding of blood flow from the heart by putting your finger on the left ventricle in **Figure 10.9** and tracing the path blood follows as it flows through the body. In your head, describe what is happening to the blood. Here's a starting line you can use: "The blood that leaves the left ventricle is rich in oxygen. As it flows through the ..."

Coordinated Action

Even though the left and right ventricles are like two separate pumps, they pump at exactly the same time. That's because cardiac muscle encloses both ventricles. When it contracts, it sends blood out of both sides of the heart at the same time. Then the ventricles relax at the same time. When the ventricles are relaxed, the atria contract. This pushes the blood through the AV valves to fill the ventricles. (An **AV** valve is between the **a**trium and **v**entricle.) When the ventricles are full, they contract again. This action closes the AV valves and sends the blood out of the heart and on to the lungs and the rest of the body. Study the drawings in **Figure 10.10**. They summarize the coordinated action of the heart.

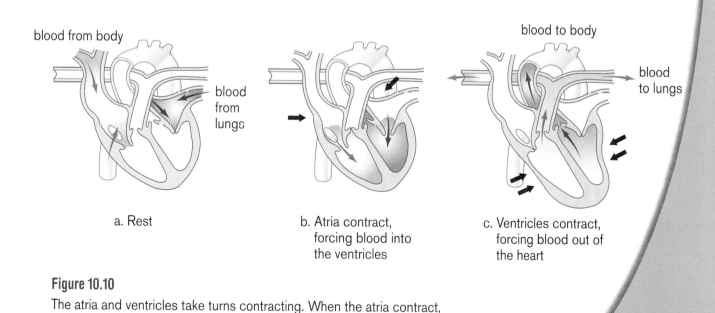

a. Rest

b. Atria contract, forcing blood into the ventricles

c. Ventricles contract, forcing blood out of the heart

Figure 10.10

The atria and ventricles take turns contracting. When the atria contract, which valves close? Which open? Why?

You may remember that nerves trigger muscle contraction. There is not a single nerve that "tells" the heart to contract. Instead, there is an area of specialized nerve tissue built into the right atrium. Its position is shown in **Figure 10.11**. It sets the pace of muscle contractions, and so is called the **pacemaker**.

The pacemaker triggers all of the muscle cells in the two atria to contract at the same time. One-tenth of a second later, the ventricles contract. An artificial pacemaker can be used if a person's own pacemaker is not working properly. It sends electrical impulses to make the cardiac muscle contract.

Contract. Relax. Contract. Relax. A heart will beat more than two and a half billion times in a 75-year lifetime. We cannot make a pump as reliable. Yet the living tissues of a healthy heart will stand up under the beating of a lifetime.

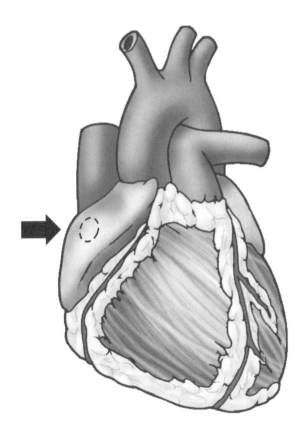

Analysis Questions

Before you answer these questions, label and color your copy of the handout, *A Double Pump*.

1. Which ventricle has thicker walls? Why?

2. How does blood in the right ventricle differ from blood in the left ventricle?

3. How many times a minute does one of your AV valves open and close? (Think about this one; it's easy.) Explain your answer.

4. Imagine that your left AV valve is not closing all the way. What will happen when your left ventricle contracts?

5. HB Connections: List at least two other systems with which your circulatory system interacts. Explain.

6. You observed a preserved heart from a sheep or from some other large mammal. You just read about human hearts. List some ways the two kinds of hearts are alike and different.

Extension Activity

Animal Kingdom: Choose an animal other than a mammal and do some research to learn about its heart. If you choose something such as a fish or an earthworm, your teacher might be able to obtain an animal for you to dissect. Look closely. How will you know the heart when you see it?

Figure 10.11

The arrow is pointing to the location of the pacemaker in the right atrium.

All Hearts
Are Not Alike

Many animals have single-chambered hearts. Vertebrates, however, have more complex hearts. A fish heart has two chambers, but all birds and mammals have four-chambered hearts.

A blue whale has the largest heart of all, which isn't surprising since it's the largest animal in the world. A blue whale can grow to be 30 meters (100 feet) long. It takes a big strong heart to pump blood to the very tip of the whale's tail. A blue whale's heart is about 1.8 meters (6 feet) long and it weighs about 540 kilograms (1,200 pounds). If you stood by a blue whale's heart, would you be able to see over it? If you looked in the heart, how many chambers would you see?

Blood Flow

When blood leaves the heart, it goes through a series of blood vessels in a specific sequence. It does not flow freely through the tissues; it stays in vessels. In some ways, it's like traveling in a car; you can drive on one-lane streets or on large, multilane highways, but you're always on a road.

In order to learn the pattern of blood flow, imagine that you are a red blood cell in the left ventricle of the heart. You are about to take a 60-second trip through the circulatory system. On your way, you will travel through a series of blood vessels and pass through the heart again. At the end of 60 seconds, you will be back in the left ventricle. Be aware of the "sights" you pass on your trip! (Ignore the numbers in parentheses until later.)

To the Lower Part of the Body

We have so many blood vessels that a drawing of the human circulatory system would look like a solid tangle of vessels of varied sizes. **Figure 10.12** is a highly simplified piece of art that shows the blood supply to one organ, a muscle by the knee. Keep referring to this illustration while you read this section.

When the heart contracts, blood spurts out of the **left ventricle** (1) through the **aorta** (2). This blood is rich in oxygen. On its "journey," it will deliver this oxygen to all of the tissues in the body. Almost immediately, the aorta branches, giving off **arteries** (3) that carry blood to the upper part of the body. Imagine that you are following the aorta itself (2) to the lower part of the body.

Arteries branch off the aorta to deliver oxygen-rich blood to all the tissues and organs it passes. For example, some smaller arteries branch off to supply blood to the wall of the stomach (4). Others go to the intestines (5), while yet others go to the kidneys, bladder, and other body parts.

As the arteries continue to branch, they form smaller and smaller vessels. The very tiniest vessels are called **capillaries** (6). When blood reaches the capillaries, it is as far from the heart as it is going to get.

From the capillaries, the blood will start the return trip to the heart through veins. Tiny **veins** (7) drain the oxygen-poor blood from the capillary beds. (See **Figure 10.13**.) The small veins empty into slightly larger ones. These veins come together to form even larger veins. The veins from the lower part of the body drain into the **lower vena cava** (8) and the vena cava enters the **right atrium** (9) of the heart.

To the Upper Part of the Body

When the blood spurts out of the left ventricle (1) through the aorta (2), some of it enters the arteries (3) that go to the arms and head. It flows through smaller and smaller arteries until it reaches the capillaries. There it gives off oxygen and begins the return trip to the heart through veins. The large vein that drains the upper part of the body is called the **upper vena cava** (8). Like the lower vena cava, it enters the right atrium (9).

Figure 10.12

The circulatory path to the left knee.

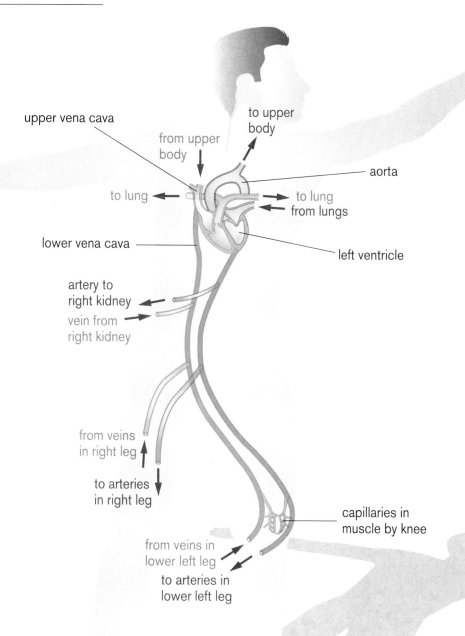

upper vena cava

from upper body

to upper body

aorta

to lung

to lung
from lungs

lower vena cava

left ventricle

artery to right kidney

vein from right kidney

from veins in right leg

to arteries in right leg

capillaries in muscle by knee

from veins in lower left leg

to arteries in lower left leg

Topic: Circulatory System
Go to: www.scilinks.org
Code: MSLS3e263

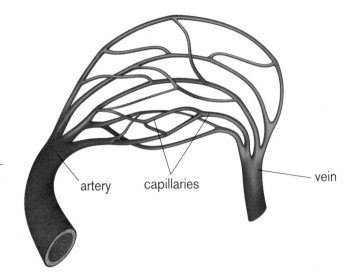

Figure 10.13

Capillaries link arteries and veins.

artery   capillaries   vein

To the Lungs

Blood in the right atrium contains very little oxygen and lots of carbon dioxide. It must go through the lungs to get rid of the carbon dioxide and pick up more oxygen.

The blood flows from the right atrium into the **right ventricle** (10). When the heart contracts, the blood is forced out into the **pulmonary artery** (11). The pulmonary artery branches, carrying blood to both lungs. Once in the lungs, the arteries branch again and again until eventually they form a network of tiny **capillaries** (12). The blood in these capillaries gets rid of the carbon dioxide and picks up oxygen. It is now ready to return to the heart.

Small veins drain the blood from the capillary beds. As always, small veins empty into larger veins. These veins eventually join to form the **pulmonary veins** (13). The pulmonary veins return blood to the left side of the heart, into the **left atrium** (14). The newly oxygenated blood now flows into the left ventricle, and will leave the heart through the aorta again.

> It helps to remember that
> - *cardiac* refers to the heart,
> - *pulmonary* refers to the lungs.

Every minute of every day, throughout your life, blood flows, following this circulatory pathway. Its route never changes. It is the same for every human on Earth.

A Handout for Review

You have just read the section, "Blood Flow." If you are lucky, you understand it perfectly. However, most people need time to think about this process before they really understand it. Therefore, read "Blood Flow" again, but this time fill in the handout. Read the directions on the handout. When you work, use a pencil in case you want to change your mind.

Analysis Questions

1. Which vessel carries blood:
   a. to the lungs from the right ventricle?
   b. from the left ventricle to the entire body?
   c. from the lungs to the left atrium?
   d. from the body to the right atrium?

2. Remember: The heart is like two pumps. Think about the pump on the left.
   a. What is its main function?
   b. Which parts of the circulatory system are part of the left pump pathway?

3. Now think about the pump on the right.
   a. What is its main function?
   b. Which parts of the circulatory system are part of the right pump pathway?

4. Imagine that you are studying blood. For your study, you need blood rich in oxygen. Which of these structures carry oxygen-rich blood?

right atrium	left atrium
right ventricle	left ventricle
pulmonary artery	coronary artery
aorta	pulmonary vein
tibial vein	

5. The coronary arteries supply the heart with blood. You may have seen these if you dissected a preserved heart. (See also **Figure 10.14**.)

   a. How do you think blood gets into a coronary artery? Where does it come from?

   b. Where does the blood in a coronary artery go?

   c. Why do you think you need coronary arteries?

6. Why do you think doctors worry when fat starts to build up *inside* the aorta?

Figure 10.14

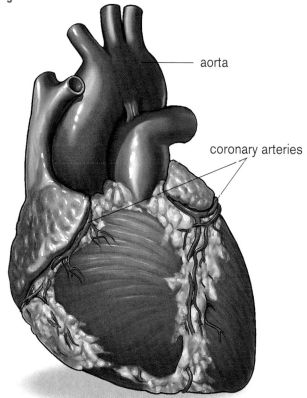

aorta

coronary arteries

Extension Activities

1. **Going Further:** Use a preserved heart to trace the flow of blood through the vessels and heart. (It may be difficult to identify the larger vessels because they are often damaged when the heart is removed.)

2. **Health Application:** Every year, many babies are born with heart defects. Sometimes a baby is born with an opening between the right and left ventricles as shown in **Figure 10.15**. Find out how this affects the baby.

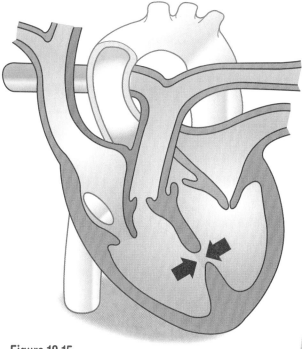

Figure 10.15

Blood Vessels

Blood vessels (arteries, veins, and capillaries) are important parts of the circulatory system. Each type of vessel has a different function, so, as you might expect, they each have a characteristic structure.

How can you tell the difference between an artery, a vein, and a capillary when you look at them through a microscope?

Why are capillaries important?

You will view microscope slides of a cross section of an artery, a cross section of a vein, and a cross section of a capillary. Your task will be to decide which vessel is which. In order to answer the focusing questions and complete this task, you will need to read about the three types of vessels.

- microscope
- prepared microscope slides, cross sections of vessels
- Handout, *Blood Vessels*
- Reading, *Three Types of Vessels,* page 268

Procedure

1. Read *Three Types of Vessels.*
2. Go to one of the microscope stations. It does not matter which station you go to first.
   a. Focus the microscope and record the magnification on your handout.
   b. Find the cross section of the most obvious vessel. Sketch the vessel in the space provided.
   c. Label the wall of the vessel and the space in the vessel.
   d. Do not identify the vessel yet.
3. Repeat step 2 at the other two stations.

4. When you have observed the three microscope slides and completed the reading, answer the analysis questions.

5. Look at your handout and decide which vessel is which. Record your decisions on the handout. (You may want to go back and look at the microscope slides again before you make your final decision.)

6. Write your conclusion.

Analysis Questions _ _ _ _ _ _ _ _ _

1. Make a chart comparing arteries, veins, and capillaries. Think about the best headings to use for the rows and columns.

2. Choose one type of blood vessel and describe its structure, and then explain how knowing about the structure helps you understand its function.

Decide whether statements 3 through 9 are true or false. If a statement is false, explain why it is false.

3. T   F   All arteries carry blood away from the heart.

4. T   F   All veins carry oxygen-poor blood.

5. T   F   Arteries and veins look alike.

6. T   F   Oxygen, carbon dioxide, and nutrients are exchanged between body cells and the blood in arteries.

7. T   F   The upper vena cava contains oxygen-poor blood.

8. T   F   Blood cells never get out of blood vessels.

9. T   F   Pulmonary arteries carry blood to the lungs.

10. You have a deep cut on your arm. Blood is gushing from it in spurts. Do you think you have cut an artery or a vein? Explain.

Conclusion

Write a few sentences to answer each of the focusing questions.

Three Types of Vessels

You will understand the circulatory system much better if you understand basic facts about arteries, veins, and capillaries. While you read, keep asking yourself, "How are these vessels alike? How are they different?"

Arteries

Arteries carry blood away from the heart. Blood surges out of the heart through two large arteries—the pulmonary artery and the aorta. These large vessels are about as big around as a garden hose. The large arteries branch to form smaller and smaller arteries.

Artery walls are strong. They have to be, because each time the heart beats, blood surges through them, pushing against their walls. If the walls were not strong, the artery might burst under the pressure.

The walls of all arteries contain a thick layer of smooth muscle and some elastic connective tissue fibers. Each time the ventricles contract, the smooth muscle fibers around the arteries contract too. This squeezing action helps move the blood through the vessels. Find the layer of smooth muscle in the cross section of an artery shown in **Figure 10.16**.

Smaller arteries have thinner walls than big ones, but the walls of even the smallest arteries are too thick for oxygen to cross. Oxygen and other small molecules can only diffuse across the walls of capillaries.

Figure 10.16

The photograph shows a cross section through an artery. On the right is a drawing of the same artery.

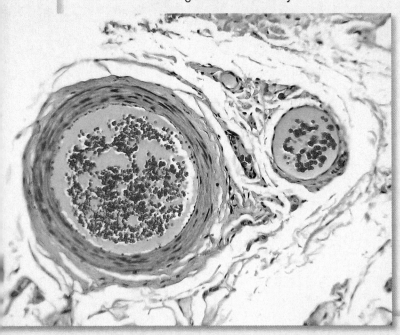
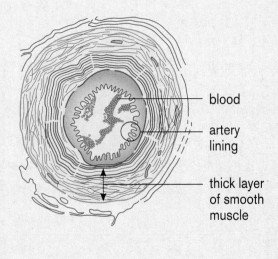

blood

artery lining

thick layer of smooth muscle

Capillaries

Capillaries are the smallest blood vessels in the body. The walls of these fragile vessels are only a single cell thick. They do not contain any muscle fibers. The vessels themselves are so fine that even tiny red blood cells must bend and twist to squeeze, in single file, through their narrow openings. (See **Figure 10.17**.)

It is impossible to sketch a capillary pathway. There are so many of these microscopic vessels that the sketch would look like a dense tangle of spider webs. Most tissues are so laced with capillaries that just about every cell is within *one-millionth of an inch* of a capillary. Oxygen and food molecules can diffuse out of the capillaries into the cells, and carbon dioxide and other waste molecules can diffuse from the cells into the capillaries. This is illustrated in **Figure 10.18**. No other blood vessels have walls that are thin enough for diffusion to take place. As you can see, these vessels are where the important functions of circulation occur.

Figure 10.17

This photograph of a capillary was taken through a microscope. Notice that blood cells must flow single file through a capillary.

Dr. Richard Kessel & Dr. Randy Kardon/Visuals Unlimited, Inc.

Figure 10.18

Small molecules can diffuse across capillary walls. What diffuses into capillaries? What diffuses out?

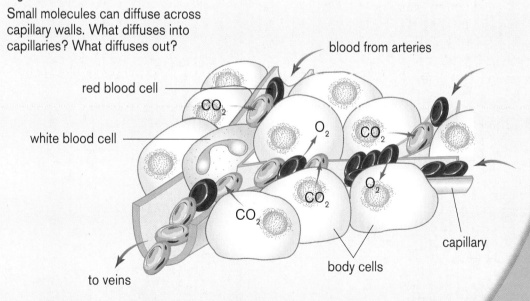

blood from arteries

red blood cell

CO_2

white blood cell

O_2

CO_2

CO_2

O_2

O_2

CO_2

capillary

to veins

body cells

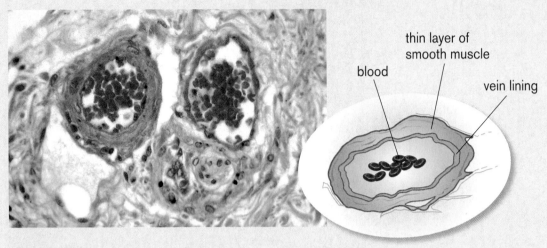

Figure 10.19

The photograph shows a cross section through a vein. On the right is a drawing of the same vein.

blood

thin layer of smooth muscle

vein lining

Veins

Very small **veins** drain blood from the capillary beds. Then, from these small veins, the blood flows into larger and larger veins on its way to the heart. The walls of veins are thinner than the walls of arteries. They contain less smooth muscle and elastic tissue. Because their walls are thinner, veins may look "collapsed" under a microscope. Find the layer of smooth muscle in the vein shown in **Figure 10.19**. How does it compare with the artery shown in **Figure 10.16**?

Blood flows fairly slowly in veins because there is not much pressure on the blood. In fact, there is so little pressure that the blood may flow backward. In order to prevent this, most veins contain one-way valves. (See **Figure 10.20**). These valves open when blood is flowing toward the heart and close when blood starts to flow backward. When the valves do not close properly, the veins swell and look like raised, bluish lines. You may have noticed them on people's legs; they are called **varicose veins**.

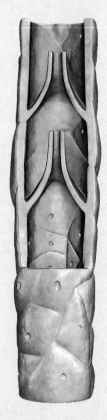

Figure 10.20
Valves in veins

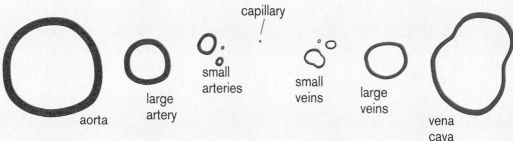

Figure 10.21

Cross sections of blood vessels

Sizes

There is quite a difference in the sizes of vessels when viewed in a cross section. The largest artery (the aorta) and the two largest veins (the venae cavae) are each about 2.5 cm (1 inch) wide. A capillary, however, is much smaller than the dot at the end of this sentence. Note the sizes of the vessels shown in **Figure 10.21**.

Who Needs a Circulatory System?

Single-celled organisms are bathed in their environment. Nutrients and oxygen can cross their cell membranes, and wastes can leave. Big organisms, however, are different. When wastes leave the cells in a penguin's pelvis, a chameleon's kidney, or even a worm's muscles, where do they go? Complex, multicellular animals, because of their size, have specialized systems to transport nutrients and gases. All of these systems have "tubing" of some kind, but not all of them have hearts. Each group of animals has its own characteristic circulatory system.

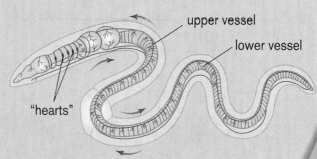

A Circulation Simulation

All body systems must work together for the body to function properly. This is especially obvious in the case of the circulatory and respiratory systems. Both systems have to be working, and in coordination, if your cells are going to get oxygen.

How can a simulation help you understand the way the circulatory and respiratory systems interact?

You will work as part of a large team to simulate the workings of the circulatory system. Once you have the system "circulating," you will be given oxygen and carbon dioxide "molecules." Use these molecules to simulate how the respiratory system interacts with the circulatory system.

Materials

- Handout, Role Cards—Blank Set (four sets per team)
- simulation packet (one per team)

Procedure
(Work in a team of students.)

Simulation 1: Moving a Blood Cell

1. Pass out the Role Cards. Each person on your team gets one card. The card tells your role in the simulation.

    a. If you are the coordinator, you are in charge of the simulation packet. If your group does not have a reader, you should also read the directions out loud for your team.

    b. If you are a structure, on the back of your card, write where blood is before it reaches your structure and where it goes when it leaves your structure. (Look back through this chapter if you need to review your facts.)

2. Work as a team. Position yourselves so you will be able to simulate the flow of blood through the circulatory system. The heart parts should sit with their backs together. Other structures should sit at least one arm length apart.

3. When you think you are in the correct position, sit down and have the coordinator take a "Blood Cell" out of the packet.

   a. Give the Blood Cell to the Left Ventricle. Have him or her pass the Blood Cell to the next part of the circulatory system. The "structure" who receives the blood should announce the name of his or her structure and pass the blood cell to the next structure.

   b. Continue until the "blood" returns to the Left Ventricle. If your team runs into a problem, rearrange yourselves until your system "flows" smoothly.

   c. When your team thinks you have a perfect system, have the coordinator ask your teacher to observe it in action.

4. When your teacher approves your circulatory system, get a new Role Card. Make sure you get a new role. Write notes on your card as you did before.

5. Again, get in position, sit down, and make sure you have a working circulatory system. You should be able to do this much more quickly, since you now have a better understanding of how the simulation is supposed to work.

6. When you think the system is working well, have your teacher observe it in action.

Simulation 2: Lots of Blood Cells

1. When your teacher approves your circulatory system, get another new Role Card. Again, write notes on your card like you did before.

2. Get in position, sit down, and make sure you have a working circulatory system. (Can you do this in 2 minutes or less?)

3. Have the Coordinator pass out the materials. Each person who is part of the circulatory system gets a "Blood Cell."

4. The pacemaker should announce the beat. Every time the heart beats, pass your Blood Cell to the next structure in the system. (Do you remember where the pacemaker is located? That's how to decide who should announce the beat.)

5. When you think your system is working perfectly, ask your teacher to observe it in action.

Simulation 3: Oxygen-Rich or Oxygen-Poor?

1. When your teacher approves your circulatory system, have the coordinator make sure that each part of the circulatory system has a "Blood Cell." (The lung cells and quadriceps cells don't get a blood cell.)

2. Oxygen and carbon dioxide are carried in the blood cells. Have the coordinator take the "Gas Molecules" out of the packet.

   a. Each part of the circulatory system should take one gas molecule—oxygen (red) if the blood in the structure is oxygen-rich, carbon dioxide (blue) if the blood in the structure is oxygen-poor. Put the "gas molecule" in the "cell."

   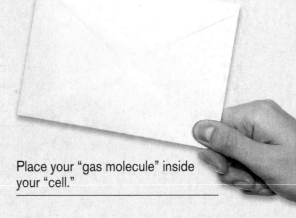

   Place your "gas molecule" inside your "cell."

   b. The lung cells get the extra oxygen molecules (red).

   c. The quadriceps cells get the extra carbon dioxide molecules (blue).

3. Have the pacemaker announce the beat. Every time the heart beats, pass your Blood Cell to the next structure in the system. Pay attention to the gas molecules. Exchange carbon dioxides and oxygens when appropriate.

4. When you think the system is working well, have your teacher observe it in action.

Simulation 4: Perfectly Coordinated?

1. Get new role cards and do the simulation one more time. If you understand what's going on, you should be able to get into position quickly and get your systems working.

2. When your system is working perfectly, ask your teacher to observe it in action.

Analysis Questions _ _ _ _ _ _ _ _

1. Explain at least three ways that this simulation illustrates how the circulatory and respiratory systems work together.

2. Explain at least two ways that this simulation does *not* illustrate how the real circulatory and respiratory systems interact.

3. What is one change in the rules that you would suggest to make this a better simulation?

Conclusion

Write a paragraph that answers the focusing question.

Extension Activity

Design and Do: Write a plan that explains how to change the simulation so it will show how nutrients are absorbed into the blood and circulate throughout the body. Include a way to show how the nutrients are carried to cells such as those in the quadriceps and how wastes are returned to the blood. Check your plan with your teacher and, if your teacher approves, try it with the class.

A Vital Body Fluid

You have studied the path blood follows as it travels through the heart and blood vessels. In order to understand the purpose of the circulatory system, you also need to know about blood itself.

When you look at blood under a microscope, you are looking at a liquid tissue, and like all tissue, it is made up of cells—red blood cells, white blood cells, and platelets. The liquid that the cells are in is called plasma. The makeup of blood is illustrated in **Figure 10.22**.

Plasma

Plasma makes up a little more than half of the blood, about 55%. This yellowish fluid is mostly water, but it also contains thousands of other substances. These include vitamins, glucose, fats, enzymes, hormones, and a variety of proteins. Two groups of proteins are especially important. One group helps blood clot. The other proteins are antibodies for fighting germs.

When you go for a medical checkup, the physician may suggest a blood test to check substances in plasma. For instance, blood may be drawn to check your blood sugar level or to analyze the fats in your blood.

Red Blood Cells

Most of the cells in blood are **red blood cells**. In 1 cubic millimeter of blood, there should be about 5 million red blood cells. They are small disk-shaped cells with depressions on both sides. (See **Figure 10.23**.) We need lots of these cells because they carry oxygen and carbon dioxide. These gases do not float around in the blood cells. Instead, they bind to large molecules in the cells called **hemoglobin**. Red blood cells could not transport much oxygen if they did not contain hemoglobin.

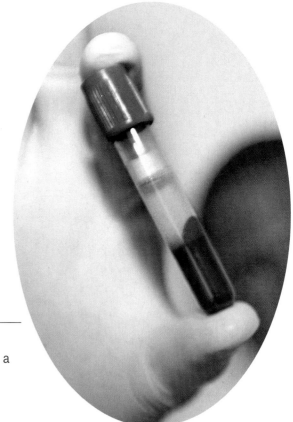

Figure 10.22

The cells can be separated from the plasma in a sample of blood.

A red blood cell is a package of hemoglobin molecules. As small as a red blood cell is, it contains about 250 *million* molecules of hemoglobin. If you could "unzip" the red cell membrane and look inside, you would see thousands of hemoglobin molecules and not much else. A mature red blood cell does not even have a nucleus. For this reason, a red blood cell is a good example of a highly specialized cell.

Blood is pumped out of the heart with such force that it takes only about a minute for a blood cell to travel throughout the body and return to the heart. This puts a lot of wear and tear on a cell. Because it does not have a nucleus, a red blood cell cannot repair itself when it is damaged. Thus, the average red blood cell lives only about 120 days.

White Blood Cells

Although **white blood cells** are called blood cells, they actually are found throughout the body. These cells can squeeze out through the little spaces in the capillary walls. Once they are in the tissues, they can do their jobs. There are several different kinds of white blood cells, each one having a specific job. All white blood cells fight infection. Some do so by producing antibodies; others engulf bacteria and destroy them. Still other white cells destroy harmful substances and break down fragments of old cells. (See **Figure 10.23**.)

There are usually about 8,000 white blood cells in a cubic millimeter of blood. If there are more, it could mean that the person has an infection.

Platelets

The third type of "cell" is the **platelet**. Platelets are not really cells, they are actually pieces of cell membrane. They help stop bleeding when a blood vessel is injured by plugging the hole. (See **Figure 10.24**.) If the injury is serious, platelets are important in causing a blood clot to form. A person with a low platelet count could have a bleeding problem.

Figure 10.24
These platelets have been magnified 14,000 times. Platelets are much smaller than red and white blood cells.

More Than a "Red Liquid"

As you can see, blood is more than a "red liquid." It is actually a yellow liquid that contains many substances important for life. It looks red because of the hemoglobin in red blood cells. Oxygen-rich blood is bright red; oxygen-poor blood is a purplish color. You probably have never seen blood when it is low in oxygen. Do you know why?

Analysis Questions

1. Compare and contrast red and white blood cells.

   a. List at least two ways they are alike.

   b. List at least two ways they are different.

2. What are platelets?

3. Why do you need hemoglobin in your red blood cells?

4. Describe three characteristics of red blood cells that make them well designed for their job.

5. What do white blood cells do?

6. Why does your body produce more white blood cells when you have an infection?

7. Challenge: Compare microscope slides of blood from humans, frogs, fish, and birds.

   a. Make a data table to summarize your observations.

   b. How are the blood smears alike?

   c. In what ways do they differ?

Extension Activity

Math Connection: In 1 cubic millimeter of blood, there are about 5,000,000 red cells. (See **Figure 10.25.**) What size box would it take to hold this many red blood cells if the cells were the size of marbles?

Figure 10.25

Heart Trouble? Not Me!

How healthy is your circulatory system? It's something to think about because heart disease is the leading cause of death in the United States. Take this short self-inventory to find out if you are at risk for heart disease. Read each of these statements and keep track of the ones that are true for you:

- I exercise for 30 minutes at least three times a week.
- Neither of my parents has high blood pressure.
- I am not overweight.
- I try not to eat high-fat foods.
- I do not smoke.
- Neither of my parents has had a heart attack.
- I am usually a calm person.

How many items are true for you? The more that are, the healthier your heart is. Usually, when students read this list, they ask questions such as the ones that follow.

What is heart disease anyway?

Any medical problem that involves the heart or blood vessels is often referred to as heart disease. Two common examples are heart attacks and high blood pressure.

My dad has high blood pressure. What does that mean?

Blood pressure refers to how hard the blood "pushes" against the walls of the vessels. A blood pressure reading consists of two numbers. For example, your blood pressure may be 115 over 70. The larger number measures the pressure your blood puts on your arteries when it spurts out of your heart. The smaller number tells how much pressure is on the arteries between beats. The cartoon in **Figure 10.26** may help you understand what the numbers mean.

Why is high blood pressure bad?

When blood pressure is high, the heart has to work too hard to pump blood through the body. Very high blood pressure can even cause a blood vessel to burst. If this happens in the brain, it can cause a stroke.

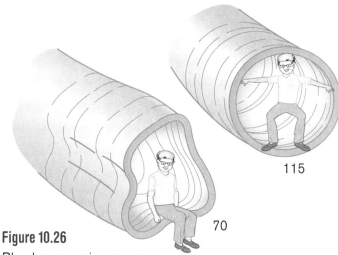

115

70

Figure 10.26
Blood pressure is a measure of the pressure that blood puts on your arteries.

Includes *sci*LINKS®

Topic: Cardiovascular Disease
Go to: www.scilinks.org
Code: MSLS3e279

What causes high blood pressure?

Sometimes there is no apparent reason for having high blood pressure, but it can result from other health problems. For example, people who have conditions such as diabetes or kidney disease often have high blood pressure. Emotions such as anger and pain or stressful situations can increase blood pressure. Also, some people inherit genes that make them more likely to develop high blood pressure.

Why is being overweight bad for the heart?

People who are overweight are more likely to have other problems that can lead to heart disease. For instance, they may develop diabetes or high blood pressure. Both of these conditions are hard on the heart. Heavier people also have more fat circulating in their blood.

Fat in the blood doesn't sound good, but how can that hurt the heart?

Some fats, such as cholesterol, can form deposits on the walls of the arteries. When this happens, the insides of the arteries become narrower. (See **Figure 10.27**.) Less blood can flow through the arteries. The parts of the body that receive blood from these arteries will not get enough blood.

Can I check my genes?

No, but a heart "gene screen" would be nice. You can be aware of what goes on in your family. Listen and ask questions about the health of your relatives. High blood pressure, heart attacks, high cholesterol or triglyceride levels, and strokes—these are all clues to family heart problems.

What can I do to prevent heart disease?

Look over the list at the beginning of this lesson. Which items on the list are under your control? For example, do you exercise for 30 minutes at least three times a week? When you exercise, you build up your cardiac muscle. A strong heart can pump more blood so it doesn't have to pump as often.

Think about it. How can you change your lifestyle in order to lower your risk for heart disease? Also, talk to your doctor. He or she will tell you if you need to take any special precautions to keep your heart healthy.

My parents are careful about what they cook because they both have "high triglycerides." What does that mean?

Your parents probably had their blood tested and found that they had high levels of one type of fat (triglycerides) circulating in their blood. By cutting down on certain fats in their diet, they may be able to lower their levels of this fat. You're lucky. Your folks' healthy cooking is good for your heart, too!

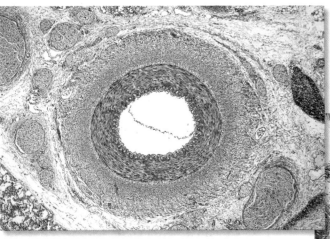

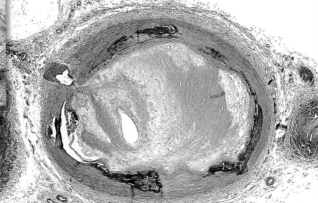

Figure 10.27
These are cross sections through two arteries. The one on the left is healthy. The one on the right is partly clogged by fat.

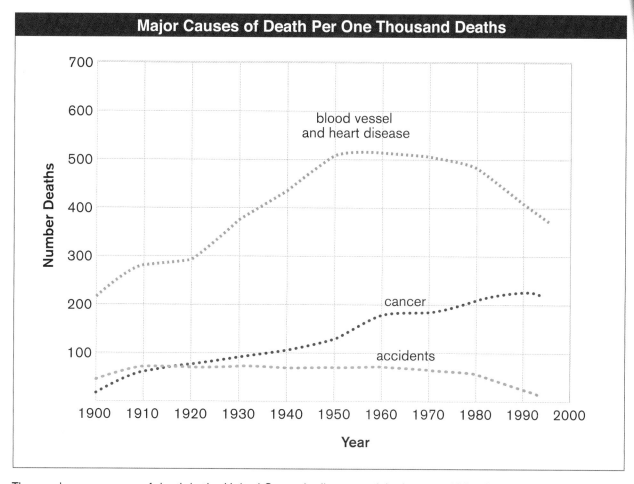

Major Causes of Death Per One Thousand Deaths

blood vessel and heart disease

cancer

accidents

The number one cause of death in the United States is diseases of the heart and blood vessels.

Me? I don't have to worry about heart disease!

Yes, you do. Studies have shown that fat can start accumulating in blood vessels of children. You may be at risk for heart disease, since both of your parents have high triglycerides. In fact, the genes you inherited from your parents are important indicators of whether or not you will get heart disease.

Summary Chart

Make a chart summarizing the information presented in this reading. List the problems that can affect the circulatory system, the causes of the problems, and how the problems can be prevented. Indicate whether or not each problem is contagious.

Extension Activities

1. **Health Application:** Look up other health problems that can affect the circulatory system to add to your chart. Examples of problems include anemia, leukemia, shock, heart attack, arteriosclerosis (hardening of the arteries), heart failure, and hemophilia. You may know of others you want to research.

2. **Careers:** Call the laboratory at a hospital and talk to the medical technologists who do blood tests. Find out what tests they do and why. You might also want to ask them to describe the most interesting and boring aspects of their jobs. Find out what training is needed to become a medical technologist.

Removing Wastes

You inhale about 2,000 liters of air every day. You eat quantities of fat, protein, and carbohydrate. You drink water, milk, and other liquids. These things are necessary for your survival. You must have them. And yet, you can't keep taking things in without getting rid of something. If you did, you'd soon burst. How do you get rid of all the waste products? How do you get rid of the things you don't need?

Carbohydrates and Fats

Carbohydrates (fiber, starch, and sugar) make up the bulk of your food. As you learned, your digestive system can't use *fiber*. Fiber travels through the large intestine and out the anus.

fiber (in) ⟶ fiber (out)

Starch is a big molecule that has to be digested; it is broken down into sugar molecules. *Fat* molecules are also digested and changed into sugar molecules. The *sugar*

molecules diffuse out of the small intestine and into the blood where they are delivered to all the cells in the body.

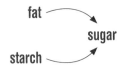

When you breathe, you inhale the oxygen your body needs to survive. Like sugar, oxygen, too, enters the blood and is delivered to every cell in the body. During cellular respiration, cells use sugar and oxygen to produce energy. Along with the energy, carbon dioxide and water are produced.

sugar + oxygen → energy = carbon dioxide + water
$(C_6H_{12}O_6)$ (O_2) (CO_2) (H_2O)

You get rid of the carbon dioxide through your respiratory system, and the water enters the blood vessels. Cells use some of the energy to stay alive, and some of the energy becomes heat. The heat warms the blood, and as the blood flows through the capillaries in skin, heat is lost to the air.

Water

You need water. About two-thirds of your body weight is water. Without water, blood wouldn't flow, lungs could not absorb oxygen, and cells would die. We need lots of extra water to dilute the wastes that are produced by the cells. If the wastes aren't diluted, they can be poisonous.

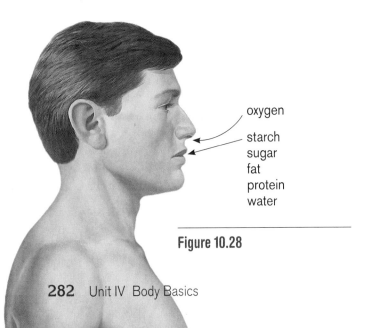

oxygen
starch
sugar
fat
protein
water

Figure 10.28

Most of our water comes from the liquids we drink and the food we eat. In addition, our cells make some water as a result of cellular respiration. We lose some water each time we exhale, some leaves the body through the digestive tract, and some evaporates from our skin. We still have excess water to get rid of, however.

Protein

The digestive system breaks down protein into amino acids. The amino acids circulate in the blood to every cell in the body. The cells use some of the amino acids, but they cannot store the extra ones. Extra amino acids are broken down.

Like fats and carbohydrates, amino acids are made up of carbon, hydrogen, and oxygen. They also contain nitrogen. When an amino acid is broken down, the carbon, hydrogen, and oxygen are used to make sugar. The nitrogen can combine with hydrogen to make ammonia, but ammonia is poisonous, so our systems convert extra nitrogen into urea. **Urea** is a waste product formed when proteins are used by the body. This waste product enters the blood, but how does it get out of the body?

protein ⟶ amino acids ⟶ sugar + urea

Needed: A Waste Disposal System

As you can see, blood accumulates extra water and urea. We have to get rid of both these substances. This function is taken care of by another system, the **urinary system**.

The story of the urinary system starts with the **kidneys**. They filter the blood, removing the urea and excess water. The kidneys are two bean-shaped organs, located just above the waist on either side of the spine. They are protected by the ribs and a cushion of fat. Each kidney is about 10 centimeters long. A kidney has a very complex structure. If you could cut

one open and examine it with a high power microscope, you would see a mass of tiny tubes and blood vessels.

Every day, arteries deliver about 170 liters (42 gallons) of blood to the kidneys. Once in the kidneys, the artery branches, forming smaller and smaller arteries until finally they become a thick mass of capillaries. The tubules in the kidney are closely associated with the capillaries. As blood flows through these capillaries, water, urea, and other small waste molecules diffuse out of the capillaries into the tubules. The blood that leaves the kidney no longer contains wastes; only cells and useful molecules remain.

The waste fluid left in the tubules is 95% water; the other 5% includes small amounts of calcium, other salts, and urea. This fluid is called **urine**. It flows down the tubules, which eventually empty into the collecting space within the kidney.

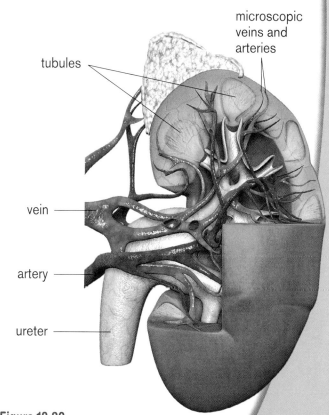

Figure 10.29
The kidney is a mass of veins, arteries, and tubules.

Figure 10.30

Urine is 95% water. It is usually a pale yellow, but it may look darker if a person has not been drinking enough water. A slight pink color may indicate that there is blood in the urine or that the person has been eating beets.

Figure 10.31

The urinary system

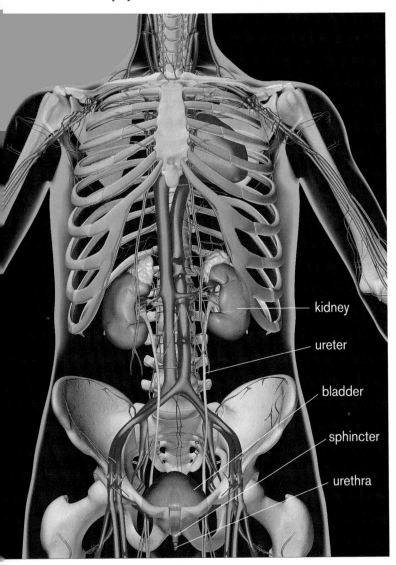

kidney

ureter

bladder

sphincter

urethra

Urine continuously trickles out of the kidneys through two long, narrow tubes called **ureters**. These tubes carry urine to the bladder. The **bladder** is a muscular pouch that stores urine. Smooth muscle fibers in the walls of the bladder expand as it fills and contract when it empties. An adult's bladder can stretch to hold about 500 mL (2 cups) of urine. Locate the ureters and bladder in **Figure 10.31**.

A ringlike **sphincter** muscle prevents urine from leaking out of the bladder. This muscle is important for bladder control. When the bladder is full, nerves send messages to the brain that the bladder needs to be emptied. Young children cannot control their sphincter muscles, so they may have "accidents." As they mature, they develop better control of this muscle. Urine leaves the body through a muscular tube called the **urethra**.

This waste removal process goes on every minute of every day. If both kidneys stop functioning, wastes build up and the person will die within a few weeks.

As you can see, the body has several ways of getting rid of waste products. We use the respiratory system to get rid of carbon dioxide. Water and some salt are lost through the sweat glands in our skin. Fiber and other indigestible parts of our food are eliminated through the digestive tract. Urea, diluted by water, is removed by the urinary system. If any one of these systems is not working, serious health problems, or even death, can result.

Analysis Questions

1. Fill in the handout, *Urinary System*. It will help you learn the parts of the system and see the relationship between the circulatory and urinary systems.

2. Think about urine.
   a. Where is it produced?
   b. What does it contain?

3. Make a table that summarizes what happens to substances that enter your body.
   a. In the first column, list the substances that you take into your body every day.
   b. In the second column, describe what happens to each substance once it is in the body.
   c. In the third column, list the wastes that are produced.
   d. In the fourth column, explain how your body gets rid of each waste.

4. HB Connections: Explain how body systems work together to get rid of wastes. In your description, make sure to mention at least five body systems.

Extension Activity

Health Application: Ask a doctor or the school nurse what they look for when they do a urine test (urinalysis). See if you can have a copy of your last urinalysis results.

Fertilizer!

Different groups of animals have evolved different ways of getting rid of nitrogen. Birds, for example, do not produce urine. Urine contains lots of water and water is heavy so it would weigh down the bird. Instead of urea, birds produce a semi-solid substance called uric acid. Where birds gather, uric acid can accumulate, creating quite a build-up. Sometimes these uric acid deposits (called guano) are harvested and sold as fertilizer. Thus, the nitrogen is returned to the soil and can be used again by plants and soil microorganisms.

Station Test

How well do you understand the circulatory system? You started this chapter by studying the heart. You investigated how long it took for your heart to return to a normal pace after you stopped exercising, and you looked at a preserved heart. Then you studied the circulatory pathway and you studied blood. Now it's time to find out what you learned.

What do you think was the most important thing you learned by studying this chapter?

Think about this question while you take a station test. What's a station test? You will go from station to station and answer questions. At some of the stations you may find a preserved heart, a diagram, a microscope, or something else you will need to use to answer the question.

- pencil and paper

Procedure

1. Listen for your teacher to tell you which station to go to first.
   a. You will need to take a pencil and sheet of paper with you.
   b. Sit down when you get to a station. Do NOT touch anything.
2. Your teacher will tell you how long you will have to answer the question. Write down the number of the station and answer the question. Do not change stations until your teacher calls time.
3. Each time you get to a new station, sit down and answer the question. If you begin at any station other than station 1, when you finish the last station, go back to station 1 and complete all the stations.

Conclusion

How do you think you did on the test? What questions were easy for you to answer? Write a paragraph in answer to the focusing question.

Body Controls

Chapter 11

The Nervous System

What does this photograph tell you about the brain?

11-1

Brain Characteristics

The brain may well be the single most important organ in the human body. Although it looks quite simple, it is incredibly complex. Some people question whether we will ever totally understand how the brain works. How would you describe a brain? What do you know about its structure, and the tissues it's made up of?

In this chapter, you will study the nervous system. In each of the activities in this chapter, there will be some mention of the brain. It will help you understand the concepts of the brain if you have a good understanding of its structure. Which parts of a brain can you locate and name?

How does a brain compare with other organs you have observed?

There are several parts to this activity. Follow the procedure and the directions in the observation guide while you inspect a preserved sheep brain. Then read about the human brain while you color and label a diagram.

Materials

- preserved sheep brain
- dissecting pan
- paper towels
- Handout, *Observation Guide*
- colored pencils
- soap
- Reading, *The Human Brain*, page 293

Procedure   (Work in a group with two or three other students.)

1. As a group, answer the focusing questions.
2. Read the rest of this procedure so you know what to do. As a group, decide how you will share responsibilities. You may want to assign roles to make sure you finish on time.
3. Follow the directions in the *Observation Guide*.
   a. Handle the preserved brains carefully! Brain tissue is fragile.

b. As you finish each step, record your observations in the space provided.

4. Read *The Human Brain*. While you read, label and color the diagram of the brain according to the italic directions in the reading.

5. Answer the analysis questions and write your conclusion.

Analysis Questions — — — — — — — — —

1. On average, how much does a human brain weigh?

2. Name the three main parts of the brain, and an important function of each part.

3. Which part of the brain is most important for controlling each of the following?

   a. seeing

   b. breathing

   c. solving a problem

   d. sneezing

   e. running

4. Which tissues did you *not* find on the sheep brain? Why do you think these tissues are not needed?

5. List at least two examples of things that are worn to protect the brain or spinal cord during specific sports activities.

Conclusion

Now that you have observed a brain, write several sentences to answer the focusing questions.

Extension Activity

Going Further: Many scientists are studying the way the brain works. One area of research is about left and right brain functions. What does it mean to have a "dominant right brain"? Write a one-page summary of this idea. Do you think your right or left brain is dominant?

Alike Yet Different

bird

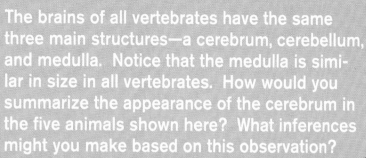

The brains of all vertebrates have the same three main structures—a cerebrum, cerebellum, and medulla. Notice that the medulla is similar in size in all vertebrates. How would you summarize the appearance of the cerebrum in the five animals shown here? What inferences might you make based on this observation?

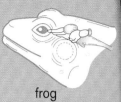

frog

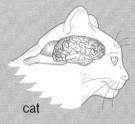

cat

shark

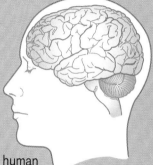

human

□ cerebrum

□ cerebellum
□ medulla

Figure 11.1

The Human Brain

A human brain weighs about 1.3 kg (2.5 to 3 pounds), making it one of the largest organs in the body. It has a complex structure made up of three main parts—the cerebrum, cerebellum, and medulla, as well as many smaller parts.

The Learning Center

The largest, most obvious, structure of the brain is the **cerebrum**. (Find it on **Figure 11.2**.) It is the structure most people think of when they hear the word *brain*. The cerebrum is made up of a right and left half. Its thin outer surface contains billions of nerve cells. Even though this layer is only a few millimeters thick, it is involved in many of the things we do. Different areas of this layer have different functions. The front of the cerebrum is the "brainy" part of the brain. It controls many aspects of thinking, such as decision making, planning, and learning.

Figure 11.2 Compare this drawing of a human brain with the preserved sheep brain you observed.

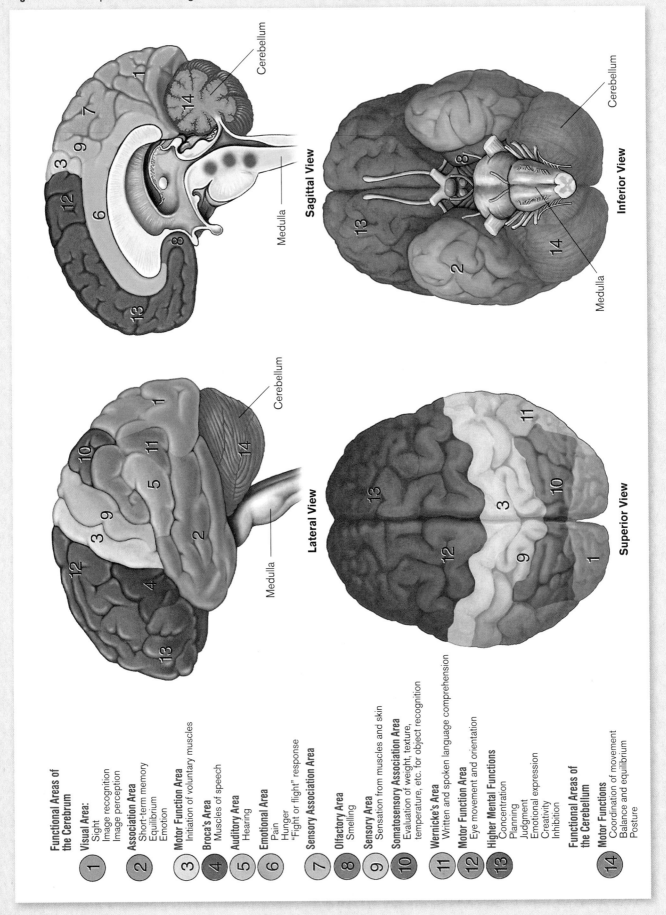

Sagittal View

Inferior View

Lateral View

Superior View

Functional Areas of the Cerebrum

1. **Visual Area:**
   Sight
   Image recognition
   Image perception

2. **Association Area**
   Short-term memory
   Equilibrium
   Emotion

3. **Motor Function Area**
   Initiation of voluntary muscles

4. **Broca's Area**
   Muscles of speech

5. **Auditory Area**
   Hearing

6. **Emotional Area**
   Pain
   Hunger
   "Fight or flight" response

7. **Sensory Association Area**

8. **Olfactory Area**
   Smelling

9. **Sensory Area**
   Sensation from muscles and skin

10. **Somatosensory Association Area**
    Evaluation of weight, texture, temperature, etc. for object recognition

11. **Wernicke's Area**
    Written and spoken language comprehension

12. **Motor Function Area**
    Eye movement and orientation

13. **Higher Mental Functions**
    Concentration
    Planning
    Judgment
    Emotional expression
    Creativity
    Inhibition

Functional Areas of the Cerebellum

14. **Motor Functions**
    Coordination of movement
    Balance and equilibrium
    Posture

The very back of the cerebrum receives messages from the optic nerve and is responsible for vision. Other areas of the cerebrum are responsible for hearing, taste, and touch sensations. Find these areas on **Figure 11.2**.

Color the cerebrum blue. Label the areas that are involved with thinking, hearing, and seeing.

Muscle Coordination

The cerebellum is located under the cerebrum near the base of the brain. (Find it on **Figure 11.2**.) The **cerebellum** coordinates muscle movement. When you decide to move, your cerebrum starts the activity, but the cerebellum quickly takes over.

Learning to ski, playing a piccolo, or riding a unicycle requires intense concentration—that's the *cerebrum's* job. Once the motions are learned, they become more automatic and can be regulated by the *cerebellum*. Think about it. How much thought do you give to walking? If you decide to skip backward, could you do it automatically, or would you have to concentrate on your movements?

Color the cerebellum green.

Basic Body Control

Right below the cerebellum is the **medulla**. It connects the brain with the spinal cord and looks like a slightly enlarged part of the spinal cord. (Find it on **Figure 11.2**.) Even though the medulla is only a few centimeters long, it is extremely important. It monitors basic body functions such as heart rate, blood pressure, and body temperature. Reflex actions such as coughing, swallowing, blinking, and vomiting are also controlled by the medulla.

The medulla controls all of the automatic functions that keep us alive. If the medulla is injured, death may result.

The medulla narrows to form the **spinal cord**. It is made of nerves that carry messages to or from the brain. If the spinal cord is cut, paralysis from the point of the injury and down results. That's because the brain is no longer able to communicate with the nerves below the injury. No amount of willpower can move muscles when the nerves are severely damaged.

Color the medulla red. Label the spinal cord and color it purple.

Protection

Because the brain is so fragile and important, it is well protected. Without the support of the bony **cranium**, the brain would collapse. That's because

the brain does not contain any tissues that provide support. Although the cranium is a protective case for the brain, it is not enough. Every bump on the head would jar the brain if it weren't for the **fluid** that surrounds it. A tough **membrane** keeps this fluid from leaking out. The spinal cord is also protected by bone—the vertebral column.

Label the membrane and color it yellow. Shade the space that contains the fluid a pale yellow. Label the cranium and color it brown.

Injuries to the brain or spinal cord often result in death or paralysis. Special protective devices have been designed to prevent such damage. For example, if you are involved in sporting activities, you may have worn protective helmets, pads, or back supports, designed to prevent injuries to the head, neck, and/or back.

This reading is just an introduction to the amazing structure we call the brain. Throughout this chapter you will learn more about how the brain works with the other parts of the nervous system, and allows us to do all the things we do.

The Nervous System

Although the brain is extremely important for regulating and coordinating body functions, it cannot act alone. It works with other organs as part of a system, specifically, the nervous system. In addition to the brain, the **nervous system** is made up of the spinal cord, nerves, and receptor organs. This system has two basic functions; coordinating what goes on inside the body and controlling how the body responds to the environment.

Nerve Cells

The nervous system is made up of billions of nerve cells. The brain alone contains an estimated 100 billion nerve cells. The spinal cord and nerves are bundles of nerve cells.

Nerve cells are specialized to carry messages from one part of the body to another. As you might expect, these cells do not look like other cells. In general, they are long and skinny with many projections. (See **Figure 11.3**.)

A message starts at one end of a nerve cell, moves through the cell body, and continues down to the other end. The message can travel in only one direction. Some cells carry messages from an organ to the brain; other cells carry messages from the brain to an organ. Many nerve cells act as "connectors" and pass messages on to other nerve cells.

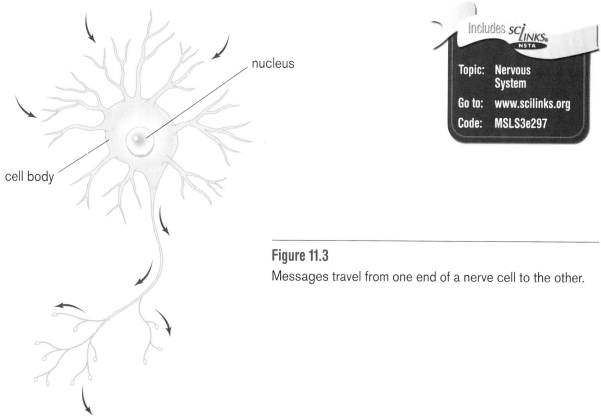

nucleus

cell body

Includes SC*LINKS*
NSTA

Topic: Nervous System
Go to: www.scilinks.org
Code: MSLS3e297

Figure 11.3
Messages travel from one end of a nerve cell to the other.

Senses and Sense Organs

The brain does not suddenly "decide" to do something. It responds to a message it receives. The messages originate in receptor cells. **Receptor cells** are specialized nerve cells. Each type of receptor cell receives a particular kind of signal.

Some receptors monitor how things are working inside the body. For example, some receptors are sensitive to the level of carbon dioxide circulating in the blood. Based on the messages the brain receives about this gas, your breathing rate may speed up or slow down. Other receptors monitor things such as blood pressure, blood sugar level, and even the amount a muscle cell has been stretched. Depending on the signals the brain receives, you may reach for a snack, get a glass of water, or shift position in response to a tingling in the foot.

Other receptors are located in **sense organs**. These receptors allow you to take in information about the environment. Most people are aware of the five basic senses: smell, taste, touch, vision, and hearing. There's a different type of receptor for each type of environmental signal. The general shapes of some of these receptors are shown in **Figure 11.4**. The skin contains the specialized receptors that can detect various types of touch including pain, pressure, and temperature. Receptors in the eye detect light, and those in the ear are sensitive to air waves. The tongue and the nose are sense organs that detect chemical substances, allowing you to taste and smell.

Nerves

Nerves, like nerve cells, can only carry messages in one direction. Some nerves carry messages *to* the spinal cord and brain that help you sense the environment. Other nerves carry messages *away* from the brain and spinal cord; which allows you to respond to the environment. The part of the nerve that attaches to the spinal cord is fairly large. It branches again and again until it forms a network of tiny nerves. These small nerves reach all parts of the body.

Figure 11.4

Each type of receptor cell is specialized to receive a particular type of signal.

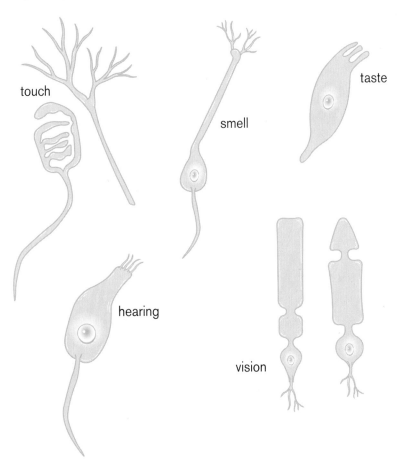

touch

smell

taste

hearing

vision

Figure 11.5

A network of nerves reaches all parts of the body.

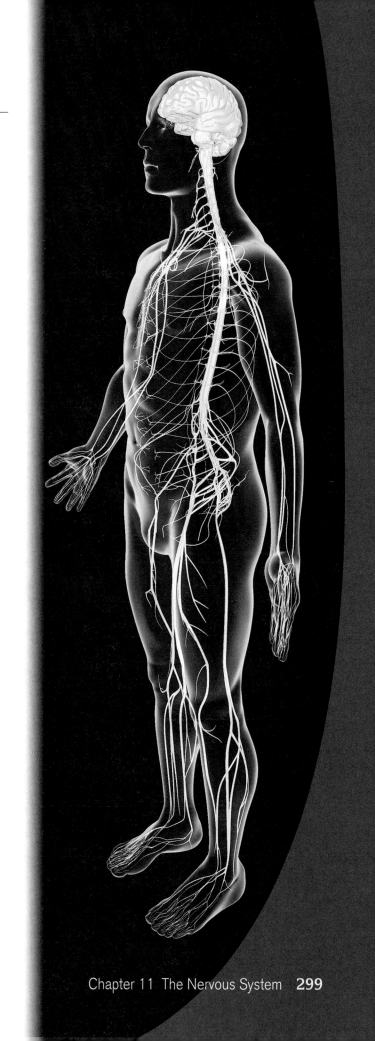

Responding to the Environment

The nervous system evolved to improve an organism's chance of survival. It allows an organism to do things such as detect danger, seek shelter, and locate food. You, too, rely on your nervous system for survival. You are constantly responding to the environment even though you may not realize that's what you are doing. For example, if it's cool outside, you may put on a sweater. Even this simple reaction involves the action of many nerves. Thousands of receptors in your skin send messages to your spinal cord and then to the brain that your skin temperature is dropping. When your brain receives the message, you can then decide to respond. Messages will travel from your brain along nerves to the muscles in your shoulders, chest, back, arms, and hands, allowing you to position the sweater so you can put your arms in the sleeves and put it on.

It takes at least a minute to put on a sweater. Sometimes you need to react much more quickly in order to avoid danger. For example, if you touch a hot pan, you would quickly pull your hand away. Nerves from your hand would send pain messages to your spinal cord. In the spinal cord, the message is immediately relayed to another nerve that tells your muscles to contract. Your hand jerks away from the pan. This is called a **reflex** response. The message does not need to go to the brain for you to react.

You can understand that it takes the coordinated action of the entire nervous system for you to respond to your environment.

Figure 11.6

Reflex reactions allow you to react quickly. The message goes from the hand along a nerve to the spinal cord and along another nerve back to the hand.

Analysis Questions

1. What are the main parts of the nervous system?

2. What is the main function of the nervous system?

3. What is a receptor cell?

4. What is the difference between a nerve and a nerve cell?

5. What is meant by the term *reflex*?

6. List at least three examples of reflexes.

7. Consider what you know about the circulatory and nervous systems.

   a. List at least two ways these systems are similar.

   b. List at least two ways these systems are different from each other.

8. Challenge: Observe a microscope slide of nerve cells.

   a. Sketch what you see. (Check the rules for making a drawing through a microscope.)

   b. Describe at least three observations that would help you recognize nerve cells if you were looking at unlabeled microscope slides.

   c. Estimate the length of one nerve cell. Be prepared to explain how you made your estimation.

Extension Activity

Careers: Physical therapists help people who have suffered spinal cord injuries. If there is a physical rehabilitation center near you, call and ask to talk with a physical therapist. Decide in advance what questions you will ask. Here are some questions you will want to include.

1. How do most spinal cord injuries occur?

2. What does therapy typically involve?

3. What training is required to become a physical therapist?

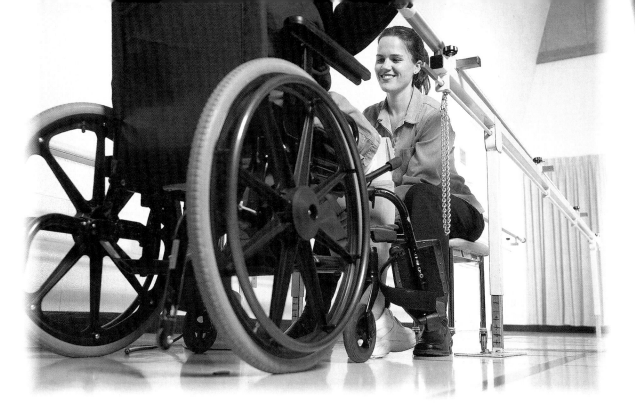

This physical therapist is working with a person whose spinal cord has been injured.

Rapid Responses

A nervous system allows animals to respond rapidly to changes in the environment. Animals such as sponges don't have a nervous system, but they do have nerve cells that allow them to respond to environmental changes.

Even worms have a main nerve cord with a "swelling" at one end that is sometimes called a brain. Many animals have a central nerve cord that carries messages from one end of the animal's body to the other. In vertebrates, this nerve cord is called the spinal cord and it runs down the middle of the back. It is protected by vertebrae.

Hydra

nerve net

Worm

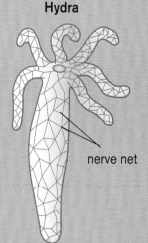

brain   nerve cord

Seeing Things

You are probably very familiar with your sense of sight. You constantly use it to take in information about your environment. And, as with so many familiar things, you probably take your sight for granted. In this lesson you will study two aspects of vision—distance vision and side vision. Your **distance vision** lets you focus on the head of a pin or on a building miles away. You also have **side vision**. This allows you to see an object even if it is not directly in front of you.

What can you learn about sight by measuring distance vision and side vision?

Find the answer to this question by testing your own vision and analyzing the data you collect.

- yellow and blue colored pencils
- Handout, *Seeing Things*
- The other materials you need are set up at stations around the room.

Procedure   *(Work with a partner.)*

1. Go to one of the stations and take turns doing the measurements. When you both have finished at one station, go to the next. It does not matter which station you do first.

2. When you have been to both stations, answer the analysis questions.

Station: Distance Vision

Materials

- eye chart
- line on floor, 20 feet from chart

1. Stand facing the eye chart with your toes just touching the line. If you wear glasses, decide whether you want to test your vision with them on or off.
2. Close and cover one eye.
   a. Start with the top line on the chart and read the letters.
   b. If you miss one letter, that's okay; go on to the next line. When you miss two or more letters, record the numbers by that line. (For example: 20/60, 20/200.)
3. Close and cover the other eye. Again, record the numbers given for the smallest line you can read.

Figure 11.7

Station: Side Vision

Materials

- two metersticks
- four objects of different colors
- focus object

You will need to do this test four times. First you will test the vision of your right and left eyes on the left side. Then you will test the vision of both eyes on your right side. Use a different colored object each time you do a test.

1. Place your chin on the tape on the table and look straight ahead at the focus object.

2. Start by taking measurements using the left meterstick.

   a. Have your partner stand on your left side.

   b. Close and cover your left eye. Keep your right eye straight ahead. It will be tempting to peek to see what your partner is doing. Avoid the temptation!

   c. Your partner will choose one of the colored objects. Do not look to see which object was chosen.

   d. Have your partner slowly move the object up the left meterstick.

      • Tell your partner when you first notice movement. Your partner should remember this distance.

      • Continue until you can correctly identify its color. Your partner should note the distance.

      • Record both distances on your handout. (Make sure you record your data in the correct spaces.)

   e. Close and cover your right eye and repeat the procedure. (Remind your partner to pick a different object.)

3. Repeat the procedure using the right meterstick. Record the distances where you first see the object and where you identify its color.

4. When both you and your partner have taken measurements, return to your seats and follow these directions to plot your data on the diagram at the bottom of your handout. ***Note: Plot data for movement only, not for color.***

   a. Use the yellow colored pencil to plot the data for your *right* eye.

- Along the left meterstick, put a dot to indicate when you first saw movement with your right eye. Draw a line from the right eye to this dot.

- Along the right meterstick, put a dot to indicate when you first saw movement with your right eye. Draw a line from the right eye to this dot.

- Shade the space between these two lines to indicate your field of vision.

b. Use the blue colored pencil and repeat these steps for the data with your *left* eye.

Analysis Questions

1. Read the information about distance vision in the box at the end of this lesson. What does it mean if a person has:

   a. 20/20 vision?

   b. 20/150 vision?

2. Why is it called "nearsighted" if a person has a distance vision of 20/100?

3. Think of at least three possible reasons for error in your distance vision measurements.

4. In the side vision test, which could you see first, color or movement?

5. Read the information about side vision in the box at the end of this lesson, and then think about some animals with side vision that you are familiar with.

   a. What are three examples of animals you think would have a wide range of vision with a narrow overlap?

   b. What do you think is an advantage to having a wide range?

   c. What are three examples of animals you think would have a narrow range of vision but a wide overlap?

   d. What do you think is an advantage to having a narrow range but a wide overlap?

6. List examples of things that would be difficult to do without side vision.

7. List at least three possible reasons for error in your side vision measurements.

Conclusion

Write a short paragraph to answer the focusing question.

Extension Activity

Careers: Show your data to an ophthalmologist (eye doctor). Find out if your measurements seem reasonable. Ask him or her to explain the other tests that are done during an eye exam.

Vision Measurements

Distance Vision

If you have normal distance vision, you can read the line labeled "20/20." Suppose the smallest letters you can read are on the line labeled "20/70." This means that you must stand 20 feet from the chart in order to read the line that a person with normal vision can read at 70 feet. If you must stand closer to read than other people, you are said to be nearsighted. A person who is so nearsighted that the best he or she can see with glasses is 20/200 is considered legally blind.

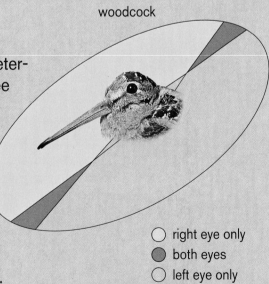

cardinal

woodcock

○ right eye only
● both eyes
○ left eye only

Side Vision

The position of an animal's eyes on its head determines how much side vision it has. You can see that the cardinal, with eyes slightly off to the side of its head, has a wide field of vision with only a very narrow area of overlap. The overlap is the area that animals see with both eyes. This binocular vision provides better depth perception. The woodcock's eyes are set very far back, giving it a full range of view. It even has areas of overlap both in front and in back of its head.

Eyes and Vision

Eyes allow you to take in information about the environment, but in order to do so, conditions have to be "right." That is, if you are viewing an object, it has to be positioned where you can see it. You have a limited range of vision, so the smaller an object, the closer you must be to it. You also have a limited field of vision; you cannot see things that are too far to the side or behind you. To see, you also need light. Why do you need light to see? In this reading, you will find out why you cannot see without light, and you will learn about the structure and function of a mammalian eye.

External Structure of an Eye

When you look at your eye in the mirror, you are seeing only a small part of this important sense organ. You are seeing the entryway into the eyeball. The black circular structure is called the **pupil**. It is the opening that lets light into the eyeball. The colored part of the eye, called the **iris**, is made up of connective tissue and smooth muscle fibers. These muscle fibers contract and relax to control the size of the pupil, which in turn controls the amount of light that enters the eye. The bigger the pupil, the more light it lets into the eye.

Internal Structure of an Eye

Figure 11.8 is a drawing of an eye as it looks in a cross section. As you read, be aware of the structures in bold type, and find each of them on the drawing.

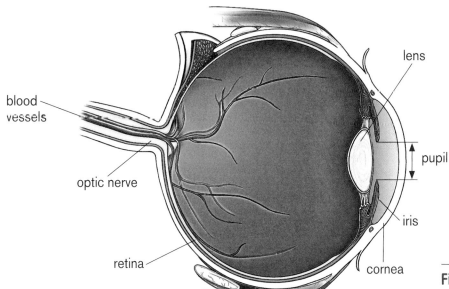

blood vessels

optic nerve

retina

lens

pupil

iris

cornea

Figure 11.8
This drawing shows what a human eyeball looks like in cross section.

The **cornea** is the tough outer layer that protects the front of the eye. It is transparent so light can pass through it. The cornea covers the **iris** and the **pupil**. Find these structures in **Figure 11.8**. Directly behind the pupil is the **lens**. It is a small disk made of clear living cells. The lens is held in place by muscles. The muscle fibers contract and relax, causing the lens to flatten or bulge. Depending on the shape of the lens, you can focus on objects that are near or far away.

There are fluid-filled spaces in front of and behind the lens. The fluid bathes the cells and helps maintain the shape of the eyeball. The **retina** lines the back of the eyeball. It is made up of highly specialized receptor cells that change a light image into a nerve message. This message travels along the **optic nerve** to the brain. When all of these structures are functioning correctly, you can see.

The Vision Pathway

Imagine that you are looking at a French horn and think about what has to happen in order for you to see it. (Refer to **Figure 11.9** while you read. It should help you picture what is going on.) Light rays strike the horn and some of them bounce off in your direction. Some of the rays that reach your eyeball go through the pupil and are focused when they go through the lens and come to a point on the retina. When

the receptor cells in your retina detect the light rays, they quickly send a message along the optic nerve. The optic nerve carries this message to the vision center at the back of the brain. The brain interprets the message so you can see the French horn.

Seeing Color

The receptor cells in the retina detect light. Most of them detect dim light, allowing you to see shape and motion (but not color) at night. Other cells are sensitive to specific colors of light. These are clustered together in a spot near the center of the retina. That's why you spotted motion before you saw color when you tested your side vision. In order to see color, you had to wait until the light reflecting off the object could reach the center of your retina.

Some people cannot see particular colors because they are missing the cells that detect those colors. For example, it's fairly common for males to have trouble detecting the difference between red and green. When this happens, they are said to be red-green color-blind. Their receptor cells do not detect red and green light.

Well, does it make sense? Can you understand how vision works? If not, go back, reread this section, and make a list of questions to go over with your teacher.

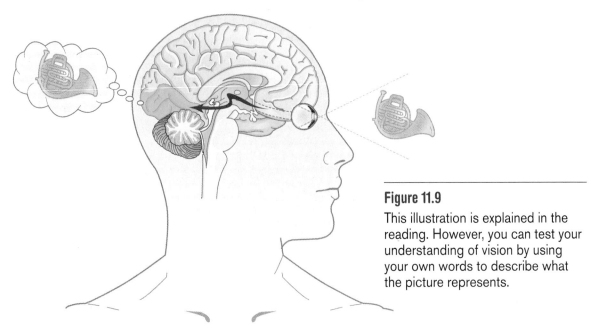

Figure 11.9

This illustration is explained in the reading. However, you can test your understanding of vision by using your own words to describe what the picture represents.

Analysis Questions

1. Make a chart that summarizes information about the structure and function of each part of an eye.

2. These structures are part of the vision pathway: brain, cornea, lens, optic nerve, pupil, retina.

   List them in the order they are used when you look at an object.

3. Why is light necessary for you to see?

4. Do you think drawing **A** or drawing **B** in **Figure 11.10** shows the way a pupil looks in bright light? Explain your answer.

5. Some people are born with damaged retinas. Why is it impossible for them to see?

6. What kinds of cells are you using when you look at an object? (Be as complete as you can.)

Extension Activity

Science Connections: When light passes through a lens, it is bent. Learn more about how lenses work. Ask a physical science teacher to help you find some lenses. Decide what is meant by the term "focal distance," and then find out how the shape of a lens affects focal distance.

A

Figure 11.10

B

Light

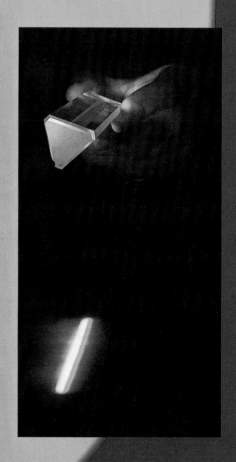

Light travels in straight lines at 186,000 miles per second. It continues going straight until it hits something such as a brick, a fingernail, or a leaf. Then some of the light is absorbed and some of it bounces off.

Sunlight contains the full spectrum of colors. When it goes through a prism, the light is "taken apart" into the familiar colors of the rainbow—red, orange, yellow, green, blue, and violet. The color of an object depends on which rays it absorbs and which it reflects. Things that look white don't absorb any rays; black objects absorb them all. A leaf looks green because it absorbs all wavelengths except green ones.

Look at your pencil. What color light rays is it absorbing? What color is bouncing off?

Eyeballs

Now that you've learned about the parts of an eyeball, do you think you can recognize them? Here's your chance to test yourself. Other mammals have eyeballs that are very much like a human's, so you can look at an eyeball from a sheep or a cow and learn about the structure of your own eye.

Which parts of an eye can you identify?

Find the answers to this question by carefully dissecting a preserved eyeball.

- preserved eyeball
- dissecting pan
- scissors
- paper towels
- Handout, *Eyeballs*

Procedure
(Work with a partner.)

Part A: Getting Ready

1. The names of the structures that make up an eye are printed in bold type in the directions that follow. As you find each structure, record your observations on the handout.
2. During the dissection, you will remove parts of the eyeball one at a time. Use a paper towel to store the parts. When you remove a part, set it on the paper towel and label it.

Part B: External Structures

1. Find the **fat** and **muscles** attached to the back and sides of the eyeball. (Remember to record your observations for fat and muscle.)
2. Locate the *top* of the eye. Try to determine how the eye was positioned in the head.

3. The **optic nerve** comes out of the back of the eyeball. It may be hidden by other tissues. If you can't find it, look at it on another team's preserved eye.

4. Feel the **cornea**. (It's the tough outer layer that covers the front of the eye.)

5. If there is a lot of fat around the eyeball, carefully trim off some of it. Put a piece of fat on the paper towel and label it.

6. Make sure you have recorded your observations for the first four structures on the data table.

Part C: Internal Structures

1. Prepare to cut open the eyeball.
   a. Choose a spot about halfway between the front and back of the eyeball. (See **Figure 11.12**.) Carefully snip at this spot with the scissors to form a short slit. Do not plunge the point of the scissors in too deeply.
   b. Cut about three-fourths of the way around the eye.

2. Gently open the eyeball. If there is a *jelly-like material* inside, carefully remove it. (In living animals, this substance is more liquid.)

3. Try to locate where you think the **optic nerve** enters the eyeball.

4. On the back surface of the eyeball you can see a whitish tissue; it's the **retina**.
   a. On your handout, describe the relationship between the retina and the optic nerve.
   b. Carefully lift out the retina and put it on your paper towel.

5. Behind the retina is a layer of dark tissue called the **tapetum**. It contains many small blood vessels. (This layer of tissue is not found in human eyes.) Remove it and put it on the paper towel.

6. Look at the front of the eye from the inside. Find the **lens**.
   a. Gently lift the lens to see how it is held in place.
   b. Remove the lens and rinse it in water. Try to look through the lens and read. Note how the letters appear.

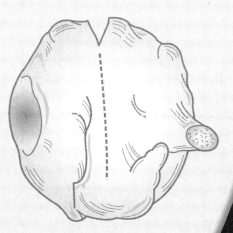

Figure 11.12

7. Now you can see the entire **iris**—it's the colored part of the eye.
   a. Lift off the iris and set it on the paper towel.
   b. The iris surrounds the opening called the **pupil**.

8. Again, find the **cornea** and describe how it looks from the inside. (You observed it from the outside in step 4 of Part B.)

9. A tough *outer tissue* encloses and protects the eyeball.

You just inspected an eyeball. Isn't it amazing that this structure allows an animal to see?

Part D: Finishing Up

1. Look at the structures on your paper towel.
   a. Make sure you have labeled them correctly.
   b. Have your teacher check the labels.

2. Clean up your equipment and work area.

3. Label the diagrams on your handout and answer the analysis questions.

Analysis Questions _ _ _ _ _ _ _

1. Which structure was the most different from what you expected? Why?

2. What do you think is the function of:
   a. the fat?
   b. the muscles?

3. One end of the optic nerve is in the eyeball. Where is the other end?

4. Cataracts are due to "clouding" of the lens. How would this affect vision?

Conclusion

Now that you have observed an eyeball, write several sentences in answer to the focusing question.

Extension Activity

Models: Make a model of an eyeball. Make sure it has at least four parts that have functions similar to a real eye, including a tapetum and a lens.

Eyes That Glow

Many mammals have eyes that seem to glow in the dark. The eyes glow only when they are reflecting light from another source such as a streetlight, headlight, or flashlight. The light enters the eye through the pupil and reaches a layer of tissue at the back of the eyeball called the tapetum. The **tapetum** acts like a mirror, increasing the amount of light in the eye so an animal can see better at night.

Figure 11.13
Leopard

Ears and Hearing

You may be wondering what causes people to lose their hearing. Scientists are constantly working to find those causes. The ear is a complex structure that is difficult to study because most of it is encased in bone. When you remove the bone, you ruin the ear.

While you read, label the diagram on your handout.

Ear Structure

Think of the ear as being divided into three general areas: the outer ear, middle ear, and inner ear. You already know something about the **outer ear** because it's what you see. The most obvious part is the cartilage structure that we call the **ear**. It funnels sound into the **ear canal**. At the end of the ear canal is the **eardrum**. It separates the outer ear from the middle ear.

Three tiny bones make up the **middle ear**. These bones are shown in **Figure 11.15**. They bridge the space between the eardrum and the inner ear.

The **inner ear** is a delicate structure that is supported and protected by the bones of the skull. Both the outer and middle parts of the ear are filled with air, but the inner ear is filled with fluid. The easiest part of the inner ear to recognize is the **cochlea** because it is shaped like a snail. This pea-sized structure contains receptor cells. When the fluid in the cochlea moves, it triggers these cells. Which cells are triggered depends on the frequency of the sound. The receptor cells send messages along the **auditory nerve** directly to the hearing area of the cerebrum. This nerve is not really part of the ear, but without it you could not hear. The cells in the cerebrum interpret the message.

Three **semicircular canals** attach to the cochlea. Even though they are located in the ear, their main function is not for hearing but for balance. Like the rest of the inner ear, they are filled with fluid. When you move, the fluid moves, triggering nerve messages that make you aware of your position. A person may have trouble walking or standing if the semicircular canals are damaged. If the fluid moves all of the time, a type of motion sickness may result.

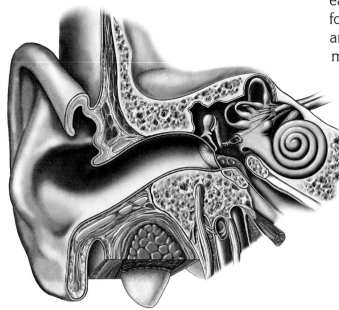

Figure 11.14

As you read, refer to this drawing to locate the parts of an ear.

Hearing Air Waves

Together, the outer ear, middle ear, inner ear, auditory nerves, and brain allow us to hear. Try to picture how this happens by imagining that someone has just blown a note on a trumpet. Sound waves emerge from the bell of the trumpet. If you could see them, they would look similar to the waves in water that appear when a pebble is thrown into a lake. We call these waves of air molecules **sound waves**. The outer ear funnels the sound waves into the middle ear. The waves cause the bones in the middle ear to vibrate. When the vibrations reach the cochlea, the fluid moves, triggering the receptor cells to send a nerve message. This message travels along the auditory nerve to the cerebrum where it is interpreted. The result? You can hear the note played by the trumpeter.

Take a break from reading to see if you are understanding how the ear works. Put the sentences at the bottom of your handout in the correct order.

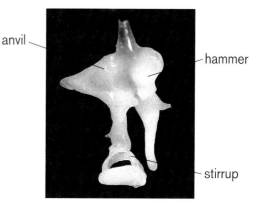

Figure 11.15
The three bones in the middle ear are the tiniest ones in the body. Their shapes give you clues to their names: hammer, anvil, and stirrup.

anvil — hammer — stirrup

Eardrums

Even though all parts of the ear are important for hearing, you are most likely to hear things about the eardrum. **Figure 11.16** shows what an eardrum would look like if you could use an otoscope to look down the ear canal. It is a thin, circular membrane, about 1 cm in diameter, that separates the outer ear from the middle ear. You can picture the eardrum as being like a piece of a balloon stretched over a ring. Like the balloon, a real eardrum can move in and out.

If you close your mouth, plug your nose, and try to breathe, you can feel your eardrums move. This happens because of changes in air pressure. This is illustrated in **Figure 11.17**

If an eardrum stretches too much, it tears. When your ears start to feel uncomfortable, you protect your eardrums from tearing without realizing that's what you are doing. You try to "make them pop" by yawning, chewing gum, or swallowing. These actions move the air in your **eustachian tubes**, making the air pressure

Figure 11.16
This is an enlarged view of an eardrum.

equal on both sides of the eardrum. The eustachian tubes extend from the middle ear into the throat. (*Label the eustachian tube on the ear diagram on your handout.*)

Infections

Sometimes, bacteria or viruses travel up an eustachian tube from the throat into the middle ear. When this happens, a middle ear infection can develop. If, however, bacteria settle in the ear canal, an outer ear infection can result. Thus, the treatment for an ear infection depends on which part of the ear is involved.

It pays to take care of your ears! Hearing is one of your most important senses for monitoring changes in the environment.

Figure 11.17
The eardrum moves when air pressure changes.

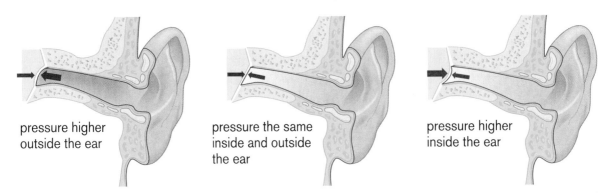

pressure higher
outside the ear

pressure the same
inside and outside
the ear

pressure higher
inside the ear

Analysis Questions

1. Make a chart that summarizes information about the structure and function of each part of an ear.

2. These structures are part of the hearing pathway: brain, ear canal, cochlea, auditory nerve, eardrum, and three tiny bones.

   List them in the order they are used when you hear a sound.

3. Check your understanding of ears and hearing by answering these questions about "ear health."

   a. *"Don't put cotton swabs, hairpins, fingernails, or anything else into your ears."* When people say this, what are they afraid you could do to your ear?

   b. Imagine that you have a cold and your ear starts to ache. Are you more likely to have an infection in your ear canal or middle ear? Explain.

   c. What kind of infection do you think could be treated with ear drops—an infection in the ear canal or in the middle ear? Explain.

4. Look at the ear diagram on your handout, and think of what has to happen in order for you to hear. Explain at least three things that could go wrong with the structure of the ear that might result in a hearing problem.

5. There must be some survival value to hearing or we would not have ears. List at least three ways hearing might have allowed your ancestors to survive.

Extension Activity

Health Application: Ask the school nurse if you can look at an eardrum with an otoscope. Find out how you can tell if someone has an infection by looking at an eardrum.

Hearing Loss

Why do you pay attention to sounds in the environment? How would your life be different if you could not hear? A world without sound is hard for most of us to imagine, but some degree of hearing loss is quite common. About one out of every thirty children has had a noticeable hearing loss.

As people grow older, hearing loss becomes more common. About one out of every four people over age 65 has lost some hearing. Most hearing problems are difficult to recognize. A young child who has never heard does not know anything is missing, and older people often lose their hearing so gradually they do not realize it is happening. Although many people have hearing problems, few people are totally deaf.

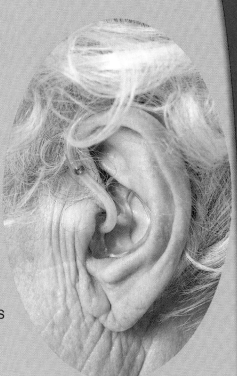

Figure 11.18

People with hearing loss often wear a hearing aid to help them hear better.

Different Pitches

Different animals can hear different pitches. Frogs can hear much lower sounds than we can hear. We can hear lower sounds than dogs, cats, and birds, but when it comes to detecting high pitch sounds, these animals have an advantage over humans. They hear high pitch sounds much better than we do. What could these animals be hearing that we can't hear?

Figure 11.19

Leopard frog

Noise Pollution

Many people lose their sense of hearing unnecessarily because they don't realize how dangerous loud sounds can be. People quickly cover their ears to protect their sense of hearing from particularly loud sounds. They do so without even thinking about it. Other sounds, even if they are not loud, can also be dangerous.

What sounds in your environment can damage hearing?

Start to answer this question by taking sound inventories at two sites, one inside and one outside. Then estimate the loudness of the sounds and read about noise pollution. This should provide you with the information you need to answer the focusing question.

- red colored pencil
- Reading, *Noise Hurts*, page 321

Procedure

1. Read the procedure so you know what you will be doing.
2. Make the data tables you will need when you take your sound inventories.
3. Take the sound inventories.
   a. For homework, use one data table to inventory the sounds in one room of your home. List all of the sounds you hear in 3 minutes.
   b. Choose an outside location and take another 3-minute sound inventory. Record these sounds on your second data table.
4. Use the information in **Figure 11.20** to estimate the decibel level for each sound on your two data tables.
   a. Record your estimates on the data tables.
   b. If you listed a sound that is not on the chart, use something that is about the same loudness to make your estimate.

Decibels

A sound may be as soft as a whisper or as loud as an explosion. The loudness of a sound is measured in decibels. This chart gives examples of the decibel levels of particular sounds.

Figure 11.20

Decibels	Sounds
0	hearing begins
10	leaves rustling, normal breathing
30	soft whisper, library
40	quiet office or room, refrigerator
50	light auto traffic - 100 feet away
60	**normal conversation**
70	dishwasher, vacuum cleaner, noisy restaurant
80	shower, motorcycle, alarm clock - 2 feet away
90	large heavy truck
100	garbage truck, lawn mower - 3 feet away, rapid transit rail noise
110	rock music
125	loud portable head stereo
140	jet engine - 75 feet away

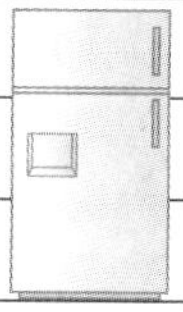

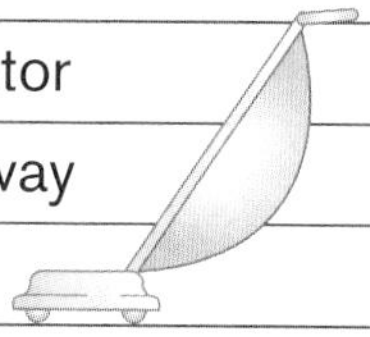

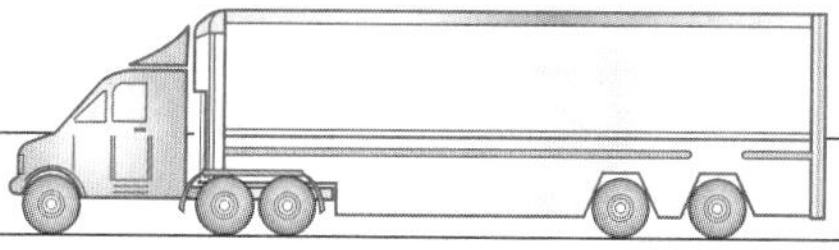

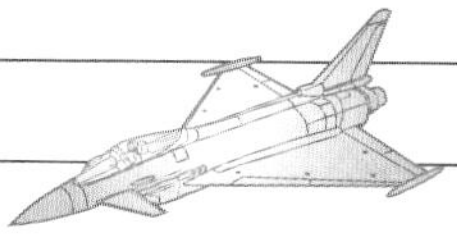

You are familiar with many units of measurements. For example, centimeters measure length and seconds measure time. When you use these measures, you can add them together to get a total. For example, 3 cm + 4 cm = 7 cm, or 5 secs + 40 secs = 45 secs.

However, you cannot add decibels to get total loudness. For instance, 60 decibels + 60 decibels does not equal 120 decibels. It equals 63 decibels. The total loudness of all the sounds at a site is about the same decibel level as the loudest sound by itself.

5. Analyze your data.

   a. Put a red star by the three loudest sounds at each site.

   b. Estimate the overall decibel level for each location and record these figures at the bottom of your table. (Read the information at the bottom of **Figure 11.20** before you do this.)

6. Read *Noise Hurts,* and then answer the analysis questions.

Analysis Questions

1. What is the lowest decibel level that can cause hearing loss?

2. Which of the loud sounds that you starred can cause permanent hearing loss?

3. Music at some rock concerts has been measured at 120 decibels.

   a. How long could you listen to the music without damaging your hearing?

   b. What could you do to protect your hearing at one of these concerts?

4. What are two things you do that are restful to your hearing? How do you know they are restful?

5. List at least four things noise can do to your body besides cause a hearing loss. For each thing you list, indicate which body systems are affected.

6. Do you think school dances should pass the "walkaway" test? Explain.

7. Describe at least three things you could do to lower the decibel level in one room of your home.

Conclusion

Write a sentence summarizing your answer to the focusing question.

Extension Activities

1. **Going Further:** Take sound inventories at three other locations in your community. Identify the noisiest sound at each location.

2. **Design and Do:** List science questions that you could answer using a sound meter. If you have access to a sound meter, and you can write a plan for your teacher to approve, use the sound meter to answer one of your questions.

"What?" Asked the Insect

Most insects cannot hear. Only locusts, cicadas, crickets, grasshoppers, and most moths have ears. They have simple ears that allow them to detect the sound of a possible mate or an enemy.

Noise Hurts

Noise has become a serious concern to many people because our environment is becoming noisier all the time. Motor vehicles, small engines, music ... many things contribute to environmental noise. Noise is a concern for at least three reasons.

- Noise can cause hearing loss.
- Noise can harm a person's health.
- Noise can make it hard to work and relax.

Hearing Loss

When you listen to music, where do you set the volume? If you turn it up high, think about the following research results. Students who were entering college were given hearing tests and the results were analyzed. One out of every three students in the

Figure 11.21

The lower the volume, the less the chance of developing a hearing problem.

study had significant hearing loss in the high frequency range from listening to loud music. Loud noises destroy the receptor cells in the cochlea that detect sounds. The cells that detect high-pitched sounds are most sensitive to noise. Once these cells are destroyed, they are gone forever. This kind of hearing loss is permanent.

Some teenagers listen to loud music even after they have been told that they are hurting their hearing. How about you? Do you have any hearing loss? You may not know if you do because a hearing loss sneaks up on you. When you realize your hearing has gone, it is too late.

Health Effects

Hearing loss is not the only problem caused by noise. Researchers have collected data that connects noise to a variety of other problems. For example, one study involved people who lived near airports where the noise level was high. Many of the people claimed that they adjusted to the noise, explaining they could "tune out" the sounds of the planes. As proof, some of them said that they sleep well at night in spite of constant air traffic. However, the data from the studies show otherwise. Even though the people thought they slept well, their heart rates jumped each time there was a loud sound. Most of the people in the study also had tense muscles and blood pressure readings that were higher than average. "Moodiness" the day after a noisy sleep was another common finding. Peoples' bodies did not actually adjust to the noise.

Figure 11.22
Airports are particularly noisy places.

This and other studies have shown that noise can cause the following health problems:

- high blood pressure
- stomachaches and ulcers
- feelings of nervousness and anxiety
- an increase in heart rate
- muscle tension

People who live in noisy areas should be aware that they are at risk for developing these noise-related health problems.

Concentration Problems

One research project found that noise makes it harder for children to learn. The researchers compared schools near the Los Angeles airport with similar schools in quiet neighborhoods. One of the findings was that children in the quiet schools were better at solving puzzles and math problems.

Did anyone do a sound inventory in or around your school? If so, how did it rate? Are your classrooms quiet enough to allow students to concentrate?

Dangerous Sounds

More studies need to be done before we understand the level of noise that causes health and concentration problems. However, there is strong evidence showing which levels cause hearing loss. In general, sounds below 60 decibels (a normal speaking level) are not harmful. Any sound above 80 decibels can cause permanent hearing loss if it continues for a long enough time. **Figure 11.23** gives you some idea of the amount of time that you can spend at specific sound levels without damaging your hearing.

Make It Safer

If you are concerned about your health and hearing, be aware of the noise level in your environment. If an area seems noisy to you, you can check the sound level by doing this "walkaway" test. Ask a friend to talk in a normal voice. You should be able to walk 8 meters (about 25 feet) away and still understand the words. If you cannot hear the words, the noise level is high enough to stress your body and possibly cause hearing damage. (This test assumes that you have normal hearing.)

Think about your own environment. Which areas are particularly noisy? What noises can you eliminate (or at least turn down)?

Figure 11.23

The times shown in this chart were established by OSHA (the Occupational Safety and Health Administration—a government agency).

Decibel Levels	Description of impact of sound on the listener	Permanent hearing loss can result if the sound continues for longer than this amount of time.
0	barely noticeable as a sound	–
20	softest sound most people hear	–
60	**normal speaking level**	–
80	annoying	20 hours
90	poor conditions for working	8 hours
100	sounds very loud	2 hours 15 minutes
110	louder yet	35 minutes
120	uncomfortably loud	7 minutes
130	strong "tingling" in ear	1 minute
140	sharp pain in ear	< 1 minute

Learning and Memory

Your eyes and ears transmit information to your brain. To a large extent, what you do with the information is up to you. You can learn from much of the information. Have you given any thought to how you learn? What kinds of things do you learn best? What is going on in your brain when you learn?

How do your senses affect your learning?

You will begin to answer this question when you participate in a demonstration about learning. This lesson also includes a reading that provides information about how the brain works. Together, these two tasks should help you answer the focusing question.

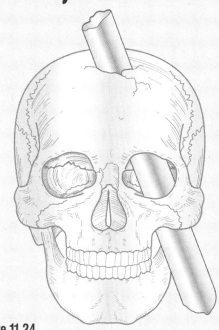

Figure 11.24

Phineas Gage suffered a serious head injury, and he survived! Your teacher will read the story of this historic accident to you.

- pencil and paper
- Reading, *Brain Activity*, page 328

Procedure

1. Participate in the demonstration about learning.

   a. Listen to the rules while your teacher explains the demonstration. If it doesn't sound like a fair test to you, explain your concerns. As a class, agree on what you need to do to make the test fair.

b. Listen closely while your teacher reads aloud a story about Phineas Gage. It's a true story about a historic event. The story is about 5 minutes long. While the teacher reads, sit quietly; do not take notes or talk.

2. Let the story of Phineas Gage "rest" in your memory while you read *Brain Activity*.

3. Answer analysis questions 1 through 4.

4. Conclude the demonstration

a. Take a test on the story of Phineas Gage. Do your best, but don't worry if you don't know all the answers. The test is intentionally a little difficult. If you don't know the answer to a question, leave it blank—don't guess.

b. Individual data is not very useful in this activity so analyze the class data. Look for patterns and then try to explain them.

5. Answer the rest of the analysis questions and write your conclusion.

Analysis Questions _ _ _ _ _ _ _ _

1. Why do you think your brain needs oxygen when you are sleeping?

2. Where does the brain get the sugar and oxygen it needs?

3. Look at **Figure 11.26** (page 330) and count the number of pathways in:

a. the first sketch (2 extensions per cell)

b. the second sketch (3 extensions per cell)

c. the third sketch (4 extensions per cell)

4. List at least three things you did today that were good for your brain.

5. Consider the results of the test on the story of Phineas Gage.

a. What was the average score for the group that listened to the story?

b. What was the average score for the group that listened and saw pictures?

c. Do you think it was a fair test? Why or why not?

6. Use information from the reading to explain the results of the demonstration.

7. HB Connections: How does the nervous system interact with other body systems? Think of at least one interaction for each of the body systems you have studied so far.

Conclusion

Write several sentences to answer the focusing question.

Extension Activities

1. **Try It:** Draw a nerve network consisting of six cells in which each cell has five endings. Compare your drawing with the ones in **Figure 11.26**. How does the additional ending affect the number of possible pathways in the network?

2. **Research:** Read about children who were raised in extremely impoverished environments. Some of these children survived on their own in nature; others were locked in rooms without any stimulation. Perhaps one of the most famous of these children, Genie, was rescued from her horrible situation in 1970 at the age of 13. Find out how these conditions affected the brain development of the children.

Brain Activity

Did you know that your brain cells need a steady supply of sugar and oxygen? Even as you read this your brain is "burning" sugar. Day and night, brains are voracious users of energy. And like all cells, brain cells need sugar and oxygen to produce energy. If the oxygen supply is shut off for even 10 seconds, unconsciousness results. Even muscle cells seldom need that much energy. Why does the brain need so much energy? Because it's constantly working, and working cells need energy.

Always Working

Millions of signals are flashing through your brain at any given moment. You don't feel overloaded because you ignore most of the usual signals. You "tune out" things such as the hum of a light bulb, the feeling of clothes on your skin, and the color of the walls in the room. Signals that relay information about body functions such as blood pressure and breathing rate are taken care of automatically. In addition to these routines tasks, you constantly use the thinking parts of your brain.

Humans have the most complex brains and are capable of more complex learning and thinking than any other animal. For example, the brain allows us to store memories of past events and to learn from the experiences of other people without having to experience the same things ourselves. A complex language may also be unique to humans. Language allows us to communicate our experiences and thoughts. Language is actually a series of smaller tasks. Which of these language tasks did your brain have to do when you listened to the story of Phineas Gage?

- associate words with things or ideas
- choose the right words to express a thought
- recognize spoken words
- put words together so they make sense
- think in words
- recognize written words
- put together sentences so they make sense

These language tasks are all coordinated by the cells in the surface area of the cerebrum. Different regions of the cerebrum control different aspects of language. **Figure 11.25** shows three views of the brain; each view shows brain activity for a specific language task.

Includes *sci*LINKS®
NSTA

Topic: Learning/Memory
Go to: www.scilinks.org
Code: MSLS3e328

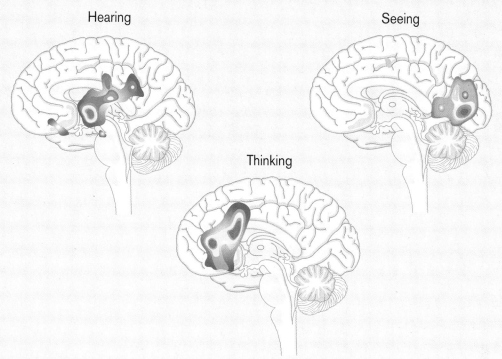

Hearing

Seeing

Thinking

Figure 11.25

These scans show which regions of the brain are active when the brain is performing specific tasks. The first scan was taken when a person was listening. The second shows the active area when a person is looking at something, and the third shows thinking. Red marks the areas of greatest activity, then yellow, and green. Blue is the area of least activity.

Nerve Cell Networks

Most of the nerve cells in the surface layer of the cerebrum are not involved with either the senses or with movement. Instead, they are part of elaborate networks that process information. The complexity of the networks corresponds with the number of extensions on the nerve cells. The more extensions, the more complex the network. This is illustrated in **Figure 11.26**. The "cells" in the most complex network in **Figure 11.26** have four extensions. Nerve cells in the cerebrum average about *one thousand* extensions per cell, and remember, there are about 100 billion cells in the brain. The number of possible pathways is huge. When a message comes into the brain, it can speed along any one of these pathways. Depending on the pathway it uses, the end result can differ. Some paths may result in movement, some in sensation, some in emotions, and some in thoughts. The same incoming message can also trigger lots of thoughts by traveling many different pathways.

The complexity of the nerve network makes the brain capable of astounding feats. But this comes with a cost. Unlike most other body cells, nerve cells in the brain are not replaced if they are damaged. In most tissues, new cells

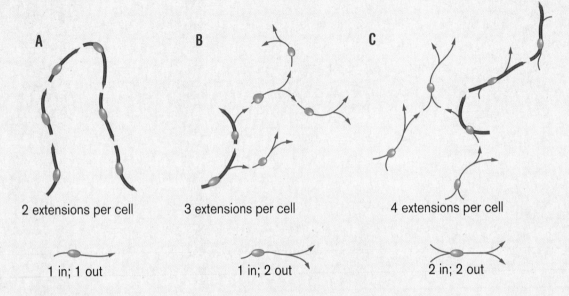

A 2 extensions per cell **B** 3 extensions per cell **C** 4 extensions per cell

1 in; 1 out   1 in; 2 out   2 in; 2 out

Figure 11.26

This illustration shows how the number of extensions on a nerve cell affects the complexity of a network. In each of these short networks there are 6 nerve cells. In *A* there are two extensions per nerve cell; *B* shows three extensions per cell; and *C* shows four. One pathway has been outlined in red in each network. Count the total number of pathways in each network.

can replace ones that die. That doesn't happen in the brain because all one thousand or so projections from that cell would have to make all the right connections. When nerve cells die, they are gone forever.

Brain Exercise

Because nerve cells in the brain do not reproduce, that means you were born with nearly all the brain nerve cells you will ever have. That does not mean you're "stuck" with what you've got. Research is showing that you can improve your brain by using it. Think about the effect of physical activity on muscles. The more you use your skeletal muscles, the stronger they become. Similarly, the more you use your nerve cells, the more you learn, and the more extensions they will develop. More extensions mean more complex networks, which may mean improved thinking ability.

Most of this research has been done on animals. Typically, laboratory animals are raised in different types of environments and then their brain cells are compared. Drawing *A* in **Figure 11.27** shows what a nerve cell from the brain of a rat might look like if it were raised in a large cage with other rats and a variety of colorful "toys." Drawing *B* shows how the nerve cell might look if

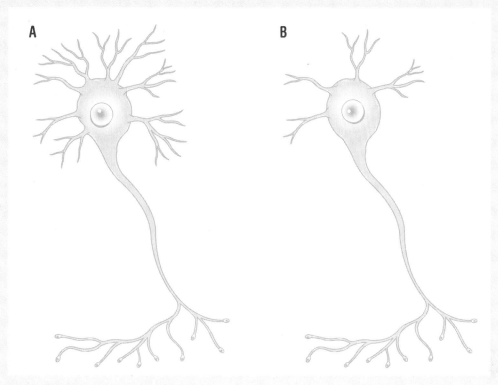

Figure 11.27

These drawings illustrate the differences in nerve cells from the brains of rats that have been raised in different environments. Drawing *A* represents a cell from a rat that was provided with lots of things that would get its brain "working." Drawing *B* represents a cell from a rat that was raised in a boring environment. Count the number of extensions on each cell.

the rat were raised alone in a cramped cage without anything to explore. At first glance, both cells look similar, but as soon as you count the extensions, you can see some of the differences. Some studies on people show similar effects.

As you grow older, some brain cells die naturally, and once they do, they're lost forever. Fortunately, brains have a surplus of cells so you can afford to lose some without losing your thinking capacity. Disease can also take its toll. Few infectious diseases destroy brain cells but sometimes, for reasons not yet understood, the brain cells begin to deteriorate. Alzheimer's disease is the best known example of this type of condition.

Some people destroy their brain cells by the choices they make. Anyone who takes mind-altering drugs is tinkering with the structure and function of his or her brain cells. You will learn more about the effects of some of these drugs in the next chapter.

Chapter 12

Decision Making and Drugs

SAY NO TO DRUGS

Do you think anti-drug slogans such as this are effective?

Sixteen Substances

Think of all the different things you eat, drink, and swallow. It doesn't matter if it's a mango, cough syrup, or a vitamin tablet, your system has to break it down; otherwise, these substances would be in your body forever. In this activity, you will focus on the many types of substances people take into their bodies. How can these substances be grouped?

What characteristics are most useful when sorting substances?

Keep this focusing question in mind while you devise a way to sort 16 substances into groups. Consider what you learned about sorting in previous activities. Sorting helps you identify patterns, and it helps you become aware of what you do and do not know about the things you are sorting.

- envelope
- Handout, *Sixteen Substances* (substance cards)
- red colored pencil

Procedure *(Work in a group with one to three other students.)*

Part A: Main Groups

1. Get a set of substance cards and look over the names of the 16 substances in the set.
2. Talk about characteristics of the substances. Answer analysis question 1.
3. Choose *one* characteristic and use it to sort the substances into groups.
4. Check your groups. If you think of a characteristic that might work better to sort the substances, try it. Continue resorting your substances until you arrive at a set of groups that you like.
5. On a sheet of paper, write headings for each of your groups and list the substances under the appropriate headings.
6. Answer analysis questions 2 through 4.

Part B: Subgroups

1. Look at the substances in your largest group.
   a. Think of several characteristics that could be used to sort this group into two or more smaller subgroups.
   b. Use one of the characteristics to sort the substances in this group. (In this way, you will be making subgroups.)
2. On the back of your paper, write the name of the group and the subgroups.
   a. List the substances in the appropriate subgroup.
   b. Write a sentence explaining which characteristic you used to make the subgroups.
3. Repeat these steps for any of the main groups that have three or more substances. (You can use a different characteristic to sort the substances in each of the main groups.)
4. Discuss each of the substances and decide whether or not it is a drug. If you think it is a drug, circle it in red.
5. Put the substance cards back in the envelope and return it to your teacher.
6. Answer analysis questions 5 and 6, and write your conclusion.

Analysis Questions _ _ _ _ _ _ _ _

1. Think about characteristics of substances.
   a. List at least three examples of characteristics of water.
   b. List at least three examples of characteristics of coffee.
   c. List at least three examples of characteristics of Ibuprofen.
2. What characteristic did you use to sort the substances into main groups?
3. What problems did you have when you put the substances into groups?
4. List at least four other characteristics you could have used to sort the substances.
5. What characteristic did you use when you divided your largest group into subgroups?
6. What criteria did you use to decide whether or not a substance is a drug?

Conclusion

Write complete sentences to answer the focusing question.

What's a Drug?

In broad terms, any substance that can affect the normal activities of cells or body systems is considered to be a **drug**. From this definition, you can see that lots of things can be called drugs, even foods. Usually, the word *drug* is more narrowly defined to mean either a substance used as medicine to treat a disease or an illegal substance that affects the mind and is addictive. Even this definition can be difficult to interpret. For example, what about bran and vitamin C—are they drugs? In a sense, they are because they can be used to treat medical problems. However, they are also natural components of food.

Medicinal Drugs

For centuries, people have used the roots, bark, leaves, and flowers of various plants to treat specific illnesses. In the last hundred years or so, scientists learned how to isolate particular chemicals from plants, making it easier to control dosage and minimize side effects. Now, scientists can create many of these same chemicals in the laboratory. These chemicals are used as medicinal drugs to treat particular health problems. There are drugs for a wide variety of health problems, from controlling blood pressure, to killing microorganisms, to slowing the contraction of smooth muscles in the intestines.

Particular drugs are used for specific reasons, and every drug has side effects. For example, people often take aspirin to relieve pain, but it also irritates the lining of the stomach. Many antibiotics not only kill microorganisms that cause infection, but also the good

bacteria in the intestine, resulting in diarrhea. Although medications for asthma make it easier to breathe, they can also cause gum disease.

Many medicinal drugs target the nervous system. There's a long list of drugs that can be used to relieve pain. Other drugs affect the mind. These drugs are important in the treatment of medical problems such as seizures, severe depression, and other mood and personality disorders.

Even though drugs are important medically, they can also be dangerous. They can make existing health problems worse, affect other body systems, or even cause death. For this reason, the government regulates the sales of drugs. Some can be sold off the shelves of a store. Others, however, need a prescription from a physician. The physician will make sure that the dosage is correct and will watch for dangerous side effects.

Figure 12.1

Some prescription drugs are useful for treating nervous system problems. However, these drugs can be dangerous if misused.

Illegal Drugs

Some substances do more harm than good, and cannot be bought or sold legally. Cocaine, LSD, marijuana, heroin, and PCP are all examples of illegal drugs that affect the nervous system. All of these drugs have a similar characteristic.

- They distort reality. They make it impossible to think clearly, make wise decisions, or even sort out emotions. Vision and other senses may be distorted or dulled. Reaction time is often slowed and it may become difficult to coordinate muscle movement.

- Illegal drugs can make an existing mental condition worse. For instance, a person who is depressed may become suicidal under the influence of one of these drugs.

- These drugs can trigger other medical problems. For example, a person with a heart problem may go into heart failure as a result of using some of these drugs.

- Long-term use of illegal drugs may damage brain cells. Because human nerve cells cannot divide, they must last a lifetime. When some of them are damaged, they cannot be replaced.

Legal Status

Many things are considered when deciding on the legal status of a drug. Here are three of the main considerations:

1. whether or not it is addicting
2. if it has any medical uses
3. how serious the side effects are

The benefits and risks are weighed against each other.

a. If there are many benefits and the risks are slight, the drug will be sold in the stores without a prescription.

b. If the drug has definite benefits, but there are some risks, the drug is sold by prescription only.

c. If the risks far outweigh the benefits, the drug is made illegal.

- Most concerning of all, these drugs are addicting. When a person is addicted, he or she needs a steady supply of the drug. An addict is a slave to the drug. The user spends much of his or her time thinking about the drug, getting high, and in the company of other users.

- These drugs are even riskier when taken by children or teens. They interfere with the normal physical and social development that leads to adulthood. Many of these drugs affect the hormones that direct growth and sexual development. Frequent drug use, even in small doses, may affect friendships and family relationships.

- Illegal drugs ruin lives and cost society billions of dollars in crime and lost income.

Analysis Questions

1. What is a drug?
2. Think about medicinal drugs.
    a. Write the name of one medicinal drug.
    b. Explain why people usually take this drug. (To treat what health problem?)
    c. Which body systems are most directly affected by this drug?

3. List at least five characteristics of illegal drugs.
4. Decide if you agree or disagree with each of the following statements. Write a statement to explain your decisions.
    a. Teens know more about illegal drugs than adults do.
    b. Society thinks it's worse if a teen uses illegal drugs than if an adult does.
    c. Illegal drugs are more harmful to teens than to adults.
    d. An addiction is the same thing as a habit.

Extension Activities

1. **Health Application:** Read the package labels for some of the over-the-counter medications you have taken. Notice the intended use of each medication as well as the side effects. How many body systems are affected by each drug?
2. **History Connection:** Research the patent medicines that were widely sold in the 19th century. Some claimed to cure ulcers, scurvy, cankers, and dozens of other problems. Many people bought these "miracle drugs" out of desperation, hoping for cures that did not exist.

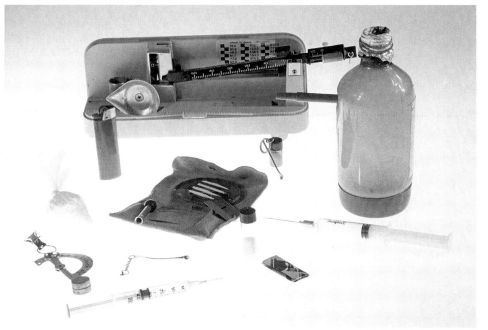

Figure 12.2
Illegal drugs come in a variety of forms. Would you recognize these as illegal?

Decision Making

One of the most complex tasks the human brain can do is make a decision. Decision making involves analytical thinking, and that's something you have to learn how to do. Of course, not all decisions require lots of thought. We all make decisions every day, most of which are simple and routine. As you grow older, however, you are faced with more difficult decisions, and some of these decisions can shape your life.

In this chapter, you'll consider the kinds of things that affect people's decisions to use drugs such as alcohol and marijuana. Sometimes people make these decisions without giving them much thought, and then they may regret their decisions. Did you know that decision making can be broken down into a series of steps?

What are the basic steps in making a decision?

In this activity, you will use a decision-making model to solve a problem that someone might face in real life.

- Handout, *Decision Making*
- Handout, *Situation Cards*

Procedure   *(Work in a group with one or two other students.)*

1. Read the procedure and decide if you want to assign roles. (Possible roles: reader, recorder, timer.)

2. Review *Decision Making* handout.

3. Read the situation described on the card. Discuss and decide what the *problem* is, and then record it on your *Decision Making* handout.

4. Think about any relevant information or *facts* related to the problem. What scientific facts do you know about this issue? Write down the information in the second column on your handout.

5. Who are all the *people* who are involved in this situation? They might be people directly related to the decision or people indirectly impacted by the decision, such as family or friends. Write your list of involved people in the appropriate column.

6. As a group, brainstorm a list of *options* that the person in this situation has. List as many options as you can. If you stop before listing at least six, you may miss important ones. Have the recorder list the options on the handout.

7. Analyze these options. Think of the consequences for each option. Decide on the *best choice* for the person this situation. Record it on your handout.

8. What *actions* should this person take as he or she follows through with the decision. Write down each step in the last column on your handout.

9. Compare your group's handout with the one from another group that worked on the same situation.
   a. Did you identify the same problem?
   b. Which options did the other groups think of that yours did not?
   c. What are their best options?
   d. Do you agree with their best choice? Why or why not?

Homework

Analyze a decision of your own. It might be a decision you already made or one you will have to make. Use the *Decision Making* handout to help you reach your best choice.

Conclusion

Try to answer the focusing question without looking at the handout.

Remember

1. There are many options for every situation.
2. Every decision has consequences.
3. "Do nothing" can also be a decision and it, too, has consequences.
4. You have to live with your decisions.
5. A solution that satisfies one person may not satisfy you.

12-4

Know the Facts About Drugs

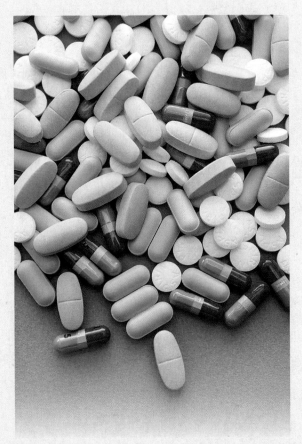

Figure 12.3

Your brain controls everything your body does. Drug and alcohol use affects a person's brain; changing the way the person thinks, his or her personality and, ability to learn. The more often someone uses drugs or alcohol, the more the person's brain may change in fundamental and long-lasting ways. These lasting changes are a key factor in drug and alcohol addiction.

How do drugs and alcohol affect a person's brain and body?

Think about what you learned about the nervous system in Chapter 11. Use this information to help you understand the processes of the nervous system and how drugs and alcohol interact with the brain and body.

- Handout, *Drug Information Cards*
- pen or pencil
- paper
- supplies for presentation

Procedure

1. Your teacher will assign you and your teammates one of the following drug categories to study. Use the drug information card and resources your teacher provides to help you with your research.

   a. alcohol

   b. nicotine (tobacco)

   c. cocaine, methamphetamines, amphetamines (speed), caffeine, Ritalin (Stimulants)

   d. marijuana

   e. inhalants

   f. ecstasy, mushrooms (Hallucinogens)

   g. anabolic steroids

2. Answer the research questions below and be prepared to present your information. You will be responsible for teaching your classmates about the drug. Use a model, role-playing, or a visual aide (poster, etc.) to help you present information about the drug.

Research Questions

1. What is the drug?
2. How does the drug affect the brain? Be detailed and specific.
3. How does the drug affect other parts of the body or body systems (think of all the systems you have studied)?
4. Is this drug legal or illegal?
5. What are other dangerous effects (physical, social, emotional, etc.) caused by using the drug?
6. What are other important facts everyone should know about the drug?

3. As your classmates present, answer the questions listed above for each drug type. You will be responsible for learning this information from your classmates.

4. Answer the analysis questions.

Analysis Questions _ _ _ _ _ _ _ _ _

1. List at least four reasons why you think a teenager might try drugs or alcohol. Circle the reason you think is most convincing.

2. List at least four reasons why you think a teenager should not try drugs or alcohol. Circle the reason you think is most convincing.

3. In most states, it is illegal for people to drink alcohol until they are 21. Do you think that is fair? Explain your answer.

4. Do you think the marijuana laws are fair? Explain.

Conclusion

Write several sentences answering the focusing question.

Alcohol, Tobacco, and Drug Questions

It is important to know the facts about drugs and alcohol, but it is more important to be able to apply what you know. People often say things about drugs and alcohol that may or may not be true, or they may ask for information about what drugs and alcohol do to the body. If someone asks you a question about drugs or alcohol, would you be able to answer it? In this lesson, you will respond to people who have questions about drugs or alcohol.

Procedure

(Work with a partner or by yourself.)

1. Read all six letters and choose three to answer.
2. Plan your answers.
   a. Refer back to the last lesson if you want to check your facts.
   b. It will probably take several sentences to answer each question.
3. Underline the facts about drugs or alcohol you included in your answers.

Letter 1

I've heard that marijuana can help a person do better in school. They say you can concentrate better when you're high and that you do better on tests if you smoke a little pot first. It's also supposed to make it easier to get through boring classes. Is all this true? Could marijuana help me in school?

Letter 2

I went to a big party last Saturday. We talked, listened to music, ate, and drank soft drinks. Everything was okay until some kids showed up with a six-pack of beer. Most of us ignored them, but Joe and couple other kids drank some beer. About a half hour later, Joe threw up. Why did he do that? My mom doesn't throw up when she drinks beer.

Letter 3

I heard someone at school say she tried marijuana once and it wasn't any big deal. In fact, she didn't feel any different. She can't understand what all the fuss is about. Is it true? Why do people use marijuana if it doesn't have any effect?

Letter 4

This morning Mom and Dad were talking about a party they went to last night. Mom said Mrs. Hawkins shouldn't have driven home after the party because she'd had too much to drink. Dad told her it was all right because Mrs. Hawkins had had two cups of strong black coffee before she left. Is dad right? Does black coffee help sober up a person?

Letter 5

My sister told me about a party she went to last night. She's worried because one guy smoked two joints and then drove off with a carload of people to go to another party. Her friends told her that it's not dangerous to drive after smoking pot, that it's worse to drive after drinking alcohol. Who's right? Is it safer to drive after using marijuana, than after drinking alcohol?

Letter 6

Trev just called. He's going with Josh to a big party. I think they're going to try cocaine. Josh's sister uses cocaine, and she's offered some to Josh and Trev in the past. Trev says it won't hurt to try it once. Is that true? Is it okay for him to try just it just this one time?

Includes sciLINKS
NSTA

Topic: Alcoholism
Go to: www.scilinks.org
Code: MSLS3e344

Alcoholism

Alcohol addiction is a costly social problem. It can ruin the lives of not only the alcoholic, but also the lives of his or her family. People with alcohol problems can get help from Alcoholics Anonymous. Their families and friends can get help from Al-Anon; children of alcoholics can call Alateen.

Tough Situations

Cocaine, LSD, heroin, crack, speed, and inhalants. The list of illegal drugs is long. You could take the time to study each of these drugs, or you could remember these four basic facts.

1. They affect the nervous system.
2. They alter a person's thinking, personality, and ability to learn.
3. They are addicting.
4. They are illegal.

These drugs are dangerous because they are often unpredictable. It is frightening to think that some teens try these drugs without realizing what they are getting into. Sometimes, if they had given their decision a little more thought, they would have said, "No, thanks." instead of, "Sure, why not?"

What kinds of things make it difficult to know what to do in a situation that involves drugs?

In this activity you will read about four situations and then use the *Decision Making* handout to analyze one of them. Keep the focusing question in mind while you work.

Materials

- Reading, *Four Situations*, page 347
- Handout, *Decision Making* (two per student)

Procedure
(Work with a partner.)

1. Review the *Decision Making* handout.
2. Read the four situations and then, with your partner, agree on one to analyze. Record the problem on the handout.
3. Think about any relevant information or facts related to the problem. What scientific facts do you know about this issue? Write down the information in the second column on your handout.

Includes *sci*LINKS.
NSTA

Topic: Teen Drug Prevention
Go to: www.scilinks.org
Code: MSLS3e345

4. Who are all the people who are involved in this situation? They might be people directly related to the decision or people indirectly impacted by the decision, such as family or friends. Write your list of involved people in the appropriate column.

5. With your partner, brainstorm a list of the options you would have if you really were in this situation. Record the options on your own handout. List at least five options, but more if you can.

6. Analyze each option. (Think of several negative and positive consequences.) Identify your best choice. (You don't have to agree.)

7. Now, imagine that you are your parents. Circle the option you think they would choose.

8. What actions would you take if you were really in this situation? Write down each step in the last column on your handout.

9. Answer the analysis questions after you have completed the homework assignment.

Homework

Take home the blank handout and a copy of the situations. Ask your parents or some other adult to work through the same situation you did and arrive at their best choices. Then answer the analysis questions.

Analysis Questions _ _ _ _ _ _ _ _ _ _

1. On a scale from 1 to 5, rate how well you and your parents agreed on how to handle the situation, a 5 meaning you agreed completely.

2. Compare the consequences you listed in class with the ones your parents mentioned. What did they think of that you did not?

3. What did you learn from this activity?

Conclusion

Write an answer to the focusing question after you have listened to the class discussion about all the situations.

Extension Activity

Health Application: Write a situation involving drugs that you think is realistic for your school. Discuss the situation with at least four people—a friend, a parent, a teacher, and a counselor or health care provider. Compare the "best choices" to resolve the situation identified by these people.

Baby-Sitter's Dilemma

You're baby-sitting for the Haley's; they are friends of your parents. Mr. Haley is supposed to drive you home. (You live about three miles away.) When they get home, you suspect that Mr. Haley has been drinking alcohol so you watch him closely. He doesn't seem to be walking very steadily, and you think he's talking more slowly than usual. You know that a person who's been drinking should not drive. What will you do?

Everyone's Doing It

You got invited to a party. You were really excited because all of the popular kids will be there. When you arrive, you realize that there are no adults at the house. One of the popular kids approaches you and starts to talk about something called "X". You really want to be cool and fit in, but are unsure when the kid offers you some. What will you do?

Hallway Conversation

You're on your way to class when you overhear someone say, "I got it from my brother. It makes you feel really good." Glancing over, you notice three kids; one of them is putting something in his pocket. You catch another part of the conversation, "It's not a drug; it's just a pill." Your school has been involved in an anti-drug program. You remember one of the slogans, "Drugs are everybody's problem." What will you do now?

Worried Friend

You're really worried about one of your friends because she uses marijuana. You've told her that smoking pot isn't smart, but she doesn't seem to care. She says she likes the way it makes her feel and it doesn't hurt anyone. You aren't so sure about that. "And besides," she says, "it is kinda legal now." She's not doing very well at school and she's gotten in trouble a couple of times, which isn't like her. What will you do now?

Life Cycles

Chapter 13

Body Changes

Besides height, what other clues can you use to estimate the age of a person?

The First Ten Years

You've been growing older every day of your life, but have you paid attention to what's been happening to you? Think about when you turned 10 years old; try to remember what you were like and how you looked. Now picture what you must have been like as a newborn. Obviously, you have changed quite a bit. In this activity, you will consider some of the changes that occur during the first ten years of life.

What do you think is a good definition of the word *growth*?

Each stage of life can be described by particular characteristics. Position 30 human characteristics on a time line and then reconsider the answer to the focusing question.

Materials

- four sheets of paper
- scissors
- Handout, *30 Characteristics*
- red and blue markers
- tape

Procedure *(Work in a group with two or three other students.)*

1. As a group, answer the focusing question. Spend a few minutes on this task so you feel comfortable with your definition.

2. Prepare a time line by taping together the four sheets of paper and labeling them as shown in **Figure 13.1**.

Figure 13.1

3. Cut apart the 30 characteristics.

   a. Set aside the four blanks to use later.

   b. Choose one of the characteristics, and with the other students in your group, decide when you think a child first meets this characteristic.

   c. Discuss evidence for your decision.

   d. When you all agree where the characteristic should be positioned, tape it on the time line.

   e. Do this for the other 25 characteristics.

4. Think of at least four more characteristics to add to the time line. Write one characteristic on each of the blanks. Attach the characteristics where they belong on the time line.

5. Put a red or blue box around each characteristic to indicate whether it is a physical characteristic or an ability.

   red = physical characteristic (a basic structure or function of the body)

   blue = ability (something a child learns to do)

6. Answer the analysis questions.

Analysis Questions _ _ _ _ _ _ _ _ _

1. Which characteristic was the easiest to place on the time line? Why?

2. Which characteristic was the most difficult to place on the time line? Why?

3. List at least three examples of the types of evidence you used when you decided where to place the characteristics on the time line.

4. Which changes do you think occur at exactly the same age for everyone?

5. Now think about the characteristics in terms of body systems. Find a characteristic that involves each system listed here. Try to find a different characteristic for each system.

   a. skeletal system

   b. muscular system

   c. digestive system

   d. circulatory system

   e. respiratory system

   f. nervous system

6. Which system do you think changes the most as a person grows? Why?

Figure 13.2

7. Look at the three children pictured in **Figure 13.2**.

   a. Write down how old you think the child in the first photograph is, and list at least three characteristics that helped you arrive at your decision.

   b. Record the same information for the second child.

    c. Record the same information for the third child.

Conclusion

Now that you have thought about growth, write several sentences in answer to the focusing question.

Extension Activity

Math Connection: If you have records showing your own growth for the first ten years of your life, use the measurements to make a graph. Decide which year you grew the most and which year you grew the least.

What Is Puberty?

Children grow a lot during the first ten years of their lives, and they continue to grow until they are 16, 18, or even 20. Sometime during these years, a person's body changes from that of a child to that of an adult. This time of change is called **puberty**. While some of the changes are obvious, you are probably unaware of some of the subtle things that are going on.

How are the changes that take place during puberty like the changes that take place during the first ten years of life?

Keep this question in mind while you work with a small group of other students and read about puberty.

- Reading, *Puberty: A Time of Change*, page 357

Procedure

Part A: Defining Puberty

(Work in a group with two or three other students.)

1. Choose one person in your group to be the recorder.
2. Review the rules for working in a group.
   a. Only one person talks at a time.
   b. All members of a group participate.
   c. Respect each other's ideas by not laughing or judging.
3. Work quickly and quietly. You will have 5 minutes to complete this part of the activity.
4. As a group, brainstorm a list of changes that occur during puberty. Have the recorder write down all of the things that your group mentions. (When brainstorming, some ideas may sound strange but that's all right. An idea that sounds funny or wrong may turn out to be the best of all.)

5. Be prepared to use your list to help create a class list of changes that occur during puberty.

Part B: Learning the Facts

(Work by yourself to complete this part of the activity.)

1. On your own paper, make a chart with three columns like the one shown here.

Changes that Take Place during Puberty		
Boys Only	Girls Only	Both Boys and Girls

2. Look at the class list of changes.
   a. Copy each item from the class list onto the correct column of your chart.
   b. If you are not sure in which column something belongs, write it in pencil at the bottom of the page. You will be able to change your answers later.

3. Read *Puberty: A Time of Change*. While you read, check your chart.
   a. If you find you have written something in the wrong column, erase it and rewrite it in the correct column.
   b. Add any changes you haven't already listed on your chart.

4. Answer the analysis questions.

Analysis Questions _ _ _ _ _ _ _ _ _

1. Decide which of the following statements you think is true and write a paragraph to defend your choice. Include at least three reasons for your decision.
   a. Puberty is a time of more change for boys than for girls.
   b. Puberty is a time of more change for girls than for boys.
   c. Both girls and boys experience about the same amount of change during puberty.

2. Many children study the reproductive system in elementary school.

    a. If you were helping the people who work on curriculum, at what grade level would you suggest the reproductive system be taught?

    b. What do you think are the three most important topics that the students at that grade level should be taught?

Conclusion

Answer the focusing question.

Extension Activity

Going Further: Talk to your parents about puberty. Find out if anyone told them what changes to expect during puberty. Ask them if they think it was easier to grow up when they did or if they think it is easier for you now.

Puberty: A Time of Change

Human growth is fairly predictable and continual; we all grow from infants to children, then go on through our teen years and into adulthood. Each of these life stages can be described by its own set of body changes.

The Early Stages

An *infant* grows at an astonishing rate. From birth to 5 months of age a baby's weight usually doubles, and by 1 year its birth weight has tripled.

Between the ages of 2 and about 6, a *child* continues to grow and develop, but somewhat more slowly. Then, around age 6, growth slows down and may even appear to stop during the last few years of childhood. This seems to be a resting time, allowing the body to prepare for what comes next.

Now comes another stage of fairly rapid growth called **puberty**. Puberty marks the beginning of the changes that lead to physical maturity. Boys and girls vary widely in the ages at which they enter puberty. Sooner or later, however, in their own time and at their own rate, boys acquire the physical characteristics of an adult male and girls acquire those of an adult female. Some of the changes are easy to see; some are not.

Naturally, there are many things you want to know about these changes. You may already have picked up bits and pieces of information and, perhaps, some pieces of misinformation about the growing-up process. This chapter should help you get the facts straight.

Growing Taller

One of the first signs of puberty is often a growth spurt. For a girl, this usually begins between ages 8 and 13. This is about two years earlier than for a boy. A boy's growth spurt generally begins anytime between the ages of 10 and 15.

After this initial spurt, growth continues—but at a slower rate—until a girl is about 16 to 18 and a boy is about 18 to 20. By the time growth is complete, boys often end up being taller and heavier than girls. Look at **Figure 13.3**. What is the normal range of heights for 10-year-old girls? For 10-year-old boys? How do the two compare?

It is important to remember that there is no such thing as an "average" person. Rate of growth varies from person to person. Whenever your own growth spurt begins is right for you. You may have noticed gains in height and weight as early as age 8 or 9. This indicates that physically you were maturing early. If so, you may be as tall as you are going to be by age 14 or 15. If you start your growth spurt later, you may keep growing until you are in your early twenties.

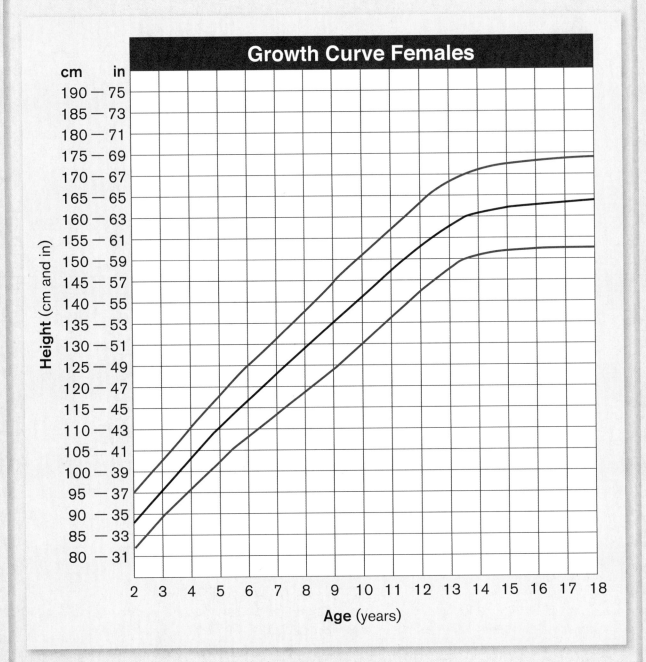

Figure 13.3

When you plot height versus age, you get growth curves such as these. The top line marks the upper limit of height for 90% of people at the given ages and the bottom line marks the lower limit. Ten percent of people will be above or below this range. If you are concerned about your height, it might be reassuring to talk to a doctor or to the school nurse.

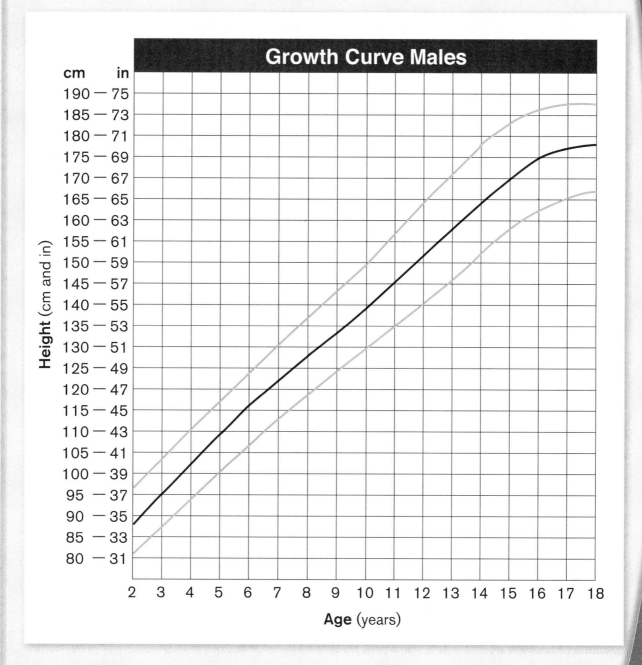

Growth Curve Males

Height (cm and in) — vertical axis
Age (years) — horizontal axis

Topic: Puberty
Go to: www.scilinks.org
Code: MSLS3e359

To some extent, you can predict your final height by looking at your parents. If shortness runs in the family, you are likely to be short. Likewise, tall parents usually produce tall children. However, once again, there are no hard and fast rules. Short parents may have tall offspring, and tall parents may have one or more children who are short. If one parent is tall and the other short, their children may be short, tall, or somewhere in between.

Growing Tissues, Growing Needs

Increases in height and weight are not the only changes that occur during puberty; all of the body's organs and tissues grow, too. For instance, the heart and skeletal muscles grow larger and stronger. This means that your strength and endurance are likely to increase. The bones grow along with the muscles, and because the bones in the hands and feet grow faster than other bones, you may feel awkward for a while. But, by the time you reach your adult height, all parts of your body will be in proportion.

Growth requires energy and the only way that humans can get energy is from food. A growing body requires plenty of nutrients—including proteins, fats, carbohydrates, and vitamins. If you always feel hungry, it may be because you are growing and your body needs the nutrients. When a person is growing, it is especially important to eat well and to choose nutrient-rich foods such as milk, vegetables, fruit, poultry, fish, cheese, red meat, and eggs. As you might expect, when you eat more food, you are likely to gain weight. This is normal. However, if you start to gain too much weight, watch the snacks you choose. High-calorie foods, such as cookies, soft drinks, and candy, may satisfy your appetite, but at the same time they can result in an unwelcome amount of extra fat.

Body Shape

Some fat is important. In addition to providing insulation, fat helps give the body its shape. During puberty body fat is redistributed and some of it may be lost. Girls generally lose less body fat than do boys, giving girls a more rounded appearance. Some of this fatty tissue accumulates in the developing breasts around the milk ducts.

A child's breasts are fairly flat with slightly raised nipples. During puberty, hormones trigger breast development, causing the nipples to stand out a little farther and the underlying tissue to swell slightly. At this time, the breasts may be tender and sensitive. This occurs in both boys and girls. However, in boys, the swelling goes down after the hormones balance out. In girls, breasts continue to develop, gradually filling out over the next three or four years. No one can say for sure when a girl's breasts will start to develop. As with puberty in general, it may begin when she is 8 or not until she is 16 or older. The age at which breast development starts has nothing to do with final bust size. An early starter may end up with either large or small breasts; the same is true for girls who develop later.

Changes in the skeletal system also affect body shape. Boys' shoulders widen, giving them an adult male form. Girls develop wider hips because their pelvises broaden. In both boys and girls, facial proportions change. The round face of a child gradually transforms to the somewhat longer face of an adult. (See **Figure 13.4**.)

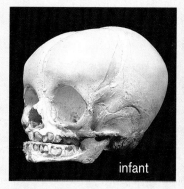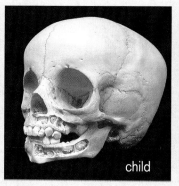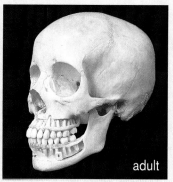

infant   child   adult

Figure 13.4

A child's face does not look like an adult's. By looking at these skulls, can you describe some reasons why?

Voice Changes

There's one change that takes place during puberty that you can hear but not see. That's the sound of the voice. Children have clear, high voices, but during puberty the vocal cords lengthen and the larynx grows to a lower position in the throat. Boys may notice that when they talk, their voices "crack" and end up on an embarrassingly high note. This transition does not last long, and when it is over, the voice is much deeper. Girls' voices also become lower and richer, but the difference is not as noticeable as in boys.

Hair and Skin

The growth of body hair is another change that marks the approach of sexual maturation. One of the first areas where hair appears for both boys and girls is around the genitals. This is called pubic hair. Hair also grows under the arms and more hair appears on the arms as well as on the legs. This hair is generally more obvious in boys than in girls.

Many boys anxiously await the appearance of hair on their faces. The first facial hairs are fine and sparse; gradually, they become coarser and thicker. Soon after the appearance of facial hair, boys may notice hair on the chest and sometimes on the back. The amount of hair varies from one male to the next. "Hairiness" has nothing to do with masculinity. Men with lots of body hair are just as masculine as men with very little body hair. Ironically, as some males approach the end of puberty and while their body hair is still developing, they may start to go bald.

Girls may also notice a change in facial hair. They may notice a bit more "fuzz" growing on the upper lip. A light growth softens facial features but, if the fuzz is bothersome, a doctor can suggest ways to remove it.

Another consequence of puberty is that the sweat glands become more active. Sweat glands are located all over the body, but they are especially numerous under the arms. The perspiration these glands produce is odorless, but the bacteria that live on the skin may cause it to have an unpleasant odor. Most of this odor can be minimized by washing daily with soap and water. You may also want to experiment with deodorants and antiperspirants. Some people find that certain brands are more effective for preventing odor than others.

In addition to sweat glands, oil glands also become more active during puberty. If the oil clogs the tiny duct that leads from the gland to the surface of the skin, it can cause a pimple or acne to form. (See **Figure 13.5**.) Pimples are especially common on the face, shoulders, chest, and back because that's where oil glands are the most numerous. Just about everyone gets at least a few pimples, but many teens are plagued by them.

The best thing to do for pimples is to clean your face with soap and water, and you may need to wash your hair every day if it is oily. If your skin problem seems to be severe, a dermatologist (skin doctor) may be able to help you. Prompt treatment can reduce your risk for long-term scarring.

So Many Changes!

Teens often ask when they will start, or stop, developing. In general, a person's growth pattern is influenced to a great extent by heredity. If your parents developed early, you are likely to develop early, too. If your parents developed later, you will probably follow that pattern.

Did you realize that all of these things are going on in your body? It's impressive to think our bodies control and coordinate so many changes. All of us eventually experience all of these changes and, in so doing, we become physically mature men and women.

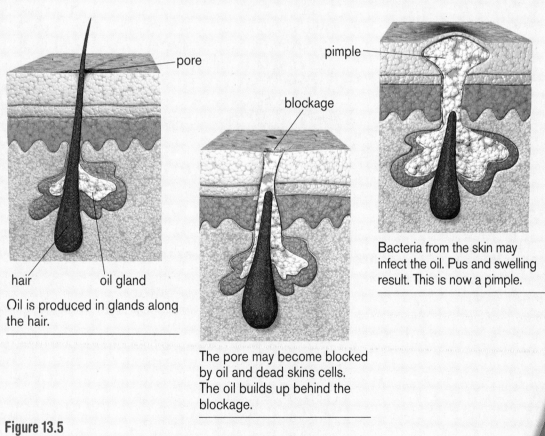

pore

hair   oil gland

Oil is produced in glands along the hair.

blockage

The pore may become blocked by oil and dead skins cells. The oil builds up behind the blockage.

pimple

Bacteria from the skin may infect the oil. Pus and swelling result. This is now a pimple.

Figure 13.5

Study each of these drawings to see if you can understand how acne develops.

A Pink Puberty?

It's often difficult, if not impossible, to judge the age of an animal by looking at it. However, flamingos are an exception. Flamingo chicks are gray and fluffy. They can fly when they are 3 months old and are almost fully grown after a year. However, they do not turn pink until they are 4 years old. After they have turned pink, they can reproduce. Is turning pink a sign that a flamingo is going through puberty? What do you think? What animals do you think go through a stage that we might call puberty?

13-3

Feelings, Friends, and Family

Some of the changes that take place during puberty are not physical changes. Feelings, thinking processes, and ideas change, too. Sometimes teens get frustrated when they try to deal with these kinds of changes. They are faced with new kinds of problems and have more complicated decisions to make. They may get upset because they don't know the best way to solve these problems without getting in trouble.

> Are middle-level students prepared to make their own decisions and deal with their own problems?

You will have a problem to solve that involves getting along with other people. Your task will be to use the decision-making model to solve the problem twice—once with other students and once with adults. In order to answer the focusing question and complete this task, you need to understand a little more about some of the nonphysical changes that take place during puberty.

- Reading, *Changes, They Aren't All Physical*, page 368
- *The Problems*, page 371
- two Handouts, *Decision Making*
- red pencil or marker

Procedure

Part A: Small Group Work
(Work in a group with one or two other students.)

1. Read *Changes, They Aren't All Physical*. (Decide if you want to read it to yourselves or if you want to take turns reading the sections aloud.)

2. As a group, answer analysis questions 1 through 4 before you go on to the next step.
3. Read the problem on the *The Problems* page that you will analyze.
4. Assign roles.

   One person should be the timekeeper and make sure the group stays on task.

   Each of you should take turns sharing ideas and opinions.

5. Fill out each column on your handout while carefully and thoughtfully thinking about each step.

Part B: Homework

1. Take your handout and a copy of the problems home with you.
2. Ask your parents or another adult to think about the problem with you.
3. Show them the handout you and your group filled out and explain the choice your group made.
4. Answer analysis questions 5 and 6, and write your conclusion.

Analysis Questions

1. What is one thing in the reading you think is probably true of you and most of your friends?

2. What changes do you think should have been mentioned that weren't?

3. Think of the advice given in the reading.

    a. What do you agree with?

    b. What do you disagree with? Why?

4. Changes such as the ones described in the reading take place because of changes in which body system?

5. How well did you and your parents agree on the best way to handle the problem?

    We agreed completely.   5   4   3   2   1   We didn't agree at all.

6. What did you learn from this lesson?

Conclusion

Write a paragraph that answers the focusing question. Use evidence from this activity to support your decision.

Extension Activities

1. **Going Further:** Are you having a problem getting along with someone? If so, use the decision-making model to try to solve it. You may discover that you have more options than you realized. Remember, if one of your choices does not work, you can go back and try another.

2. **Careers:** Talk to the psychologist or counselor at your school. Find out why they like their job. Ask them to describe what special training they had to help middle school students with their problems.

Changes, They Aren't All Physical

As children grow into adults, they realize that they feel differently about their lives, their interests, and the people they know. This is normal; it's part of growing up. You may be experiencing these changes right now and that can be confusing.

Feelings

You may have noticed that your feelings about the things you usually enjoy are changing. For instance, you may still like some of the games and books from your childhood, but something about them, or perhaps it's about you, has changed. They may not give you the same satisfaction they did before.

You may be one of those people who experiences mood swings as your feelings and attitudes change. One minute you may be cheerful and full of plans, then suddenly you may find yourself feeling depressed and angry with the world. No one knows for sure why feelings can change so quickly. Hormones are often blamed for the changes, but hormones probably are not the whole answer. As some researchers point out, lots of changes are going on during the early teen years and they take getting used to.

For instance, you may have noticed that teachers expect more from you than they did a few years ago. You are expected to take better notes, read more difficult assignments, write more thoughtful answers, and solve more complex problems. Some students are comfortable accepting these new challenges; others are not. It can be tough to take on new roles. Sometimes they are enjoyable, but other times they feel like a burden. These conflicting feelings can contribute to mood swings.

Moodiness is part of the bigger issue of behavior—the way people act. Look around you. Think about the some of the students in your class. What do you think of their behaviors? You may decide that some are childish, or stuck up, or "cool," or kind, or whatever. One reason for this wide range of behaviors is because everyone is at a different stage of puberty. Some students act childish because they still are children. Others act "strange" or "weird" because they think that is the way teenagers should act, because they are trying to impress someone, or perhaps because they don't know any other way to behave. You judge people by their behavior and you, in turn, are judged by your own behavior.

Friends

During puberty, some people may make new friends and lose touch with old ones. This is normal. As interests change, friends may change, too. One day you may even realize that your best friend has a new "special" friend—one of the opposite sex. At first your feelings may be hurt, but then you may wonder if perhaps you too should find someone special or start to date. Actually, it is not at all important to date at this age. It is much more important to learn how to get along with other people and how to be a friend—a friend to people of both sexes.

When kids date one special person, they limit themselves. They do not have a chance to meet lots of other people or to discover their own likes, dislikes, and values. These are the years you should be doing things with lots of friends—you could go swimming, watch a movie, play softball, or do other things you enjoy, but you should do so with a *group* of friends. It is often easier to talk to people and to meet new people when you do things in groups.

Dating can wait until you are older. Your parents probably have some ideas about dating. Talk to them about it. It is much better to discuss this issue in advance. That way, when you do want to date, you will already understand each other's expectations. The mention of parents brings up a whole new issue, that of family.

Family

Puberty often marks the beginning of a gradual change in your relationship with your **parents**. At times they may not seem to understand you, or you may think they are being unreasonable. Actually, your parents are probably feeling as if they are going through a kind of puberty of their own right along with you. They are watching their child develop into an adult. It is not any easier for them than it is for you. They realize that you want more independence and the freedom to make your own decisions. However, you and your parents may not agree on exactly how much independence you should have or on which decisions you should be making. Even small differences in opinion can lead to major conflicts.

> The word **parents** is used a lot in this unit. Since not all kids live in families with two "birth parents," you should substitute the word parents for whatever fits your situation. Perhaps you live in a single parent family, or with foster parents, grandparents, some other relative, or a friend of the family. When you read the word parents, think of the adults you are living with.

Many of these conflicts can be solved if you and your parents sit down and talk with each other, realizing in advance that you may not get your own way. Let your parents explain how they feel, and you tell how you feel. It doesn't do any good to get angry and yell. Talk about any differences that may come up. If you still don't think you are communicating, talk to some other adult who is not as personally involved in the situation. He or she may be able to offer some helpful suggestions.

It is also important to realize that independence does not come all at once; you have to earn it. You should show your parents that, when making decisions, you are considering more options and trying to make wise choices. This will help them acknowledge that you are growing up. Gradually they will let you assume more responsibilities until one day you realize that you are making all of your own decisions—like an adult!

The Problems

Getting His Attention

It seems like my friends talk with boys all the time. How do I get boys to notice me? Some girls flirt and giggle a lot. I don't. I try to be myself but the boys ignore me. There's a boy in my science class who seems really nice, but he never looks at me. I don't even think he knows my name. How can I get him to talk to me?

Katie

Basketball, Not Girls

Lately, my friends are more interested in girls than in basketball. We used to go to the gym every Friday. Not anymore. Most of my friends want to hang around with girls instead. My friend Jon wants to go to the shopping mall to meet some girls this Friday instead of playing basketball. I feel like I'm losing him as a friend. What should I do?

Carlos

Too Young

All my girlfriends are allowed to go on dates, but my parents say I'm too young. I think they're unreasonable and old-fashioned, and I've told them so. The other day I told them I was going out with my girlfriends, and then we met some guys and went out in a group. We are supposed to do this again Saturday. I don't like to lie to my parents, but I really want to go. What should I do?

Gena

Tongue-tied

I'd like to talk to a girl in my class. In fact, I'd really like to eat lunch with her in the cafeteria, but I don't know what to say. She doesn't seem to know I exist. Beside, she always has her friends with her, and I don't want to talk to them too. When I do say something, it sounds dumb. What should I do?

Matt

13-4

The Reproductive System: External Changes

Some of the most significant changes that take place during puberty are those that involve the **reproductive system**. This is the system that enables humans to reproduce. It is essential for life. If we could not reproduce, the human race would come to an end.

Children do not need a functioning reproductive system. It is much more important that other body systems, such as the digestive, respiratory, and skeletal systems, develop to keep pace with a growing body. During puberty, however, the reproductive system "wakes up." This chapter is about the obvious changes that result when the reproductive system begins to mature.

While going through puberty, both boys and girls grow taller, worry about being normal, and experience many similar changes and feelings. However, some changes are unique to boys and some are unique to girls. In this reading, you will learn about some of the unique changes that involve the reproductive structures on the outside of the body; these structures are called the **genitals** or **genitalia**.

Boy to Man: Changing Anatomy

The two most obvious structures of a boy's genitalia are the penis and the scrotum. During puberty, the **penis** grows a little longer and thicker; exactly how long and how thick varies from one male to the next. At some time or another, almost every male worries about the size of his penis. It is important to remember that, because puberty starts at different times for different people, some boys' penises will be bigger than other boys', even though the boys are the same age. That's okay. In fact, the size range is normal. Feet, hands, and noses are different sizes; so are penises. By the end of puberty, size differences are not nearly as great. Penis size has nothing to do with manliness, strength, or anything else; it does have to do with the genes a boy inherits.

The appearance of a penis varies depending on whether or not a boy has been circumcised. At birth, the end of the penis, often referred to as the **glans,** is enclosed in a sheath of skin called the **foreskin**. The surgical removal of the foreskin is called **circumcision**. Soon after birth, the parents usually decide whether or not they want their son to be circumcised. Their decision may be based on their religious beliefs, on the family's customs, or on their understanding of long-term health effects. Either way is okay. Whether or not there is a foreskin does not affect the functioning of the penis.

The penis functions in two body systems. As part of the urinary system, it carries urine from the bladder to the outside of the body, and as part of the reproductive system, it transports sperm. Both urine and semen pass through a small opening at the end of the penis called the **urethral opening**.

Behind the penis is a loose sac of thin skin called the **scrotum**. It contains the **testes** or **testicles**. They are an important part of the reproductive system because they produce sperm. Muscle fibers in the scrotum contract and relax, changing the position of the testes. When the surrounding air is cold, the muscles contract, bringing the testes closer to the body. In hot weather, the muscles relax so the testes hang a little farther from the body and will stay cooler. When this happens, the testes may seem larger but there really is no change in their size. By changing the position of the testes, these muscles help keep the temperature of the testes about 2°C lower than body temperature. This lower temperature is necessary for the production of sperm.

One of the early signs of puberty is often an increase in the size of the testes and, therefore, of the scrotum, too. When this happens, the skin on the scrotum becomes darker and more wrinkled. As the scrotum grows, one side may hang a little lower than the other. This probably makes movements such as walking more comfortable, since the testes won't rub against each other. Like other body parts, the size of the scrotum varies slightly from one person to another.

During puberty, **pubic hair** appears around the base of the penis. At first, it may look like a pale fuzz. Over time, the hairs become coarser, darker, and curlier.

All these changes take place in every boy as he goes through puberty. The exact timing varies from one boy to another, but eventually all boys become adult males.

Girl to Woman: Changing Anatomy

A girl's genitals are not as obvious as a boy's. Perhaps that's why many girls are not aware of the changes that take place as they go through puberty. A female's genitals are called the **vulva**. The vulva is made up of several parts. The most obvious parts are two pairs of fleshy folds of tissue—the inner and outer labia.

The **outer labia** are a pair of fat-filled folds that enclose and protect the other structures.

In girls, the inner surfaces of these folds are smooth but, as puberty progresses, small, slightly raised bumps may appear. These bumps are oil glands that keep the area moist so the skin does not get irritated.

Just inside the outer labia are thinner folds of tissue called the **inner labia**. In childhood, they may be fairly small; they grow during puberty. They, too, have oil glands that produce oil. As a girl goes through puberty, the labia become fleshier and darker in color.

Where the two inner labia meet in front, there is a firm bump of tissue called the **clitoris**. In grown women, it is about the size of the head of a wooden match. Although the clitoris has no reproductive function, it contains many nerve endings and is sensitive to touch.

Directly behind the clitoris is the **urethral opening**. As in boys, this is the opening through which urine passes on its way out of the body. However, in girls, the urethral opening is not part of the reproductive system; it functions only in the urinary system.

The **vaginal opening** is behind the urethral opening. The vagina itself is not visible, since it is inside the body. Even though this opening is larger than the urethral opening, it is not very obvious either. That's because the sides of the vagina are usually touching each other. Sometimes a membranous tissue called the **hymen** partially covers the vaginal opening. This membrane is often so thin that it is easily broken or torn. In fact, most women are unaware they ever had a hymen.

Early in puberty, **pubic hair** starts to grow in front of the vulva and on the outside of the outer labia but not on the inner labia. At first, the pubic hair is fine and sparse. As puberty progresses, it becomes coarser, thicker, and curlier.

All these changes take place in every girl as she goes through puberty. The exact timing varies from one girl to another, but eventually all girls become adult females.

Men and Women

The changes that take place during puberty are gradual, taking several years to complete. Nothing happens overnight, in a week, or even in a month. The entire process is pretty amazing if you stop to think about it. During this time, the bodies of children—boys and girls—become the bodies of adults—men and women. It's important to remember, however, that having a mature body does not mean that a person is an adult.

This reading has mentioned some of the most obvious changes, but, of course, it does not explain everything. If you notice something going on with your own body that you wonder about, or are worried about, talk to someone. Your parents, school nurse, doctor, or some other trusted adult can be helpful sources of information. You may even find that when they were young, they had some of the same questions you have now.

Includes *sci*LINKS®
NSTA

Topic:	Reproductive System
Go to:	www.scilinks.org
Code:	MSLS3e374

Analysis Questions

1. List at least two reasons you think a person should know the names of the structures that make up the genitals and how these structures change during puberty.

2. The body of an adult is different from the body of a child, but there are other differences as well.

   a. List at least three ways besides appearance that adults differ from children.

   b. How old do you think a person should be before you call him or her an adult?

Extension Activity

Animal Kingdom: In humans, the difference between a male and a female is usually fairly obvious. Read about a variety of other kinds of animals and keep track of examples where the two sexes look alike and where they look different. Can you find a pattern?

Figure 13.6

Which are the males and which are the females? The differences are not obvious in species such as these newborn penguins.

The Reproductive System: Internal Changes

Some parts of the reproductive system are visible on the *outside* of the body, but most of the parts are *inside* the body. Now that you have learned about the *external* changes that take place during puberty, you are ready to study the *internal* changes. Even though these changes are not visible, they affect the way you feel and the way your body works.

In this reading you will learn the names and locations of more of the structures that make up the reproductive system, and you will begin to learn about their functions.

Boy to Man

As you learned, one of the early signs of puberty is the growth of the scrotum. It grows because the testes inside are growing. The **testes** are paired, egg-shaped structures. They are made up of a mass of microscopic coiled tubes; they do not contain any fat or muscle tissue. Testes have two main functions— to produce sperm and to release hormones. Sperm are made in the coiled tubes; the cells around the tubes produce the male hormone **testosterone**. During puberty a testis may double in size, growing from about 2 cm to a length of 4 to 5 cm. (See **Figure 13.7**.)

Look at **Figure 13.8**, and find the tube that lies along the back side of each testis. This tube is the **epididymis**. Once sperm are produced in the testis, they travel into the epididymis where they finish maturing.

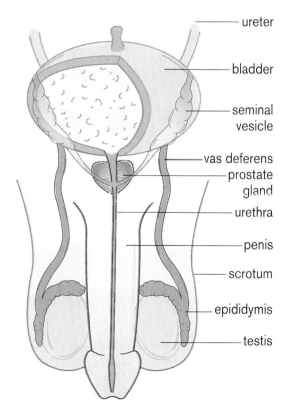

— ureter

— bladder

— seminal vesicle

— vas deferens
— prostate gland

— urethra

— penis

— scrotum

— epididymis

— testis

Figure 13.8
Male reproductive system—front view

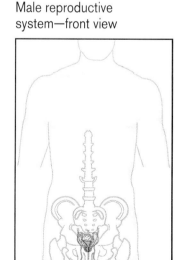

Figure 13.7
Testis size

size before puberty

adult size

Leading away from the epididymis is a larger tube, the **vas deferens**. Sometimes it is referred to simply as the vas. It extends up out of the scrotum and into the abdomen. It then loops around the bladder to join the urethra. Sperm move through the vas deferens, into the urethra, and out of the body. Follow the path of the vas deferens in **Figures 13.8** and **13.9**. (Which figure is easier for you to understand?)

The **urethra** is another one of the many tubes that make up the male reproductive system. Since the urethra drains urine from the bladder, you may have discussed it when you studied the urinary system. As part of the reproductive system, it transports sperm from the vas deferens out of the body through the penis. Even though both urine and semen pass

through the urethra, they are never there at the same time. A muscle closes off the connection between the urethra and bladder when sperm are present.

Several glands produce fluid that is released into the vas deferens. The largest of the glands, located right beneath the bladder, is the **prostate**. (See **Figure 13.8**.) In children, it is relatively small, but during puberty it starts to grow and eventually reaches 2 to 3 cm in diameter. Boys and young men usually do not need to give much thought to their prostates, but after the age of 40, men should have annual prostate checks because it is a common site of cancer. The **seminal vesicles** are a pair of glands that also produce fluid. Find them in both **Figures 13.8** and **13.9**.

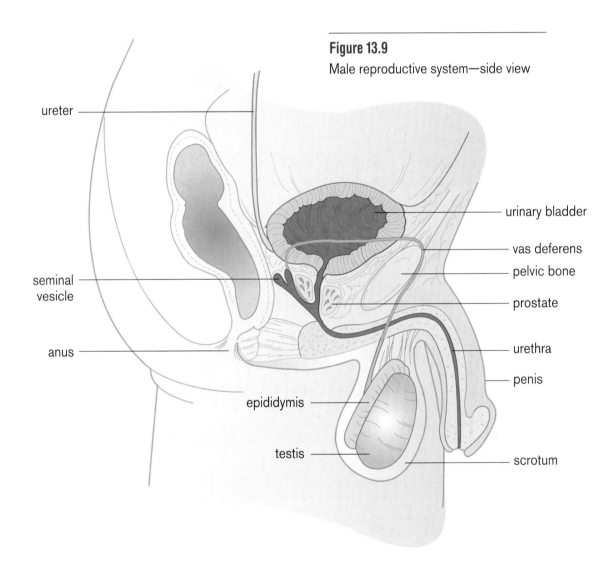

Figure 13.9
Male reproductive system—side view

ureter

seminal vesicle

anus

epididymis

testis

urinary bladder

vas deferens

pelvic bone

prostate

urethra

penis

scrotum

Girl to Woman

A girl's reproductive system has fewer "tubes" than a boy's does. The main structures of a female's reproductive system are labeled in **Figures 13.10** and **13.11**. Locate the **ovaries** in both figures. In an adult female, they are 2 to 3 cm long. In addition to releasing egg cells, they produce sex hormones such as **estrogen**. These hormones regulate many of the characteristics that we think of as being feminine. Biologically, the ovaries and testes are similar, since they have similar functions.

When an egg is released from an ovary, it enters a **fallopian tube**. These 10-cm (4-inch) long structures extend from each ovary to the uterus. They are fairly narrow, about the diameter of a pencil. The ends of the fallopian tubes flare open to guide eggs into the opening of the tube itself. The walls of the fallopian tubes are made up of smooth muscle cells. When these muscles contract, the tiny **cilia** that line the tubes wave back and forth, moving the egg along the length of the tube toward the uterus.

Cilia are tiny hairlike structures that protrude from the cell membranes. When you studied the respiratory system, you learned about the cilia that line the bronchial tubes.

To many people, the **uterus** is the most familiar part of the reproductive system because it is where a baby develops and grows. During puberty, the uterus more than doubles in size. It grows until it is about the size and shape of a pear. Like the fallopian tubes, the walls of the uterus are made up of smooth muscle cells. When a female is not pregnant, there is not much space in the uterus. But during pregnancy, the muscle fibers relax, allowing the walls of the uterus to stretch so the uterus can hold a baby—and sometimes even two or more babies!

During puberty, the lining of the uterus develops blood vessels that go through monthly changes. When these changes begin, a girl experiences her first menstrual period. This process is very complex, and you will learn more about menstruation in a later reading.

The lower part of the uterus is the **cervix**. It is a firm muscular ring that surrounds the opening to the uterus. The muscles in the cervix have to be strong because they keep the uterus closed during a pregnancy. Because of this muscle, a baby cannot "fall out" of the uterus.

The **vagina** is a flexible "tube" that extends from the uterus to the outside of the body. Like the uterus and fallopian tubes, its walls

Figure 13.10

Female reproductive system—front view

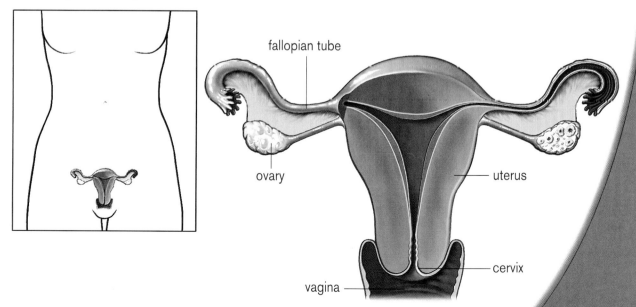

also contain smooth muscle cells. These cells can contract, relax, and stretch. Usually, the cells are relaxed so the vagina is similar to a collapsed tube. When necessary, it can stretch enough for the delivery of a baby. For this reason, it is sometimes called the birth canal.

The *urinary bladder* and the *urethra* are labeled in **Figure 13.11**. While neither one is part of the reproductive system, they are included to help you see how the different parts fit together. By now, you should realize that the female has three openings in the lower part of her body—the urethra, the vagina, and the anus—whereas, a male has two—the urethra and the anus.

In order to check the female's reproductive system for medical problems such as lumps, the physician must be able to feel the structures. This is accomplished by doing a pelvic exam. During an exam, the physician gently inserts two gloved fingers into the vagina and carefully feels for any abnormalities. It may feel a little uncomfortable, but it is not painful.

Parts of a System

This reading has described some of the structures that make up the reproductive system. As with all other systems, the parts all work together. You will learn about the functioning of the reproductive system in the next few lessons.

Figure 13.11
Female reproductive system—side view

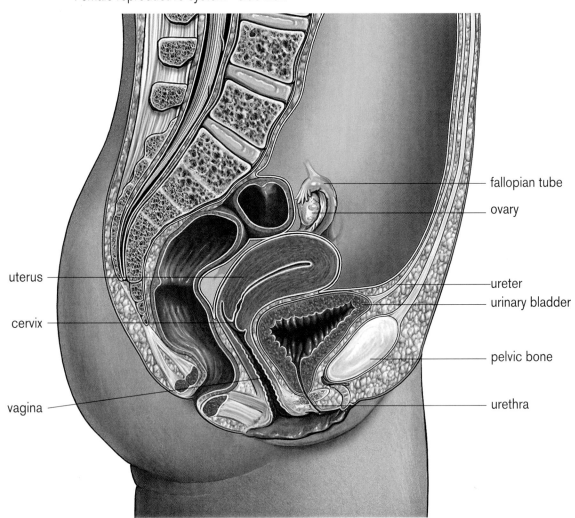

uterus

cervix

vagina

fallopian tube

ovary

ureter

urinary bladder

pelvic bone

urethra

Analysis Questions

1. Compare the testes and ovaries.
   a. List at least two ways they are alike.
   b. List at least three ways they are different.

2. Think about the changes that take place in the male reproductive system during puberty. Which structure do you think changes the most? Why did you choose this structure?

3. Think about the changes that take place in the female reproductive system during puberty. Which structure do you think changes the most? Why did you choose this structure?

Extension Activity

Plant Kingdom: Many flowers have female and male reproductive structures—ovaries and anthers. Look at several flowers and try to locate these structures.

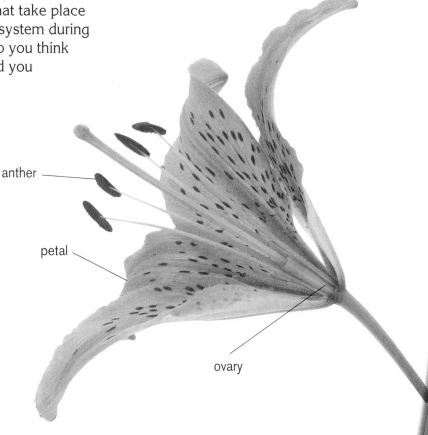

anther

petal

ovary

Uniquely Male

In addition to doing this reading, inspect a preserved bull testis. You can do so at any time during this lesson. When you look at the preserved specimen, fill in the handout.

Ten to twelve years after birth, a boy's reproductive system seems to "wake up." During puberty, the external and internal sex organs start to grow and take on new functions. The onset of these new functions can be startling and may even be scary if a boy does not know what to expect. As you read the next several pages, you will learn about these functions—about what they are and why they happen.

Erections

Males of every age, including tiny babies, experience erections. In fact, a male has his first erection before he is born. During an **erection,** the penis stiffens, becomes larger, and stands out from the body. Because the penis becomes so much firmer than it usually is, many people mistakenly assume that it contains a bone.

Actually, the penis is not made up of bone or muscle. Instead, it contains three cylinders of spongy tissue that is laced with blood vessels. This tissue runs the length of the penis. (See **Figure 13.12.**) During an erection, blood enters the vessels in the penis causing the tissues to swell and the penis to stiffen. All of this can happen almost instantly or it may take several seconds.

Erections happen for all sorts of reasons. Getting nervous or excited can cause an erection, so can pleasant, warm thoughts. Thinking about sex or about the sex organs is the most common cause of an erection. Some erections, however, have nothing to do with thoughts or feelings. They may be caused by a full bladder or by tight clothing, or they may occur for no obvious reason whatsoever. They can happen at any time of the day or night. Erections are just another normal body function like any other body function, except they are unique to males.

Figure 13.12

Cross section of a penis

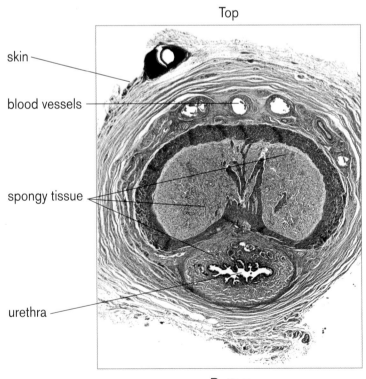

Top

skin

blood vessels

spongy tissue

urethra

Bottom

Ejaculation

Unlike erections, ejaculations do not happen in babies or young boys. A male does not ejaculate until a year or two after he has started puberty. A boy's first ejaculation is a sign that his body is developing into that of a man. During an **ejaculation,** about 3–6 mL of fluid is propelled from the end of the erect penis. (5 ml is about a teaspoonful.) This whitish fluid, called **semen,** is a mixture of sperm, nutrients, and water. Immediately after ejaculation, the erection starts to go away. The pleasurable sensation that accompanies an ejaculation is called an **orgasm**.

Many boys experience their first ejaculation as a result of masturbating. That is, a boy may be touching, stroking, or otherwise handling his erect penis when suddenly the fluid buildup inside his body is released. It is normal for boys to masturbate; it is also normal not to. For years, people thought masturbation could be dangerous. Now researchers have found that it is not harmful to the body. Even so, some people object to masturbation. You probably know what your family thinks about the subject. If you don't, ask them. They may be startled by your question, but they usually appreciate it when you take the time to discuss with them topics that can be controversial.

Wet Dreams

While erections can occur just about any time, ejaculations don't. Semen will not appear unexpectedly following an erection except occasionally when a boy is sleeping. An ejaculation that takes place during sleep is called a **wet dream**. A boy's first ejaculation is often a wet dream. Even if he has heard about wet dreams, the first time it happens is usually startling. He is likely to wake up and find that his skin, pajamas, or sheets feel wet and sticky. If he does not know what to expect, he may think he has wet the bed. However, the fluid does not look or smell like urine, and there is not nearly as much of it. Even so, many boys are embarrassed, although there is no need to be; wet dreams are another normal part of growing up. Some boys have them frequently. Other boys may never have a wet dream and that, too, is normal.

Another term that you may hear for wet dreams is *nocturnal emission*. Nocturnal means "during the night" and emission means "sending something on." So nocturnal emission refers to semen that is sent forth, or ejaculated, at night.

Semen

People often ask, "Where does semen come from?" The answer to the question is easy if you remember that semen is a mixture of sperm and fluid. Sperm are produced in the tiny tubules that make up the testes. They then spend several weeks traveling slowly through the epididymis where they finish maturing. The mature sperm move into the vas deferens where they are stored until they are released from the body.

When a male ejaculates, smooth muscles around the prostate and other parts of the reproductive system contract, forcing the sperm out of the vas deferens and through the urethra. These contractions also squeeze fluid from the prostate and other glands. This fluid mixes with the sperm to form semen. As soon as the fluid mixes with the sperm, the sperm become much more active and start to wiggle and "swim" along with the fluid as it is propelled through the urethra. The sperm pathway is illustrated in **Figure 13.13**.

A Continuing Process

Although the production of sperm and the onset of ejaculations are a sign that a boy is maturing, they do not mean he is a man yet. His body must continue to grow for several more years until he reaches physical maturity. It also takes many years of living and learning before he will be ready to take on the responsibilities of an adult.

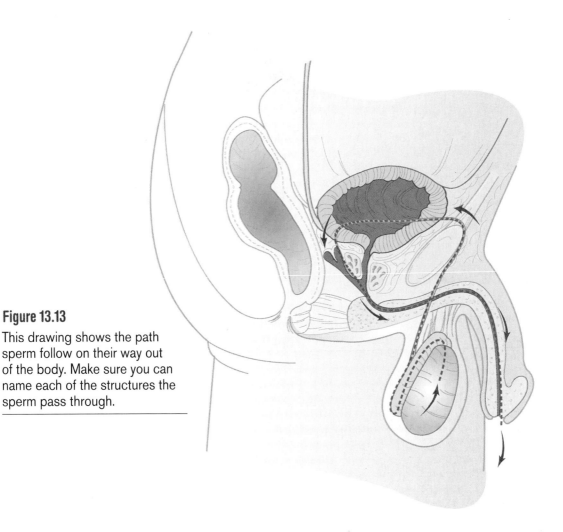

Figure 13.13

This drawing shows the path sperm follow on their way out of the body. Make sure you can name each of the structures the sperm pass through.

Analysis Questions

1. What is the difference between an erection and an ejaculation?

2. What is semen?

3. Sperm pass through these structures on the way out of the body: vas deferens, urethra, testes, epididymis.

   a. List the structures in the correct order in which sperm passes through them.

   b. Describe what happens to sperm in each structure.

4. Decide if each of these statements is true or false. If it is false, rewrite the statement so it is correct.

   a. An erection results when the penis muscle contracts.

   b. Erections are always caused by sexual thoughts.

   c. Ejaculation always follows an erection.

   d. An erection always comes before an ejaculation.

   e. Once wet dreams start, they occur about once every month.

   f. An erection is a sign of sexual maturity.

Extension Activity

Going Further: Look up semen in a biology book, medical book, or encyclopedia. Find out what each gland contributes to the semen. Show where these glands are located on a diagram of male internal anatomy.

Reproductive Systems

All species have to be able to reproduce, otherwise they would become extinct. Many plants and animals reproduce sexually; that is, two parents are required. Other plants and animals, as well as bacteria and most protists, reproduce **asexually**. That is, one parent gives rise to the offspring. There are many types of asexual reproduction. In some species, the parent splits in two; in others a new individual may grow from the body of the parent. The photograph shows young sea anemones growing off the body of the parent sea anemone. Eventually, they will detach from the parent.

13-7

Uniquely Female

In addition to reading this chapter, inspect a preserved cow ovary. You can do so at any time during this lesson. When you look at the preserved specimen, fill in the handout.

Once puberty starts, a girl becomes aware that her body is changing. She can see the changes that involve body shape and size, but those that involve functions are less noticeable. Both girls and boys have many questions about the way the female body works. This reading will answer some of the questions but probably not all of them. If you think of other questions while you are reading, jot them down so you can ask them later.

Vaginal Responses

Early in puberty, the cells that line the vagina become more active. They start to produce small quantities of fluid that keeps the vagina clean and healthy. This fluid is clear and odorless, with perhaps a whitish or a pale yellow tint.

Later in puberty, a girl may notice days when she produces more of these fluids than usual. Some of this variation may be due to her awakening sexuality. Watching a romantic movie or thinking about sex may cause more fluid production. Sexual thoughts can also trigger stronger reactions such as a pleasurable tingling or a throbbing sensation in the genitals. When these sensations reach a peak, it is called **orgasm.**

These sensations may also be a result of masturbation. That is, a girl may be touching or stroking her labia and clitoris when she notices these feelings. Not all girls masturbate. It's normal if they do, and also normal if they don't. Masturbation does not harm the body and may release sexual tension. Even so, some peowple believe that masturbating is wrong. You may know how your family feels about the subject. If you don't, ask them. They may be surprised by your questions, but they are also likely to be pleased that you want to discuss with them this controversial subject.

A Menstrual Period

One of the most significant events that tells a girl she is becoming a woman is her first menstrual period. This usually occurs some-time between the ages of 11 and 14. However, some girls may begin as early as 8 while others may not start until they are 17. Many girls start to menstruate at about the same age that their mothers started. For example, if a girl's mother started her periods when she was 15, the girl is likely to start hers around age 15 also.

When people talk about a **menstrual period** or **menstruation,** they are referring to the loss of several tablespoons of blood and mucus from the vagina. It is not released all at once; instead, it is lost a little at a time, over several days. There is then a span of 3 to 4 weeks before a girl has another period. This pattern repeats itself over and over again.

3–6 days menstrual flow	21–28 days no flow	3–6 days menstrual flow	21–28 days no flow

Menstruation is a normal body process. Most of the time it does not affect a girl's activities. During her period, she can do all of the things she normally does. Sometimes, however, girls have cramps just before or during their periods; some get cramps more often than others. Cramps may be very mild or they can be fairly uncomfortable. The discomfort usually can be relieved by applying heat with a heating pad, or by taking a mild pain reliever such as ibuprofen. Exercise helps, too. The movement probably stretches and relaxes the muscles in the lower abdomen, and this eases the cramping. Girls who have painful cramps should talk to their physicians; sometimes a stronger medication is recommended.

A lot has been said about the way periods make a person feel. The feelings are varied. Some girls and women feel more energetic or happier during their periods. Others feel quieter, more restless, or even "out of sorts with the world." In general, a period may be thought of as a signal that all is well with the body.

Changes in the Uterus and Ovaries

A menstrual period is part of a complex cycle that involves the uterus and the ovaries. All of the events are timed to support the development of a baby.

A period occurs when the spongy tissue that lines the uterus is lost. Immediately after a menstrual period, the lining is thinnest. Over the next several weeks it gradually develops a network of new blood vessels. This tissue is capable of nourishing a growing embryo. However, if there is not a pregnancy, the lining

Figure 13.14

Swimming and other exercises can be continued during menstruation.

is not needed so it breaks down and passes out of the body through the vagina. By the end of a menstrual period, the lining of the uterus is thin once again and ready to start a new cycle. The uterine lining is unique because it changes constantly—it is always building up or breaking down. (See **Figure 13.15**.)

Figure 13.15

During a menstrual cycle, the lining of the uterus gradually thickens and then is lost.

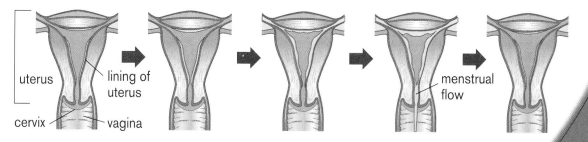

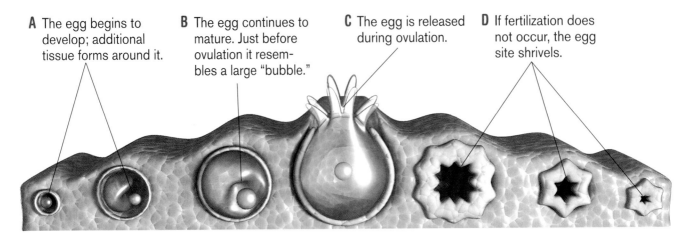

A The egg begins to develop; additional tissue forms around it.

B The egg continues to mature. Just before ovulation it resembles a large "bubble."

C The egg is released during ovulation.

D If fertilization does not occur, the egg site shrivels.

Figure 13.16

During every menstrual cycle, an egg develops in one of the ovaries.

While the lining of the uterus is developing, an egg is maturing in one of the ovaries. As the egg matures, it forms a tiny bulge on the surface of the ovary. This bulge continues to swell until it bursts, releasing the egg. The rupturing and release of an egg is called **ovulation**. Most females do not feel anything when this happens. However, some women do feel a slight twinge or cramp that could indicate ovulation. Study **Figure 13.16**; make sure you understand what it is illustrating.

After ovulation, a small pit is left on the surface of the ovary. It gradually heals until only some scar tissue remains. Meanwhile, the egg is drawn into the fallopian tube and moves along to the uterus. This journey takes several days. Since the egg cannot live for much more than a day, it usually dies somewhere along the way. Within a week or two, another egg will start to develop.

Coordination

The changes in the lining of the uterus and the development of an egg are closely coordinated. They're timed so that if a fertilized egg reaches the uterus, the lining will be rich with blood. If the egg dies, the lining breaks down. Then, as another egg matures, it must build back up again.

These changes are coordinated by hormones from the ovaries and the pituitary gland. (The pituitary gland is located on the underside of the brain. See **Figure 13.17**.) The pituitary hormones cause the egg to mature. As the egg matures, the tissues around it produce another hormone (estrogen) that causes the lining of the uterus to develop. In this way the lining is thickest when the egg arrives.

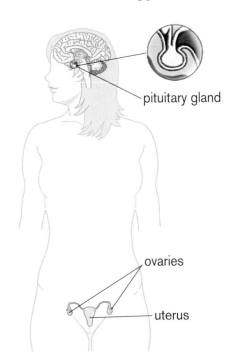

pituitary gland

ovaries

uterus

Figure 13.17

Hormones from the pituitary gland and ovaries coordinate the menstrual cycle.

Days in a Cycle

If you understand the menstrual cycle, you can predict what is happening in the uterus and ovaries on any particular day. **Figure 13.18** summarizes the daily changes that go on during a 30-day cycle. You can see that, on day 1, the lining of the uterus starts to break down and an egg starts to develop in the ovary. On day 16, the egg is mature and ovulation occurs; the lining of the uterus is developing. Two weeks later, on day 30, if the egg is not fertilized, the lining of the uterus will be lost. Study the figure. Try to describe what is happening on any given day of the menstrual cycle.

The length of the menstrual cycle varies from one individual to another. **Figure 13.18** shows a 30-day cycle. The average length of a cycle is about 28 days. Some women, however, have cycles that last about 21 days while others have cycles that last up to 40 days.

Some women are very regular; that is, their cycles are always the same length, such as every 26 days or every 30 days. Other women have cycles that vary slightly from month to month. For example, someone may have a cycle length of 27 days one time, 31 days the next, and then 29 days for the next three cycles. When a girl first starts to menstruate, her cycles are likely to be very irregular. It is not unusual for a girl to skip a cycle or two completely. It may take months or even years before a regular pattern develops.

When the Cycle Stops

The cycle continues month after month, year after year, unless a woman becomes pregnant. A pregnant woman does not have menstrual periods because the lining of the uterus is needed to nourish the growing embryo. After the baby is born, the mother's menstrual cycle starts again.

Figure 13.18

For any given day of the menstrual cycle, you can predict what is happening in the uterus and ovaries.

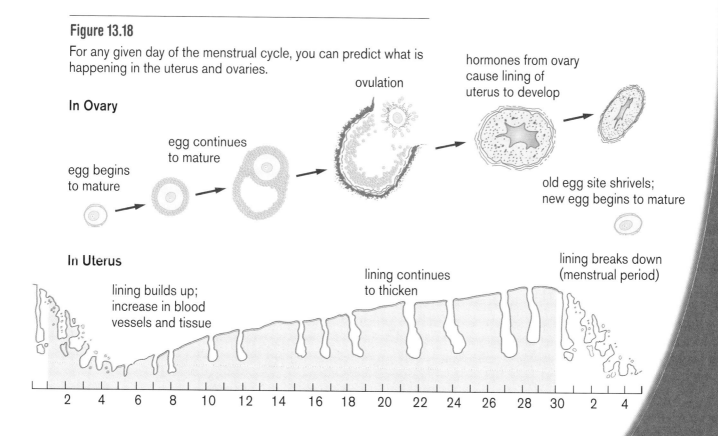

In Ovary

egg begins to mature → egg continues to mature → ovulation → hormones from ovary cause lining of uterus to develop → old egg site shrivels; new egg begins to mature

In Uterus

lining builds up; increase in blood vessels and tissue — lining continues to thicken — lining breaks down (menstrual period)

2 4 6 8 10 12 14 16 18 20 22 24 26 28 30 2 4

Eventually, between the ages of 40 and 55, a woman's menstrual cycle will come to an end. It may stop all at once or her cycle may become irregular for a while before it stops. This time in a woman's life is called **menopause**.

A Continuing Process

The whole process of growing up is pretty amazing. While the start of the menstrual cycle is a sign that a girl is maturing, it does not mean she is a woman yet. Her body must continue to grow for several more years until she reaches physical maturity. It also takes many years of living and learning before she will be ready to take on the responsibilities of an adult.

Analysis Questions

1. What is the difference between a menstrual period and a menstrual cycle?
2. Summarize the changes that take place during one menstrual cycle in:
   a. the ovary
   b. the uterus
3. Define the word ovulation.

4. Decide if each of these statements is true or false. If it is false, rewrite the statement so it is correct.
   a. A menstrual period usually lasts 3 to 6 days.
   b. A menstrual cycle usually lasts 3 to 6 days.
   c. A girl can play tennis or go swimming during her period.
   d. Periods are caused by sexual thoughts.
   e. A girl may feel weak and shaky from loss of blood during her period.
   f. A girl knows something is wrong if she gets cramps during her period.
5. Explain what is happening in the uterus and ovary on each of these days of a 30-day cycle. (Use **Figure 13.18** to help you answer this question.)
   a. day 1
   b. day 9
   c. day 16
   d. day 27
   e. day 1 (day "31")
6. If a woman menstruates every 27 days, on what day of her cycle is an egg most likely to be released from an ovary?

Estrous Versus Menstrual Cycles

Most sexually reproducing animals mate only when the female is ovulating or in "heat." This fertile time is know as *estrus*. The time from one estrus to another is called an *estrous cycle*.

While a menstrual cycle and an estrous cycle are similar, they are not the same. During a menstrual cycle, the lining of the uterus is shed if an egg is not fertilized. In an estrous cycle, the uterine lining shrinks back to its original size and there is no discharge. The bleeding that occurs in some animals right before they are in heat is caused by red blood cells that cross the wall of the vagina.

The length of an estrous cycle varies from species to species. Dogs go into estrus twice a year, cows and horses once a month, and mice, every few days. Wild animals usually come into heat so the young will be born when environmental conditions are most favorable. For instance, in wolves and deer, estrus occurs only in the fall so the young will be born in the spring.

Chapter 14
Reproduction

How are babies part of a life cycle?

Sperm, Eggs, and Fertilization

Humans, as well as most other animals and many plants, reproduce sexually. Sexual reproduction requires two kinds of cells, usually from two parents. One parent usually produces fairly large cells that provide nourishment for the future embryo; these cells are called **eggs**. The other parent often produces smaller cells that can move about; these cells are the **sperm**. Eggs and sperm are called **gametes**. Since gametes are such special cells, they deserve a closer look.

What do guinea pig sperm look like?

How can you recognize an egg cell on a microscope slide of an ovary?

Look at microscope slides of a sperm and an ovary to find the answers to these questions. This activity will challenge your microscope skills, because eggs and sperm are not easy to observe. The reading that accompanies this activity provides information about eggs and sperm and introduces the unique function of these two types of cells.

Materials

- Handout, *Sperm and Eggs*
- microscope
- microscope slides
  - guinea pig sperm
  - mammalian ovary

- Reading, *Early Stages of Reproduction*, page 393

Procedure

(Work with a partner. Start with either step 1 or step 2, depending on what your teacher tells you to do.)

1. Read *Early Stages of Reproduction*.
2. Look at the microscope slides. Fill in the handout while you work.
3. Answer the analysis questions and write your conclusion.

Analysis Questions _ _ _ _ _ _ _ _

1. Think about sperm and egg production. How many:
   a. sperm does a male produce each day?
   b. sperm does a male produce in a month?
   c. egg cells mature every month?

2. What happens to:
   a. sperm that are not ejaculated?
   b. eggs that are not fertilized?

3. Indicate whether each of the following is true of an egg, a sperm, both, or neither.
   a. a single cell
   b. contains lots of nutrients
   c. has a tail
   d. has a cell membrane
   e. is barely visible without a microscope
   f. is produced in tiny tubules
   g. contains many tiny capillaries
   h. has a nucleus

4. Define the word fertilization.

5. List five serious consequences that can result from sexual intercourse.
   a. Circle the consequence *you* think is most serious.
   b. Put a star by the consequence *your parents* would think is the most serious.

6. Do you think it would be possible for a female to get pregnant if she has had intercourse only once? Explain your answer.

7. List at least three reasons why you think some teens have sexual intercourse despite the possible serious consequences that can result.

Conclusion

Based on what you observed, answer each of the focusing questions.

Early Stages of Reproduction

In many ways, sperm and eggs are like all the body cells you've studied. They too are enclosed by cell membranes and they contain nuclei. They also have specialized functions and characteristic shapes.

There is, however, one major difference between gametes and other body cells. When an egg and sperm combine, a baby will develop. This does not happen with other types of cells. One or two skin cells cannot grow into a new individual; nor can bone cells, muscle cells, or any other cells. Only eggs and sperm have this amazing ability.

In this reading you will learn about these two kinds of cells and about the process that brings them together to form a new life.

Includes *sciLINKS* NSTA

Topic: Reproduction
Go to: www.scilinks.org
Code: MSLS3e393

Sperm

A sperm is one of the smallest human cells. As you can see in **Figure 14.1**, it looks somewhat like a tadpole with an oval head and a thin tail. The "head" contains the nucleus of the sperm cell. The tail can whip back and forth to move the sperm forward.

Sometime during puberty, the testes produce enough testosterone to trigger the production of sperm. Sperm are produced in the coiled bundles of microscopic tubules that make up a testis. If you could uncoil the tubules from one testis, they would extend for about 500 meters. This provides plenty of room for making sperm. At maturity, each testis produces about 75 million sperm a day.

When sperm leave the testis, they are not fully formed. They spend the next 50 or so days in the epididymis where they complete their development. The mature sperm then move on to the vas deferens where they are stored until they either disintegrate or are released from the body.

Figure 14.1

This photograph of a human sperm was taken through a microscope. Compare it with the labeled drawing of a sperm. (top) Dr. Dennis Kunkel/Visuals Unlimited, Inc.; (bottom) Dr. Richard Kessel & Dr. Gene Shih/Visuals Unlimited, Inc.

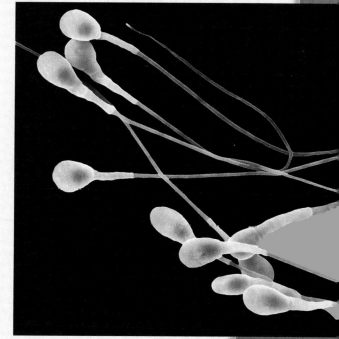

head   tail

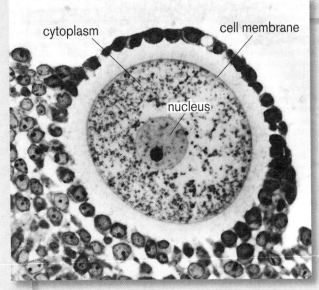

Figure 14.2

This human egg was photographed through a microscope.

Eggs

A female is born with all of the egg cells she will ever have. At birth, there are about a million of these tiny cells in her ovaries. Over the next several years, many of the egg cells die so, by puberty, only about 400,000 of the original eggs remain. Most of these will eventually disintegrate, too.

Once a girl enters puberty, one egg cell matures and is released from the ovary about every 28 days. During a woman's lifetime, only about 400 eggs ever mature. This is significantly different from the male system where millions of sperm are produced daily.

Even though an egg is thousands of times larger than a sperm, it is still very tiny. It is barely visible without a microscope. (See **Figure 14.2**.) Compared with other cells, the cytoplasm of an egg contains a large supply of nutrients. If the egg is fertilized, these nutrients will be used to nourish the embryo.

When an egg is released from the ovary, it is drawn into a fallopian tube. It takes several days for an egg to travel the length of the fallopian tube and finally enter the uterus. Since an egg lives only about 12 to 24 hours, the egg usually dies and starts to break down somewhere in the fallopian tube.

Fertilization

Sometimes the egg does not die; instead, it is joined by a sperm and a new life begins. The joining of the egg and sperm is called **fertilization**. (See **Figure 14.3**.)

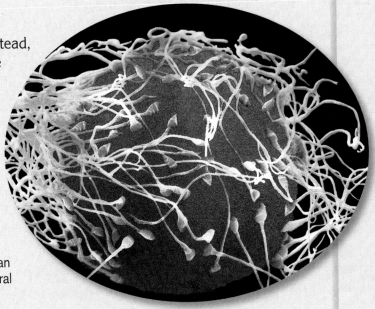

Figure 14.3

Only one of the many sperm surrounding this egg will fertilize it. How would you describe the difference in size of this human egg and sperm? Note: These are not natural colors.

Fertilization usually results from sexual intercourse. During sexual intercourse, the man's penis becomes erect and is inserted into the woman's vagina. Semen, containing about 300 million sperm, is then ejaculated close to the cervix. Many of the sperm die almost instantly. The survivors, their tails whipping rapidly, are propelled into the uterus. They move at a speed which would be about the same as that of a swimmer covering 12 meters in a second. Many more sperm die in the uterus; only a few hundred ever reach the fallopian tubes. The dead sperm cells break down and are cleaned up by white blood cells as are all other dead cells in the body.

If an egg is in one of the fallopian tubes, hundreds of sperm surround it. As soon as one sperm crosses the egg's membrane, the surface of the egg changes, making it impossible for other sperm to enter. In this way, only one sperm can fertilize an egg. A fairly predictable sequence of events follows.

The dot-sized fertilized egg divides as it continues through the fallopian tube. (See **Figure 14.4**.) It also produces a hormone that causes the lining of the uterus to become even richer with blood. If it weren't for this hormone, the lining would be lost during menstruation, along with the fertilized egg. It's almost as if the fertilized egg is sending a signal that the lining is needed because an egg is on the way. Once it reaches the uterus, the fertilized egg burrows into the spongy lining where it continues to develop into a baby.

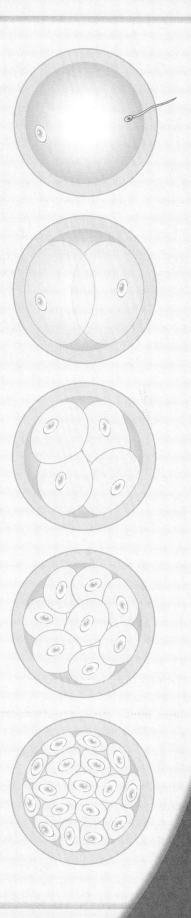

Figure 14.4
Soon after the egg and sperm join, the fertilized egg divides into two cells, then four, and so on. After several days, it is a "ball" of cells.

More Than "Sex"

Sexual intercourse was mentioned very briefly a few paragraphs ago. It is important to include a little more about this topic because no other behavior causes the same emotions or the same strong feelings. Ideally, sexual intercourse is an expression of strong love and respect between a husband and wife. Their shared love is one of the deepest emotions there is. It involves a sharing of more than just a physical attraction. In a marriage, many things are shared—a home, money, good and bad times, interests, and a deep commitment to each other. Under these circumstances, sexual intercourse is an expression of this deep love.

Under other circumstances, however, sexual intercourse can have some serious consequences. Some teenagers get carried away by their emotions and have intercourse without stopping to think about the consequences. Before you read on, close your eyes and think about the consequences. How many can you name?

Five of the most serious consequences are pregnancy, sexually transmitted disease, emotional turmoil, depression, and parental disapproval. (How many of these did you think of?) Some of these have lifelong impact. All of them are stressful and none will improve a relationship or make a person feel good about himself or herself. These are some of the reasons that teens should not be involved in sexual intercourse. **Abstinence** (not having sexual intercourse) is your best choice. There are many ways to enjoy a male-female relationship without "having sex."

Some students wonder when they are "supposed to have sex." There is not a magical age when the time is suddenly "right" for sexual intercourse. The early teen years are definitely too young. There are lots of things in addition to age that should be considered. If you have not already talked with your parents about "the right time to have sex," you should do so. They will have their opinions and reasons and will probably be happy to share them with you.

Animal Sperm

Many animals reproduce sexually. That is, the females of the species produce eggs and the males produce sperm.

All sperm look similar, but the heads of the sperm are often quite distinctive. Within the head of each sperm is the nucleus. This nucleus, along with the nucleus of the egg, determines what kind of offspring is produced.

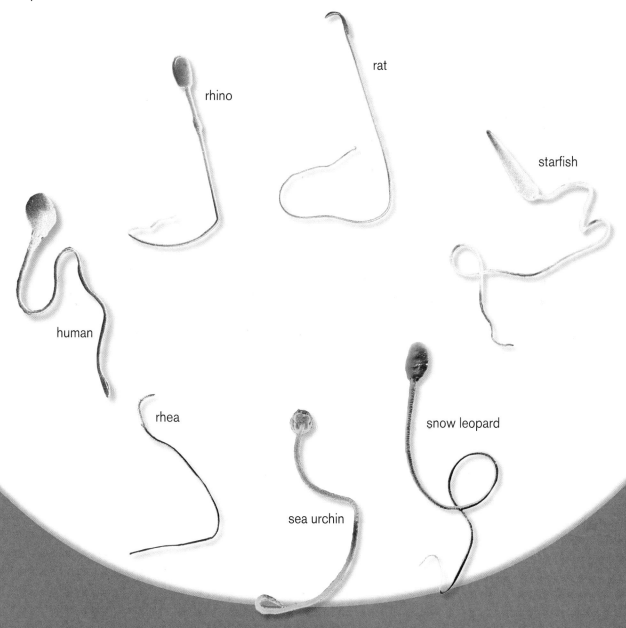

rat

rhino

starfish

human

rhea

sea urchin

snow leopard

Fetal Development

A fertilized egg soon divides to form two cells, the two cells divide to form four cells, and the four form eight cells. This continues until, after nine months, there are billion of cells that form a baby. Obviously, there is more going on than just cell division or else we would all be nothing more than huge clusters of cells. The cells also have to specialize and organize to form various body tissues and organs. The nine months of development are a time of rapid change.

When is the most critical time of development? Why?

Work in a group to research one stage of fetal development and summarize your findings on a poster which you will explain to the rest of the class. Keep the focusing question in mind while you work.

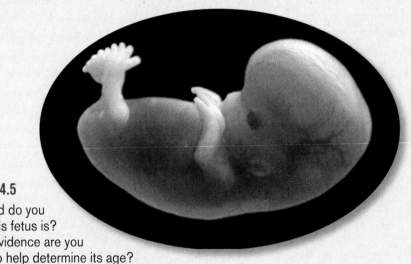

Figure 14.5

How old do you think this fetus is? What evidence are you using to help determine its age?

- resource books
- poster board or large sheet of paper
- colored markers
- glue, tape, scissors, and other materials

Procedure *(Work in a small group of students.)*

Part A: Poster Preparation

1. Record the stage of fetal development that your teacher assigns to your group, and then read these directions so you know what to do.
2. Your group will need to research one stage of fetal development. On your poster, you will need to include the following information:
   a. a title that indicates the age of the fetus
   b. the actual size of the fetus at this time of development
   c. important developmental events
   d. a status report on the heart, brain, fingers, and toes
   e. an illustration
3. Each member of your group should help with the research. Plan to do part of your research as homework.
4. Keep track of the resources you use. Record the title, author, and publication date for each of your resources on the back of your poster.

Part B: Poster Reports

1. One or two members of your group should be prepared to explain your poster to the rest of the class.
2. Take notes while you listen to the presentations by other students. You may want to make a table like the one shown here to help you organize your notes.

Age of Fetus	Size	Developmental Happenings

Analysis Questions ‒ ‒ ‒ ‒ ‒ ‒ ‒ ‒

1. What is the difference between an embryo, a fetus, and a baby?
2. What are the best sources of information about fetal development?
3. How are the changes that take place during fetal development like the changes that take place during puberty? How are they different?

Conclusion

Think about fetal development as taking place in three, three-month periods: the first three months, the second three months, and the final three months. Review your notes, look back at the posters, and write a paragraph that answers the focusing question.

Extension Activity

Animal Kingdom: Learn about the development of some other vertebrate (such as a chicken). Compare how it is similar to, and yet different from, a human.

Which Is Which?

During the first few weeks of development, all embryos look similar. Look at the embryos shown here. One is a human, one is a zebrafish, and one is a mouse. Which is which?

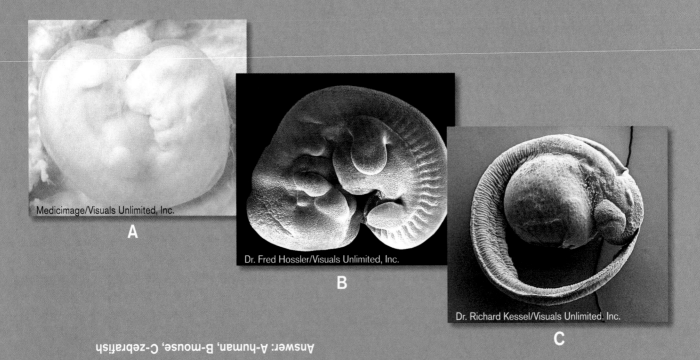

Medicimage/Visuals Unlimited, Inc.

A

Dr. Fred Hossler/Visuals Unlimited, Inc.

B

Dr. Richard Kessel/Visuals Unlimited, Inc.

C

Answer: A-human, B-mouse, C-zebrafish

Pregnancy and Birth

Pregnancy is a time of change not only for the developing baby but also for the mother. This reading describes a few of the many physical changes that the mother experiences during the nine months it takes the baby to develop.

Signs of Pregnancy

A woman cannot tell the moment she becomes pregnant; she has to wait for a "sign." This sign is often a missed menstrual period. A doctor should confirm the pregnancy by doing a physical exam and using a simple laboratory test. This is important because pregnancy is not the only reason for missing a period.

Frequent urination is another early sign of pregnancy. As the uterus grows, it presses against the bladder, producing the sensation that the bladder is full. This usually goes away but it returns again during the last few weeks of pregnancy when the uterus is much larger.

In the early stages of a pregnancy, some women are affected by morning sickness. They may be nauseated and feel like vomiting. Even though it is called "morning" sickness, this queasy feeling may last throughout the day and into the night. It seldom continues much past the sixteenth week of pregnancy.

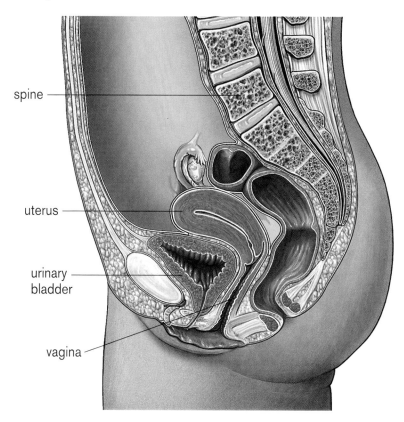

Figure 14.6
This drawing shows the size and position of the uterus in a woman who is not pregnant.

As time passes, there are other signs of pregnancy. A woman's breasts will grow and become more sensitive because the milk ducts are developing. By the fourth or fifth month, she has usually gained several extra pounds and noticed an increase in her waist size.

Early Care

One of the most important things a woman can do during her pregnancy is to see a health professional regularly. She will learn what she needs to do in order to take care of herself and the developing baby. Here are some of the things that a pregnant woman is advised to do:

- See a health professional regularly.
- Eat a balanced diet.
- Drink plenty of water.
- Get plenty of sleep.
- Exercise daily.
- Do not smoke.
- Avoid alcohol.
- Check with a health professional before taking any medication.

Did you notice that five of these recommendations address things people put in their mouths? A pregnant woman has to think before she puts anything into her body. That's because everything she eats, drinks, or smokes gets into her system and could cross the placenta and get into the fetus's system, too. A fetus needs a variety of nutrients. Drugs, alcohol, and the chemicals in cigarette smoke can interfere with fetal development.

The Placenta

The **placenta** nourishes the developing baby. It starts to form as soon as the embryo settles into the lining of the uterus. Cells from both the uterus and the fertilized egg divide and multiply to form the placenta. The entire placenta soon becomes laced with blood vessels.

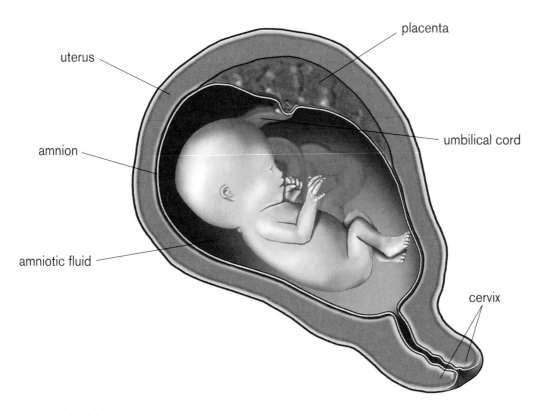

Figure 14.7

Blood vessels in the umbilical cord carry nutrients and gases from the vessels in the placenta to the growing fetus. Other vessels carry the wastes away from the fetus.

Blood vessels from the placenta enter the embryo at the site of the baby's future belly button. These vessels form the **umbilical cord**. Oxygen, glucose, calcium, and other nutrients from the mother's blood diffuse into the blood vessels that nourish the growing embryo. Carbon dioxide and other wastes from the embryo travel back through the umbilical cord and diffuse into the mother's blood. In this way, the placenta serves as the lungs, kidneys, intestines, and liver for the developing baby.

The embryo does not "hang" from its umbilical cord in the uterus. Instead it floats in a fluid-filled membranous sac called the **amnion**. The **amniotic fluid** cushions the baby from bumps and jars. The amnion and placenta continue to develop throughout pregnancy right along with the fetus. (See **Figure 14.7**.)

Labor and Delivery

The average pregnancy lasts about 265 days (38 weeks or 9 months). Its actual length usually varies by 5, 10, or even 20 days.

As the time for birth approaches, the baby usually turns upside down and its head drops into the mother's pelvis. If the mother notices this "settling," she knows that the baby is likely to be born within the next several weeks. (See **Figure 14.8**.)

The first sign that the baby is on its way is the beginning of **labor**. During labor, the smooth muscles that make up the walls of the uterus contract and relax. These contractions begin the process of birth. Right before or during labor the amnion may break, releasing the amniotic fluid through the vagina. As labor continues, the contractions get stronger and come closer and closer together. When they begin, the mother or father will usually contact the doctor. They will be asked about the timing and strength of the contractions in order to determine whether or not it is time to go to the hospital.

Meanwhile, the muscles that make up the cervix have to relax. They usually keep the uterus tightly closed; now they gradually allow the uterus to open. Eventually, the baby's head can be seen and, after what may seem like forever to the parents, the contractions push the baby out into the vagina. The muscles that make up the walls of the vagina stretch to let the baby pass through. Once the baby's head is delivered, the rest of the body follows quickly. (See **Figure 14.9**.)

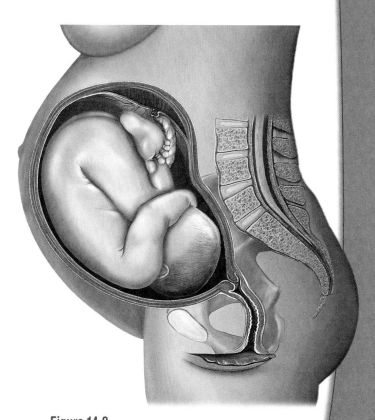

Figure 14.8
By the end of nine months, the baby takes up quite a bit of room in the mother's abdomen.

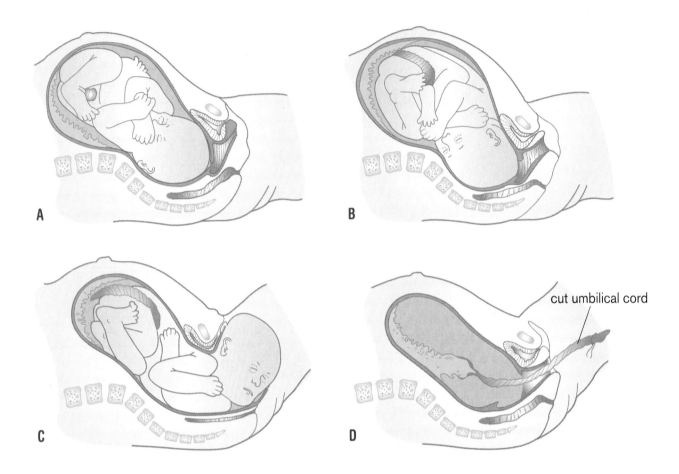

cut umbilical cord

Figure 14.9

How would you describe what is happening in each of these drawings? Which of these stages do you think would be the most uncomfortable for the baby? for the mother? for the father?

Even though the baby has been delivered, labor is not quite over. The uterus continues to contract until the placenta is also expelled.

The length of time from the beginning of labor to the delivery of the baby varies from woman to woman and from pregnancy to pregnancy. With a first child, it averages about 14 hours, although it may be much shorter (3 hours or less) or longer (24 hours or more). Labor is usually shorter for mothers who have given birth before.

Attention to the Baby

As soon as the baby is delivered, it needs attention. Since the baby's lungs are now working, the umbilical cord is no longer needed. It can be clamped and cut fairly close to the baby's body. This is not painful because there are not any nerves in the umbilical cord. Over the next several days, the stump of the cord dries and withers. When it finally falls off, it leaves the navel or "belly button" as evidence of a prebirth existence. (See **Figure 14.10**.)

Mucus is gently suctioned from the baby's mouth and nose. The baby's weight and the time of birth are recorded. The parents may then want to spend a few minutes with their baby before its checkup is finished. A few drops of antibiotic ointment or silver nitrate will be placed in the baby's eyes to prevent blindness in case the mother has an STD. A quick prick on the heel provides a few drops of blood for some laboratory tests and then the baby is ready for a warm sponge bath. Welcome to the world!

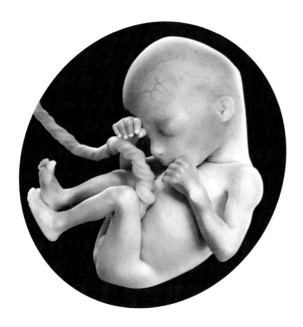

Figure 14.10

How would you describe the appearance of the umbilical cord?

they may have respiratory or heart problems. All of these problems are serious. This is why small babies have higher death rates during the first few weeks of life than do slightly heavier babies. The smaller the baby, the more likely it is to have problems. (See **Figure 14.11**.)

There are many reasons for low birth weight. One common reason is that the baby may have been born early; these babies are often referred to as being premature. Mothers who smoke or drink during pregnancy also have smaller babies; so do very young mothers or mothers who do not eat well during pregnancy. Of course, a baby may also be small at birth because its parents are small. These babies are usually not at much risk.

Small Babies

The average weight of a newborn is about 3,200 grams (7 pounds). Newborns who weigh less than 2,500 grams (5.5 pounds) are said to be *low birth weight* infants. They are more likely to have problems during the first few weeks of life than are other babies. They may not be able to control their body temperature or

Figure 14.11

An incubator can be used to help a low birth weight baby maintain its body temperature.

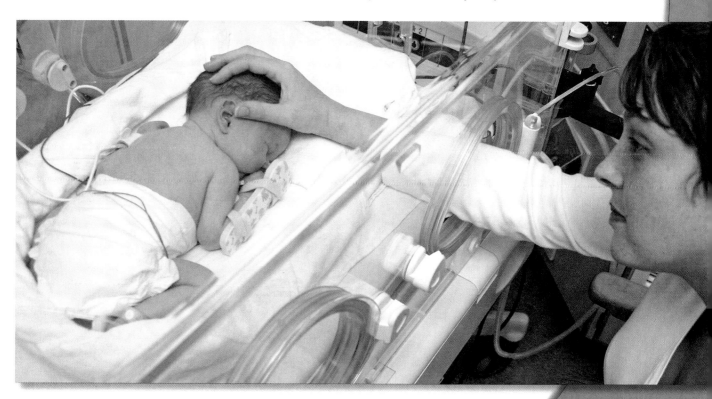

More Changes

While most newborns are healthy and most pregnancies go smoothly, problems can and do occur anywhere along the way. Sometimes problems develop either before or during labor that make a vaginal delivery risky for the mother or the baby. When this happens, the baby may be delivered through an incision in the mother's abdomen and uterus. This procedure is called a *caesarian section.*

After birth, the mother's system has to readjust to not being pregnant and the systems in the baby's body must undergo many changes to allow it to survive under totally new conditions. Suddenly, the baby has to breathe on its own, eat and digest food, maintain its own body temperature, produce antibodies—it has to take on all of the responsibilities of living. Isn't it amazing that our bodies can undergo all of these changes?

Analysis Questions

The analysis questions are supposed to help you understand the important ideas that are presented in a lesson. There aren't any analysis questions for this lesson. Instead, you decide what the important ideas are for this reading. Write one summarizing statement for each section, and make sure it includes the important ideas.

Extension Activity

Going Further: Call the newborn nursery at a local hospital. Arrange to visit the nursery or talk with someone who can answer some questions for you. Here are some questions you might want to ask. Also think of others you can add to this list.

1. What kinds of problems do you check for when examining a newborn?
2. What's the difference between a 35-week baby and a full-term baby?
3. Why do very small babies have problems?
4. Why are some babies placed in incubators?

Pregnancy: Another Point of View

You have read about an "average" pregnancy and delivery, but who is completely average? In reality, every pregnancy is unique. A pregnancy differs from woman to woman and from pregnancy to pregnancy for the same woman.

How do real pregnancies differ from the process described in this book?

What kinds of things do people recall when they describe a pregnancy?

Interview two people about a pregnancy they have experienced. Then compare what they say with what you learned in the previous lesson. Use your findings to answer the two focusing questions.

- Handout, *Pregnancy: Another Point of View*
- scissors
- Handout, *Summary of Pregnancy Interviews*

Procedure

Part A: Getting Ready

1. Read the questions on the handout.
   a. Note any questions that you think someone might be uncomfortable answering.
   b. Think about changes you might want to make to the handout.
   c. Answer analysis questions 1 through 3.
2. Consider which two adults you would like to interview. Choose people who you think will feel comfortable talking with you about their pregnancies.
3. Plan what you will say to the adults.

   Remember to tell them that this is a class assignment and explain that it is all right if they choose not to answer a question.

4. As you do the interview, record the answers on the handout. Remember to thank the person when you are done.

Part B: Summarizing the Interviews

1. Cut apart each interview question with its response. Keep the questions in order when you do so.

2. Place your questions and responses at the appropriate station. When all students have done this, everybody's responses to F-1 will be at one station, those for F-2 at another, and so on.

3. With a partner, summarize the responses for the question that your teacher assigns to you.

   a. Look for patterns.

   b. Be prepared to present your summary to the class.

4. Listen to other students summarize the responses.

   a. While you listen, take notes on the handout.

   b. Ask questions if you need further information to write a good summary.

5. After the class discussion, answer the rest of the analysis questions and write your conclusion.

Analysis Questions — — — — — — — — — —

1. Which question on the interview form do you think looks the most interesting? Why?

2. Predict how you think adults will answer:

   a. question 5.

   b. question 7.

3. Which, if any, questions do you think people might feel uncomfortable answering? Why?

4. Look at the answers to F-5 and M-5.

   a. What were the four changes listed most often by fathers?

   b. What four were listed most often by mothers?

5. Look at the answers to F-7 and M-7.

   a. On average, what age do fathers think is best for someone to have their first child? Why?

   b. On average, what age did mothers say? Why?

6. What was the most interesting thing you learned from this activity?

Conclusion

Write a paragraph to answer each of the focusing questions.

Birth Control

Students your age should not be involved in sexual intercourse, so you don't need to know how to use birth control at this time in your life. But living in our society, you hear a lot about *birth control* so you should understand what the term means.

In an earlier reading, you learned that intercourse is a caring and pleasurable way for couples to express their love for each other. However, a couple does not want a pregnancy to result each time they have intercourse. For this reason, most couples use some form of birth control.

This topic can cause some heated debates because people have different ideas about whether or not it is right to use birth control. Some people strongly object to its use, often for religious reasons. Others feel that using birth control is necessary to limit the size of their family. You may already know how your family feels about this issue. If you don't, talk to your parents. They will probably be willing to share their thoughts and beliefs with you. The following information may help you focus the discussion.

Family Planning

The desire to control family size is not new. For hundreds of years, people have tried a variety of methods to control pregnancies. One of the safest things that people tried was avoiding intercourse whenever they thought the chance for pregnancy was high. Eventually, this evolved into what we now call **natural family planning** or **rhythm**. People who use this method do not have intercourse when the woman is ovulating. This method is inexpensive and medically safe but its effectiveness varies.

Problems arise because the human body is not 100% predictable. As you learned, a woman's cycle can vary from month to month, so it can be hard to predict when ovulation will occur. It is also impossible to know exactly how long an egg will live. About one out of every four couples who use rhythm will be expecting a baby by the end of one year. For this reason, some people say that it is not a good method of birth control.

Another "method" that some people try is called **withdrawal**. This is when the penis is removed from the vagina before ejaculation. It doesn't require any special devices and it doesn't cost anything, but it is risky. In fact, it is so risky that people who use it are likely to become parents.

Contraceptives

Other methods of birth control involve some kind of chemical or a device. These methods are sometimes called **contraceptives**. They work in a variety of ways. The various methods are based on an understanding of how the reproductive system works.

Some contraceptives contain hormones that prevent egg cells from maturing so ovulation does not occur. These hormones may be in the form of a pill (the birth control pill), a shot, or a device inserted under the skin. Since hormones affect the functioning of the body, these methods require a prescription.

Other contraceptives prevent the sperm and egg from joining. They block or kill the sperm before the sperm reach the egg. Condoms, diaphragms, and spermicides are examples of this type of contraceptive. Sterilization is a more permanent way to prevent sperm and eggs from joining. When a male is sterilized, the vas deferens are cut. This procedure is called a vasectomy. In females, the fallopian tubes are cut.

Another type of contraceptive, called an intrauterine device (IUD), prevents a fertilized egg from implanting in the uterus. These devices must be placed in the uterus by a health care professional.

Contraceptives
birth control pill
condom
diaphragm
spermicide
vasectomy (male sterilization)
tubal ligation (female sterilization)
IUD (intrauterine device)

Three of the most talked about and most frequently used contraceptives are condoms, spermicides, and the various forms of hormones. They are explained briefly in the chart, *Pregnancy Prevention*.

Effectiveness

It is important to realize that none of these methods works 100% of the time. For example, from the information about condoms in the chart, you can see they are 90% effective. That means they fail 10% of the time; or, in other words, 10 out of every 100 couples who use this method for one year will end up expecting a baby. Condoms are more effective if used with a spermicide. Of course, the effectiveness of any method will go down if it is not used correctly. Keep in mind that the only 100% effective way to prevent a pregnancy is not to have intercourse.

The methods that are mentioned here have been declared safe and effective by physicians and the ***government*****. Other methods you may hear people talking about may be based on their own misconceptions. Such methods are likely to be unsafe or ineffective.

The Food and Drug Administration (FDA) is the governmental agency that monitors the safety of food and drugs and other health care products.

Pregnancy Prevention

Abstinence

- no sexual contact
- 100% effective
- avoids the possibility of sexually transmitted disease

Natural Family Planning

- timing intercourse to avoid "fertile days"
- requires careful planning and control
- 50%-80% effective
- requires help from medical professional to learn woman's cycle

Hormones (birth control pills, shots, under-the-skin inserts)

- prevent release of an egg from the ovary
- pills and shots must be taken regularly
- more than 99% effective
- require a prescription

Condoms

- covering for penis that holds sperm so they don't enter the vagina
- 98% effective if used with spermicide; about 90% effective if used alone
- provides some protection from sexually transmitted diseases (STDs)
- does not require a prescription

Spermicides

- foams, gels, and films containing chemicals that kill sperm
- are put into the vagina before intercourse
- 98% effective if used with a condom; 75%-80% if used alone
- does not require a prescription

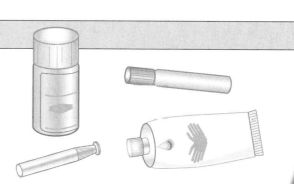

Withdrawal

- penis is removed from vagina before ejaculation
- not very effective (about 60%) because some sperm are usually released before ejaculation
- difficult for man to judge when sperm will be released

Analysis Questions

1. Make a table like the one shown here on your own paper. It should have six blank rows for you to fill in.

   a. Use the information from the reading to fill in the first three columns of the table.

   b. Use stars to fill in the rating column.

   ** = very effective

   * = effective

   x = better than nothing

2. List at least three places someone could go to get more information about one of these methods.

3. What are three reasons people may want to use birth control?

4. What are three reasons people may decide not to use any method of birth control?

5. What is the only 100% certain way not to get pregnant?

Birth Control			
Method	How Does It Prevent Pregnancy?	One Important Fact	Rating

Sexually Transmitted Diseases

Pregnancy can result from sexual intercourse. So can STDs. **STD** stands for **s**exually **t**ransmitted **d**isease. These diseases are passed from one person to another during sexual activity. Like other diseases you have studied, each STD has characteristic symptoms, each affects some body parts more than others, and each has a cause. And, most importantly, STDs can be prevented.

What are four examples of STDs?

What are typical symptoms of STDs?

Take a quiz to find out what you already know about STDs. Then read about some of the most common STDs, organize the information in a chart, and take the quiz again.

 **Materials**

- Handout, Quiz: *STDs, True or False?*
- Reading, *Basic Facts About STDs*, page 415

Procedure

1. Take the quiz.
2. Read the reading.
3. Make a chart that summarizes the information in the reading. Make sure your chart includes the following information about each STD:
   a. cause
   b. symptoms
   c. long-term effects
   d. treatment
4. Check your quiz and correct your answers.

5. Work with a partner.
   a. Review your notes about STDs. Underline the facts that you think are important for people to know.
   b. Write four questions about STDs that your teacher could use on a test. Neatly tear a piece of paper in fourths and write one question and the answer on each piece of paper.
   c. Write your names on the back of each piece of paper and give the questions to your teacher.

Conclusion

Make sure you can answer the focusing questions.

Extension Activity

Community Resources: Invite someone from the local health department to talk to your class about STDs. Ask the speaker to talk about the STDs that are the biggest problem in your community.

Basic Facts About STDs

STDs, or sexually transmitted diseases, are passed from one person to another, usually as a result of sexual activity. Millions of Americans catch STDs every year. STDs are a major health problem among teenagers who are sexually active. If untreated, STDs can result in serious consequences, maybe even death. This reading provides basic information about five of the most common STDs.

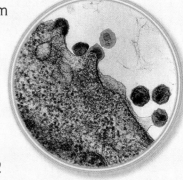

Figure 14.12

HIV is the virus that causes AIDS. The small virus particles (blue-green) are emerging from a white blood cell (shown here as purple). (magnification: 14,555 times)

Five Diseases

The most discussed STD may be **AIDS**. It is caused by a virus known as HIV. (See **Figure 14.12.**) HIV attacks the immune system, making it difficult for an infected person to fight infections. Symptoms often do not appear immediately. Early symptoms of AIDS include diarrhea, weight loss, fever, long-lasting infections, and swollen glands. The symptoms can be treated, but AIDS cannot be cured. Over time, cancer and severe pneumonia usually result in death. Fortunately, HIV is not easy to catch. The virus does not travel through the air. It is transferred in body fluids such as blood, semen, and the fluids in the vagina.

Chlamydia is caused by a bacterial infection. It is sometimes called the "silent STD" because people who have it may not have any symptoms. When there are symptoms, they include a burning sensation during urination and a discharge from the vagina or penis. Chlamydia can be cured with antibiotics. If left untreated, the long-term infection can result in constant pain in the lower abdomen and *infertility*. This is the most common *curable* STD in America; millions of new cases are reported every year.

> *infertility:* sterility, inability to have children

Gonorrhea is another example of an STD that is caused by bacteria. (See **Figure 14.13.**) Early symptoms include a burning sensation during urination in males and a yellowish discharge from the penis or vagina. The discharge has an unpleasant odor. The lucky people are the ones who have symptoms; sometimes the bacteria can thrive in the body for years without causing any noticeable symptoms. However, over the long term, gonorrhea can result in infertility and even arthritis. This STD can be cured with antibiotics.

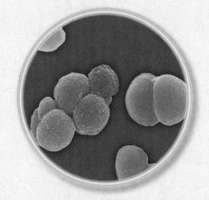

Figure 14.13

These bacteria cause gonorrhea. They are magnified about 4,250 times. Notice how much larger they are than the virus particles in Figure 14.12 (and remember: Figure 14.12 was enlarged 14,555 times).

One of the most contagious STDs is **herpes**. People who have herpes may go weeks or months without symptoms, and then there is a "flare-up." During a flare-up, the person may experience aches, pains, itching, and sores on the penis or vagina. This STD is caused by a virus so it cannot be cured with antibiotics, but drugs can be used to treat the symptoms. There are two main kinds of herpes virus. One kind is usually the cause of genital herpes and the other is the cause of "cold sores" around the mouth. However, both kinds of virus can infect either the mouth or the genitals. Millions of Americans are carriers of the herpes virus.

The **Human papilloma virus** (HPV) is the most common sexually transmitted *virus* in the United States. There are more than 40 HPV types that can infect the genital area of males and females. Most people with HPV do not develop symptoms or health problems. But sometimes, certain types of HPV can cause genital warts in males and females and other HPV types can cause cervical cancer. The types of HPV that can cause genital warts are not the same as the types that can cause cancer. A vaccine is available that protects females against the four types of HPV that cause most cervical cancers and genital warts. Ask your parents and doctor about the HPV vaccine.

Effects on Fetuses and Infants

If a mother has an STD when she is pregnant she may pass it on to her baby. Babies can catch STDs as they pass through the vagina when they are born. Depending on the type of infection, the baby may develop problems such as skin infections, blindness, mental retardation, or even death.

Prevention

STDs are no fun. Fortunately, they can be avoided. All STDs are spread by sexual activity, so one obvious way to avoid getting an STD is to avoid sexual activity. Some STDs, including AIDS, can be transmitted in body fluids. That's why drug users who share needles and people who go to tattoo parlors that don't sterilize their equipment are at a high risk for catching STDs. Medical workers who handle blood samples must be careful so they don't get infected. If people understand STDs and avoid risky behaviors, the disease rate will drop.

Changes Continue

Infancy, childhood, puberty, and adulthood. People, like all organisms, go through a predictable series of changes as they age. When people become adults, they don't stop changing; they continue to change and grow older every day of their lives. In this activity, you will consider some of the changes that occur after age 20.

What do you think is a good definition of the word *growth*?

How do the changes that occur later in life compare with the changes that take place during the first few years of life?

In lesson 13-1, *The First Ten Years,* you positioned 30 characteristics on a time line. In this lesson, you will again position 30 characteristics on a time line, but this time the time line will start at age 20 years and extend well into old age.

- four sheets of paper
- Handout, *30 Characteristics*
- red and blue markers or crayons
- scissors
- tape

Procedure   *(Work in a group with two or three other students.)*

1. As a group, answer the focusing question. Spend a few minutes on this task so you feel comfortable with your definition.

2. Prepare a time line by taping together the 4 sheets of paper and labeling them as shown in **Figure 14.14.**

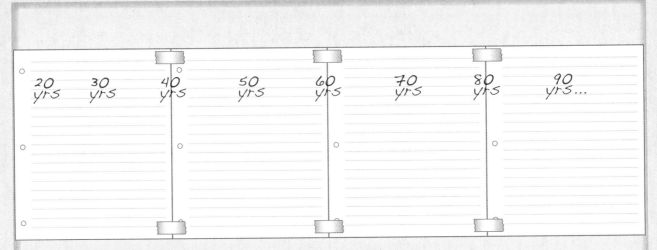

Figure 14.14

3. Cut apart the 30 characteristics.

   a. Set aside the four blank boxes to use later.

   b. Choose one of the characteristics and, with the other students in your group, decide when you think a person first meets this characteristic.

   c. Discuss evidence for your decision.

   d. When you all agree where it should be positioned, tape it on the time line.

   e. Do this for the other 25 characteristics.

4. Think of at least four more characteristics to add to the time line. Write one characteristic on each of the blanks. Attach the characteristics where they belong on the time line.

5. Put a red or blue box around each characteristic to indicate whether it is a physical characteristic or an ability.

   > red = physical characteristic (involves a structure or function of the body)

   > blue = ability (something a person does)

6. Answer the analysis questions.

Analysis Questions _ _ _ _ _ _ _ _ _ _

   1. What characteristic was the easiest to place on the time line? Why?

   2. What characteristic was the most difficult to place on the time line? Why?

   3. List at least two examples of the types of evidence you used when you decided where to place the characteristics on the time line.

   4. Which changes do you think occur at exactly the same age for everyone?

5. The ages at which any of the characteristics on the time line occur depend on decisions a person makes. For example, people decide when they want to write a will or get married.

   a. List at least six other characteristics that a person can control (at least to some extent).

   b. List at least three characteristics over which a person has no (or very little) control.

6. Which system do you think changes the most as a person grows older? Why?

Conclusion

Now that you have thought about growth, write several sentences in answer to the focusing questions.

Extension Activity

Plant and Animal Kingdoms: Do research on a particular species of plant or animal (choose something other than a mammal). Find out about the changes it undergoes as it ages. Sketch its life cycle. Two examples are shown in the reading, *Continuity of Life*.

Continuity of Life

Individual organisms die, but the species lives on. Each species has a characteristic life cycle. (The term **life cycle** refers to the series of stages that each organism passes through from birth to death. Individuals do not "cycle," but because they reproduce, the cycle can start over with each new generation.)

The life cycle of a one-celled organism may be as simple as growth and cell division. In contrast, the life cycle of multicellular organisms often starts with a fertilized egg that then divides and specializes to form a young organism. The young organism grows until it is old enough to produce eggs and/or sperm and the life cycle can continue.

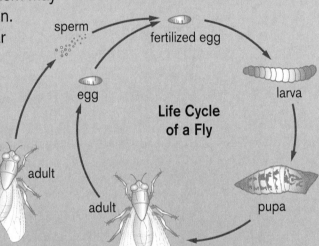

Life Cycle of a Fly

sperm — fertilized egg — larva — pupa — adult — egg — adult

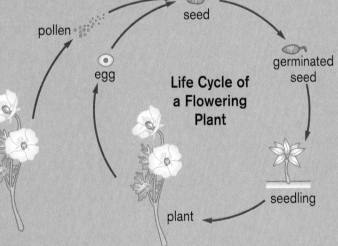

Life Cycle of a Flowering Plant

pollen — seed — egg — germinated seed — seedling — plant

Each type of organism has a distinctive life cycle. Compare the examples of the life cycle of a flowering plant with that of an insect. How are they alike? How are they different?

Includes *sci*LINKS. NSTA

Topic: Life Cycles
Go to: www.scilinks.org
Code: MSLS3e420

Reproduction and Populations

Even though all individual organisms (plant, animals, fungi, and microorganisms) eventually die, species live on because organisms produce offspring—they reproduce. The number of offspring each individual produces affects the size of the population. Among humans, the size of individual families affects the size of the total human population.

Many people are interested in predicting population size because it helps them plan for the future. You, too, can make predictions. Imagine that you want to find out what happens to the size of a population if couples have, on average, 2 children. You might do so by starting with a population of 100 adults of about the same age. If there are equal numbers of males and females, that could result in 50 couples. If each couple has 2 children, 100 children will be born. Imagine that these children grow up, marry, and have 2 children each. How many children will result?

# Adults	# Couples	# Children
100	50	100

If you do the math, you will find that the population size stays the same from generation to generation when the average family size is two. This is not the case for other family sizes.

How does *family* size affect *population* size?

Calculate the size of the population after five generations for three different family sizes. Graph your data and then use your graph to answer the focusing question.

Materials

- Handout, *Reproduction and Populations*
- graph paper

Procedure

(Work with a partner.)

1. Look at the handout—you have three charts to complete. When you fill in the charts, assume that the children in one generation grow up to become the adults of the next generation.

2. Plot the data from the three charts onto one graph. Remember to:
   a. label both axes
   b. scale the axes
   c. plot your data
   d. include a key
   e. title your graph

3. Answer the analysis questions.

4. Add the data for an average family size of two to your graph. (Remember to fix the key on your graph.)

Analysis Questions _ _ _ _ _ _ _ _ _

1. Explain what happens to the size of a population if the average family size is:
   a. one
   b. two
   c. three
   d. four

2. Use your graph to estimate the number of people in the seventh generation for an average family size of
   a. two children
   b. four children

3. The number of children in eight families is shown here. Calculate the average number of children for this group of families.

Family A: 2	Family E: 3
Family B: 0	Family F: 2
Family C: 4	Family G: 1
Family D: 1	Family H: 2

4. Number of children is not the only factor that affects population size.
   a. List at least two other things that could cause a town's population to *increase*.
   b. List at least three things that could cause the population to *decrease*.

5. Imagine that the population is growing quickly in a community.

   a. What are three advantages to a growing population?

   b. What are three disadvantages?

6. A field mouse can have 100 offspring every year. If one pair of field mice moved into a new field, after three years 250,000 mice could be living in that field. That's not likely to happen. Why?

Conclusion

Look back at the focusing question and think about what you learned by doing this activity. Write a paragraph that answers the question and provides evidence for your answer.

Extension Activity

Math Connection: Use three family sizes—2, 4, and one of your choice—and calculate the population size after 12 generations. Plot your data and explain the patterns you see.

HB:
Interacting Systems

This chapter, as well as the previous eleven chapters, has focused on the structures and functions of the body systems. The next two chapters in this book are about things that are too small for you to see-chromosomes, and genes. Before you go on to the next unit, let's review what you have learned.

This lesson is made up of three parts. During the first part, you will review and reflect on the lessons you have done and what you have studied. During the second part, you will take a test to check your understanding of important concepts. During the last part, you will think about HB one last time. You will compare what you now know about the interactions that take place within the human body with what you knew when you started Unit II.

Materials

- Handout, *The "Best" Lessons in Chapters 3-14*
- scissors

Procedure

Part A: Reflect and Review   *(Work with a partner.)*

1. Look back over each of the lessons and think about what you did and what you learned.
   a. For each chapter (3 through 14), choose the one lesson you think taught you the most. (You and your partner should agree on the choice.)
   b. Record your choice on the handout and explain the most important thing you learned from that lesson.
2. Cut along the dotted lines on your handout. Place your thoughts about each chapter at the appropriate location.

3. Summarize students' opinions.
    a. Your teacher will give you the responses for one chapter to summarize.
    b. Read all the responses.
    c. Plan a one-minute presentation to the class. In your presentation, report the three lessons in the chapter that students thought were most informative and the main idea that students learned from each lesson.

Part B: Written and Station Tests   *(Work independently.)*

1. There are two parts to the test—a written section and a station test. Your teacher will tell you which part to do first.
2. Station Test
    a. You will have 5 minutes at each station. Work quickly and carefully.
    b. Stay at your station. Do *not* touch anything unless the directions tell you to do so.
    c. Write down the number of the station and the answers to the questions.
    d. Do not change stations until your teacher calls time.
3. Written Test
    a. Read each multiple choice question carefully.
        • Select the letter that corresponds to the *best* choice.
        • Record the answer on your answer sheet. Do not write on the test.
    b. Write thoughtful answers to the short answer questions. This is your chance to show yourself and your teacher that you have learned something!

Part C: HB Wrap-Up

Look at the HB your class created at the beginning of Chapter 3.
    Make corrections. How has your thinking about the body changed?

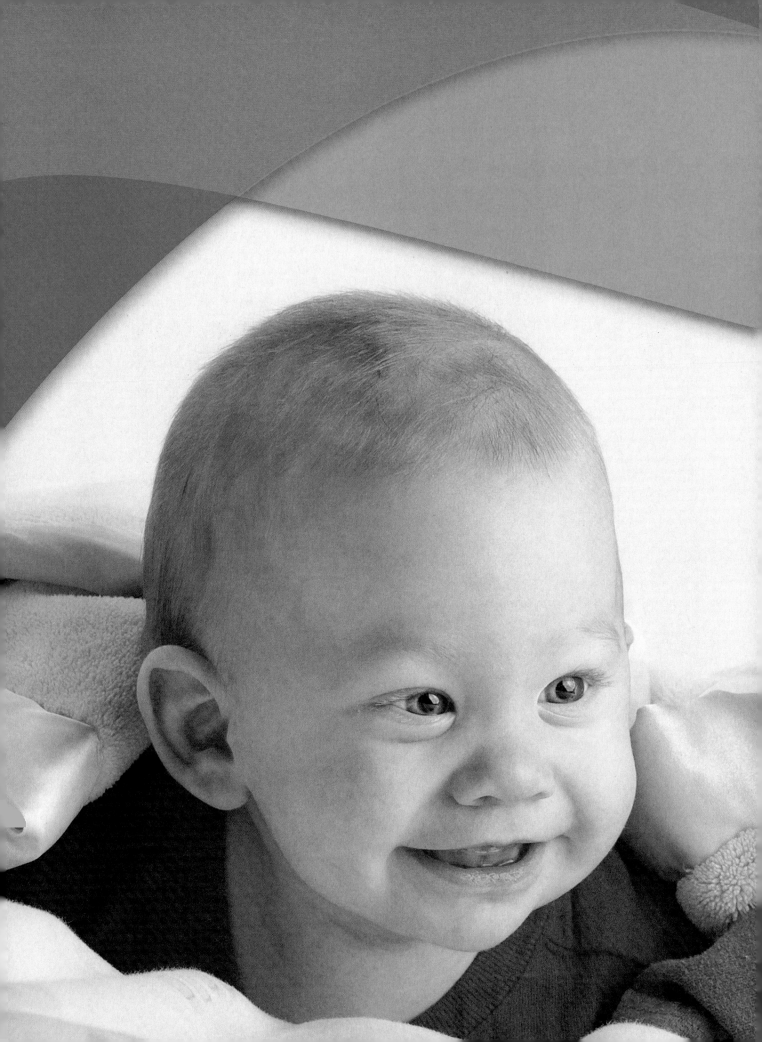

Cells and Genetics

Chapter 15

Cells and Chromosomes

Chromosomes are like an
instruction manual for a cell.

Growth and Cell Division

The body grows because the number of cells increases, not because cells grow larger. Cells never get very large. When a cell reaches a certain size, it splits in two. As you continue to grow, your cells divide into more and more cells. Cells also divide to replace cells that have died or to repair damaged tissues. The process of one cell dividing into two exact copies is called **cell division**.

How do cells make copies of themselves?

Scientists have separated cell division into different phases based on significant events happening in the cell. Several phases encompass the division of the cell's nucleus. When the nucleus of a cell divides it is called **mitosis**. During this activity, you will discover what happens to a cell during cell division—and mostly, during mitosis.

- Reading, *Grow and Divide*, page 433
- Handout, *Chromosome Models*
- scissors
- three sheets of paper
- tape or glue
- prepared microscope slide of onion root tip
- microscope

Procedure

(Work with a partner.)

Part A: Modeling Mitosis

1. Read *Grow and Divide*
   a. Answer analysis questions 1 through 3.
2. Get your set of chromosome models. In this model, your *parent* cell has four chromosomes in its nucleus.
3. Your goal: Use the chromosome models to form two identical *daughter* cells.

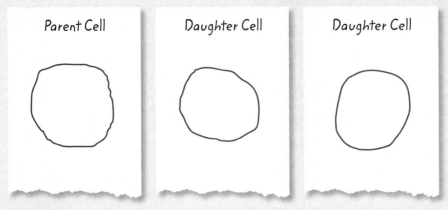

Parent Cell   Daughter Cell   Daughter Cell

Figure 15.1

4. Mark your three sheets of paper as shown in **Figure 15.1**.

5. Prepare the chromosome models.

   a. Cut out the four *parent* chromosomes and place them in your parent cell.

   b. Cut out the four *replicated* chromosomes and place them to the side.

   c. Throw away the scraps of paper. **Make sure you don't throw away any chromosomes.**

6. Imagine that your parent cell is about to undergo cell division. *(Refer to the diagram in* **Figure 15.2** *as you move your chromosomes through your model.)*

   a. The first thing the parent cell needs to do is replicate (copy) its chromosomes.

      • Select a *replicated* chromosome that matches (same size and pattern) a chromosome in the parent cell. Place it next to the matched chromosome in the nucleus.

      • Continue to place the matched replicated chromosomes into the parent cell.

      • You should now have a set of four doubled chromosomes.

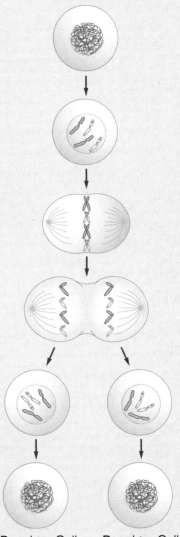

Parent Cell

Daughter Cell   Daughter Cell

Figure 15.2
A cell undergoing cell division.

7. During **mitosis,** the doubled chromosomes first line up in the middle of the cell and then move to opposite sides of the cell.

   a. Line up your doubled chromosome in the center of the cell.

   b. Move one chromosome from each doubled set to the opposite sides of the parent cell.

8. The last stage of cell division is for the cell to divide into two identical daughter cells.

   a. Move each set of chromosomes from the parent cell into each of the daughter cells.

      • You should end up with two daughter cells with exact copies of chromosomes and the same number of chromosomes as the original parent cell.

   b. When you think your two daughter cells are correct, tape or glue down the chromosomes and sign your names on the paper.

   c. Have your teacher check and initial your paper.

9. Answer analysis questions 4 and 5.

Part B: Microscope

1. Observe the cells on the prepared microscope slide.

   a. Move the slide around to find examples of cells undergoing mitosis. You may need to switch to medium or high power.

   b. See **Figure 15.3** for pictures of cells during different stages of mitosis. Use them to help you find similar cells on the microscope slide.

   c. Carefully sketch at least two examples of cells undergoing mitosis.

   d. Label the cell wall, cytoplasm, and chromosomes.

   e. Record the magnification.

2. Answer analysis questions 6 and 7.

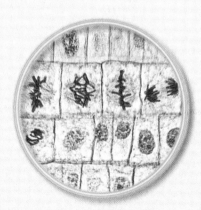

Figure 15.3

Mitosis in an onion root tip. In plants, the roots continue to grow as they search for water and nutrients. These areas of growth are good for studying cell division because at any given time, you can find cells undergoing mitosis.

Analysis Questions

1. Why is cell division important?

2. Describe, in your own words, the process of cell division, including mitosis.

3. Why do you think it is important for the chromosomes to line up in the middle of the cell before moving to opposite sides of the cell?

4. Imagine you are studying a type of cell that divides every hour. At noon, you have one cell. How many cells will you have at noon the next day?

5. How was this model similar to an actual cell undergoing cell division? How is it different?

6. List at least three things the cells you observed on the microscope slide had in common. List at least three differences.

7. Cancer is a type of disease in which a group of cells display uncontrolled cell division. Explain how a drug that disrupts mitosis could potentially be used as a cancer drug.

Conclusion

Write a paragraph that answers the focusing question.

Extension Activities

1. **Health:** Do some research on different environmental and lifestyle factors that cause cancer. Find out how pollution, tobacco, and eating habits can increase the risk of cancer. Make a list of things you can do that might prevent cancer.

2. **Careers:** With your teacher's permission, invite someone who studies cells to talk to your class. Cytologists conduct research on cells, and many lab technicians either observe or grow cells. Find out which cell types your speaker works with. Have the person also explain what can be learned by looking at these cells and what he or she likes about the job.

Grow and Divide

New cells come from cells that already exist. They do not suddenly appear from nowhere. This concept is part of the cell theory. Most, but not all, cells can divide to produce more cells. Cells divide for three reasons: so an organism can grow, to replace dead cells, and to repair damage. An organism would not live very long if cells could not divide.

Growth and Cell Division

A cell has a lot to do before it can divide. It has to make another nucleus and make new copies of all of the organelles in the cytoplasm. This takes energy and nutrients. When a cell is ready to divide, an animal cell will pucker around the middle and a plant cell will form a new cell wall in the middle of the cell. The cell separates into two almost even halves. Where there was one cell, there are now two, almost identical new cells, called **daughter cells**.

The process of producing two *daughter cells* from a *single parent* cell is called **cell division** (**Figure 15.4**). During cell division, the chromosomes in the nucleus are carefully copied. Each copied or doubled chromosome is held together at its center, forming the shape of an X. During **mitosis,** each doubled chromosome lines up in the center of the cell and then splits, with each half being pulled to the opposite side of the cell. This ensures that each daughter cell ends up with a complete, identical set of chromosomes. Finally, during the last phase of cell division, the cell divides into two identical daughter cells. In human cells, this means that each daughter cell has 46 chromosomes—just like the parent cell.

It is in this way that skin cells give rise to new skin cells and cheek cells form new cheek cells. Imagine what a hodgepodge a body would be if this were not the case. Strange combinations would result—intestine cells might form in the eyeball and muscle cells could appear in the kidney. The control center of a cell, the genetic

Parent Cell

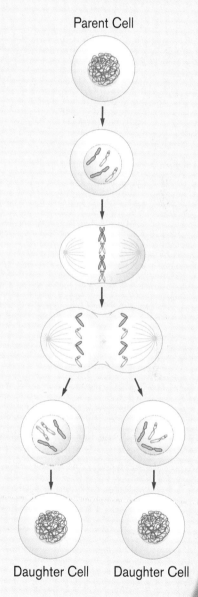

Daughter Cell   Daughter Cell

Figure 15.4
A "parent" cell dividing into two "daughter" cells.

material in the nucleus, is responsible for making sure that the right kind of cells form in the right places.

Replacement and Repair

Millions, perhaps even billions, of your cells die every minute. This is normal. Sometimes an injury kills additional cells. If you didn't replace the dead cells with new ones, you would soon be nothing but a mass of dead cells. Fortunately, in most tissues, cell division continues even after a person reaches adulthood.

How often cells divide depends on the cell type. Cells in the living layer of the epidermis must divide fairly often to replace themselves, since some skin cells live only about 8 hours. If these cells were not replaced, you would soon be left without an epidermis.

The cells that line the intestines also experience lots of wear and tear. They, too, must be replaced constantly. Each of these cells lives only about a day and a half. In total, your intestine has a new lining every 3 to 6 days. White blood cells live only about 13 days and a red blood cell lives an average of 100 days. These cells, too, need to be replaced constantly.

Cells in organs such as the kidney and lung do not divide. If they are damaged from disease or injury, they cannot be replaced.

What happens to dead cells? They self-destruct. They produce enzymes that break themselves down from the inside out. All of the molecules that are released can be recycled—they are used to build new cells. As you may remember, white blood cells are part of the "cleanup crew" that disposes of dead cells.

Out of Control

Normally, cell division is an orderly, controlled process, but sometimes it goes out of control. When this happens, the new cells may not look like the original ones, they may invade surrounding tissues, and they may not function normally. Uncontrolled cell division is the cause of cancer. Scientists who study cancer are learning a lot about cell division. They are finding out why some cells can divide and others cannot. They are also learning about the molecular signals that keep cell division under control. Because of their research, more and more kinds of cancers are becoming curable.

Aren't you glad that your body automatically replaces your dead cells? It's hard to believe that cell division is going on 24 hours of the day, no matter what you are doing.

Includes sci LINKS.
NSTA

Topic: Cancer
Go to: www.scilinks.org
Code: MSLS3e434

Single-Celled Organisms

The plants and animals you see every day are made of millions of cells. In addition to the large organisms, there are thousands of much smaller ones; some of the smallest organisms are the one-cell protists. These small organisms can move, eat, and "breathe." They also reproduce, sense what is going on in their environment, and get rid of waste.

Like other cells, a single-celled protist has a membrane, cytoplasm, a nucleus, and organelles. Some of these organisms look like shapeless masses; others have a distinctive appearance.

An Amoeba Experiment: **Amoebas** are one-celled protists that are often used in experiments to study cells. In one experiment, a scientist cut 100 amoebas in half. One-half of each amoeba had a nucleus; the other half did not. The scientist observed these half-amoebas for 40 days and recorded the results in the chart.

Day of Experiment	Number of Half-amoebas with Nuclei	Number of Half-amoebas without Nuclei
1	100	100
2	74	78
3	66	60
5	59	48
10	52	14
20	47	3
30	61	0
40	88	0

Answer these questions about the amoeba experiment:

a. On the second day of the experiment there were 74 half-amoebas with nuclei. What happened to the other 26?

b. Describe what happened to the half-amoebas with nuclei by the fortieth day.

c. Describe what happened to the half-amoebas without nuclei.

d. What do you think the scientist concluded?

Discovering DNA

For many years, no one knew what the nucleus did. Scientists knew that almost every cell had a nucleus, and they knew that the nucleus divided when cells divided, but they did not know its function. Now we know that the nucleus is the "control center" of a cell because it contains chromosomes made from molecules called **DNA,** or **d**eoxyribo**n**ucleic **a**cid. DNA is a critical molecule of life. It contains all of the information needed to operate a cell, organize tissues, and coordinate the activities of all the organ systems in a body.

How do scientists take DNA out of a living thing?

During this activity, you will have a chance to see this stringy molecule by extracting it from strawberries.

- small plastic baggies
- strawberries
- extraction buffer
- ice cold rubbing alcohol
- small graduated cylinder or beaker

- test tube
- test tube holder
- funnel
- coffee filter or paper towel
- wooden stir stick

Procedure
(Work with a partner.)

1. Get 1–3 strawberries from your teacher. Put them into a small plastic bag, sealing the bag. Make sure most of the air is out of the bag.

   With your fingers, squeeze and smash the strawberries in the bag for about 2 minutes.

2. Open the bag and add 10mL extraction buffer into the bag. (The extraction buffer contains soap and salt.) See **Figure 15.5**. Again, squeeze the air out and seal the bag.

3. Mash the strawberries and buffer together into the bag for about 60 seconds. Try not to create bubbles.

4. Open the bag and pour the solution through the filter your teacher provides. Collect the liquid (filtrate) that goes through the filter into a test tube. Try to get about 3 mL. (**Figure 15.6**)

5. Tilt the test tube and SLOWLY pour 6 mL cold isopropyl alcohol (rubbing alcohol) down the side of the test tube. The alcohol should form a layer on top of the strawberry liquid. (Don't let the alcohol and strawberry liquid mix. The DNA collects between the two layers!)

6. Let the strawberry filtrate and alcohol sit for 2 minutes.

7. Take a wooden stir stick and slowly swirl it between the two layers. Do not stir! A white substance should attach itself to the stir stick. This is the DNA from the cells of the strawberries.

8. If possible, look at the DNA under a microscope and sketch what you see.

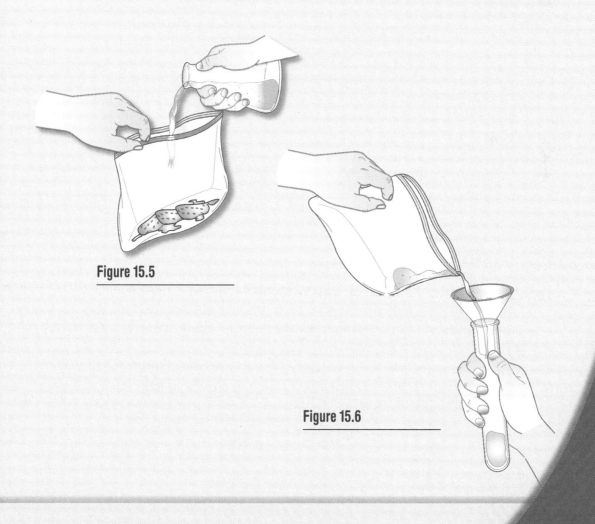

Figure 15.5

Figure 15.6

Analysis Questions _ _ _ _ _ _ _ _ _ _

1. Where in the strawberries is DNA located?

2. What do you think was the function of each step in the activity? See if you can match each step of the procedure with its function.

Steps:

   a. initial mashing of the strawberries

   b. addition of extraction buffer

   c. filtering of the strawberry solution

   d. addition of the cold isopropyl alcohol

Functions:

   1. separates the broken cell parts and proteins from the DNA

   2. causes the DNA to become a solid (precipitate) from the solution

   3. breaks up proteins and dissolves the cells' membranes

   4. breaks the strawberry into smaller clumps of cells

3. If you were to extract DNA out of another organism, do you think it would look the same? Why or why not?

4. Does your food contain DNA? How do you know?

5. You can see a strawberry with your eyes but would need a microscope to see its individual cells. That is because a strawberry is made up of millions of individual cells. Why do you think you were not able to see the DNA in individual cells but could see it after the DNA extraction?

Conclusion

Write a complete answer to the focusing question.

Extension Activities

1. **Careers:** With your teacher's permission, invite someone from a local forensic lab to talk with your class. Forensic scientists can use the DNA found in the blood, semen, skin, saliva or hair at a crime scene to help apprehend and convict criminals.

2. **Try It:** You can try these steps to purify DNA from lots of other living things. Grab some oatmeal or kiwis from the kitchen and try it again! Which foods allow you to see the most DNA?

15-3

Chromosomes and Karyotypes

Imagine you could focus on one single chromosome in a cell. If the magnification were high enough, you would see that it looks like it is made up of coils. If it were possible to unwind the coils, you would see an incredibly long molecule. This molecule is **DNA,** or **d**eoxyribo**n**ucleic **a**cid. (See **Figure 15.7**.)

Structure of a Chromosome

It would be interesting to look at your own chromosomes, but that requires careful laboratory work and more equipment than you are likely to have at school. So, instead of actually looking at your chromosomes, you'll have to imagine that you are doing so. First you clip a prepared microscope slide to the stage of your microscope and focus using the low-power objective. Slowly, you move the slide until you

find a cell, then, switching to medium power, you fine-tune the focus. Now you are ready to use an objective that can magnify 1,000 times. As you focus, the chromosomes suddenly come into view. They look like the ones pictured in **Figure 15.8**. Count them. How many chromosomes are in a cell from a human? (Be careful not to make any marks on the picture.)

A Karyotype

It's not easy to keep track of chromosomes when you're looking at them through a microscope. That's why people who study chromosomes often use karyotypes. Making a **karyotype** involves taking a picture of the chromosomes through the microscope, cutting up the photograph, and arranging the chromosomes in a particular order. **Figure 15.9**

Figure 15.7
Chromosomes are made of DNA.

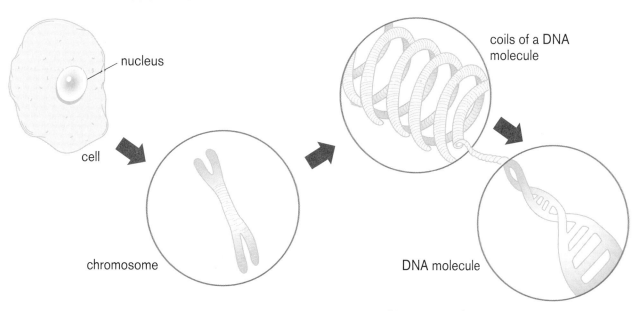

nucleus

cell

chromosome

coils of a DNA molecule

DNA molecule

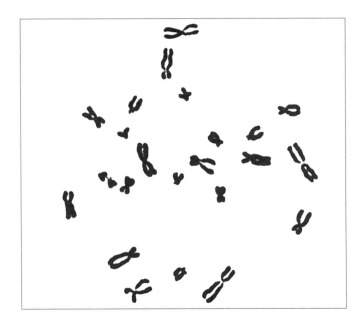

Figure 15.8
This is what human chromosomes look like when they are magnified 1,000 times. How many chromosomes are there?

Each chromosome can usually be identified by its size, the location of its narrowest point, and the "stripes" or bands that appear when the slide is stained. These characteristics are useful when making a karyotype. (Refer to **Figure 15.9** while you read this section.)

- *Size*: Chromosomes are arranged from longest to shortest.

- *Narrowest point*: Chromosomes are also arranged based on where the narrow point is located. You can see that the narrow point is near the middle of chromosomes 1, 2, and 3, but it's close to the top of chromosomes 4 and 5. (See **Figure 15.10**.)

- *Bands*: Each chromosome has a unique pattern of light and dark bands.

shows a karyotype that was made from the chromosomes in **Figure 15.8**. Count the chromosomes in the karyotype shown in **Figure 15.9**. You should find that humans have 46 chromosomes. That is, there are 46 long DNA molecules in the nucleus of every cell in the body.

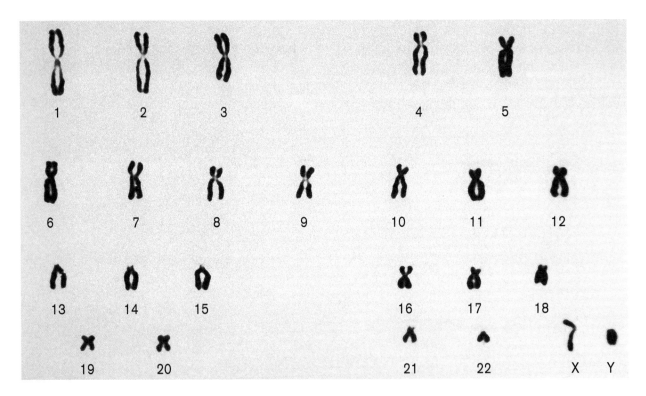

Figure 15.9
This karyotype was prepared from a cell that came from a male. Chromosomes numbered 1–22 are two chromosomes (called chromotids) attached together at their narrow point.

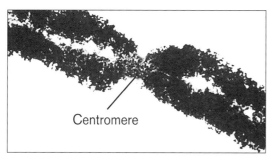

Figure 15.10

You probably noticed that some chromosomes are bent. That is not a characteristic which can be used to identify a chromosome. Since a chromosome is a long molecule, it's likely to bend, especially the longest ones.

Learn how to identify chromosome 1. First, find the two chromosomes labeled number 1 in **Figure 15.11**. Observe them closely—a number 1 is the longest chromosome; it narrows in the middle; and the top of the chromosome doesn't stain, so it is light in color. (You may also need

to look at the arrangement of the other light and dark bands if you need more clues for recognizing this chromosome.) Now try to find the two chromosome 1s in **Figure 15.8**. Observe carefully; you should be able to spot them.

Sex Chromosomes

Did you notice that most of the chromosomes in the karyotype in **Figure 15.9** are paired? That is, there are two number 1s, two number 2s, and so on, through pair number 22. Next are two chromosomes labeled X and Y. Because these two chromosomes do not match, the cells for this karyotype must have come from a boy.

Now look at **Figure 15.11**. A cell from a girl was used to prepare this karyotype. Again, there are 46 chromosomes, but this time they are all paired, even the last two chromosomes. Girls have two X chromosomes; boys have an X and a Y. The X and Y chromosomes are called the **sex chromosomes**.

Figure 15.11

This karyotype was prepared from a cell that came from a female.

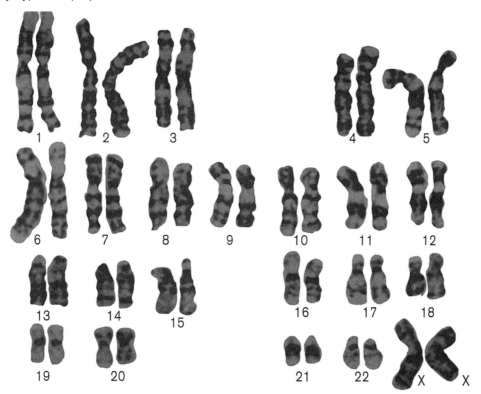

All the boys in your class have karyotypes similar to the one in **Figure 15.9**. All the girls have karyotypes that look like **Figure 15.11**. That's because all humans have the same basic set of chromosomes. Even though they look similar, the DNA contains slightly different information. How do you know this is true?

Chromosome Numbers

All organisms have chromosomes, but they don't all have the same number. Each species has a characteristic chromosome number. Humans have 46 chromosomes, so do red squirrels, and small-mouth bass (a type of fish). There probably isn't any significance to the fact that these three organisms have the same chromosome number. But it does show that chromosome number doesn't have anything to do with the size of an organism.

Glance over the chromosome numbers listed in the table, *Chromosome Numbers*. You'll notice that there doesn't seem to be any pattern to chromosome numbers. We don't know why some organisms have so many more chromosomes than others. For example, why does a canary have 80 chromosomes, but a giraffe has only 30? (If you want to be a scientist, you could tackle this question after you graduate from college.)

The study of chromosomes is fascinating and exciting because there is so much to learn. Right now, the main thing for you to remember is that all organisms have chromosomes—efficient ways to package DNA.

Chromosome Numbers		
Plants	**Mammals**	**Reptiles**
14 cucumber	40 beaver	36 boa constrictor
40 peanut	60 bison	22 American toad
24 pine trees	38 domestic cat	
48 potato	60 cattle	**Fish**
80 sugarcane	48 chimpanzee	46 small-mouth bass
	78 dog	48 bluegill
Birds	56 elephant	84 trout
80 canary	30 giraffe	
68 dove	64 horse	**Insects**
80 duck, mallard	46 human	446 butterfly
80 duck, redhead	38 tiger	32 honeybee
78 raven	32 walrus	6 mosquito
	44 killer whale	2 roundworm

Analysis Questions

1. Where are chromosomes found?

2. The chromosomes in **Figures 15.8**, **15.9**, and **15.11** were magnified 1,000 times. This line, -, is 2 mm long.

   a. Sketch how it would look if it were magnified 10 times.

   b. How long would it look if it were magnified 1,000 times?

3. What is the name of the chemical that makes up a chromosome?

4. Use the three characteristics that scientists use to make a karyotype to describe chromosome 16. (Look at the karyotypes to answer this.)

5. Look at the cell in **Figure 15.12**.

   a. How many chromosomes does it contain?

   b. How many pairs of chromosomes?

   c. Sketch the cells that will result if this one divides.

Figure 15.12

6. How many chromosomes are in a cell from a:

   a. human's liver?

   b. a needle on a pine tree?

   c. trout's bone?

   d. dog's nerve?

7. Why are chromosomes important?

Extension Activity

Math Connection: There are about 100 trillion cells in the human body. Figure out what that means in terms of chromosomes.

- Write out 100 trillion in numbers.

- Since each cell has a set of chromosomes, how many sets of chromosomes are there in the human body? How many individual chromosomes?

- If a chromosome were the size of a pencil, how much room would it take to store all of the chromosomes from one person?

DNA Structure

Chances are you've seen an illustration of DNA's double-helix structure. Often, this double-helix has been described as looking like a twisted ladder. The *sides of the ladder* are made of alternating sugar and phosphate molecules. The *rungs* of the DNA molecule are made of four chemical bases called adenine (A), thymine (T), guanine (G), and cytosine (C). The bases on one side of the molecule always pair up with specific bases on the other side. So adenine is always joined with thymine (A-T) and cytosine is always joined with guanine (C-G). (See **Figure 15.14**.) This feature is critical to the DNA molecule's ability to replicate (make an exact copy of itself) during cell division.

It is the order, or sequence, of the four chemical bases that provide the "instructions" to a cell on how to operate. You may wonder how only four bases tell a cell what to do. Think of the four bases (A, T, G, and C) as letters in an alphabet. Then imagine the letters make words (TAT, TGG, CTA) and the words make sentences (ATG TAT TGG CTA TAG). These sentences are called **genes** and tell a cell what to do by instructing it to make specific proteins.

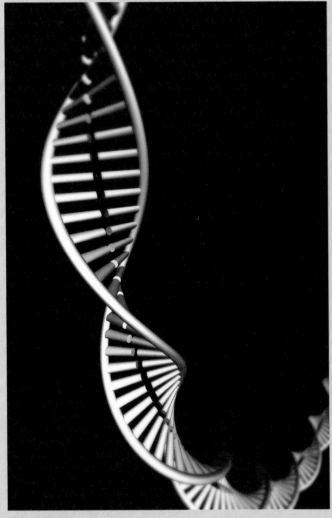

Figure 15.13
DNA Double Helix

Figure 15.14

DNA is a double helix formed by base pairs attached to a sugar-phosphate backbone.

These proteins enable the cell to perform its specific functions. For example, specific genes may instruct cells to be a muscle or to carry nerve signals.

The DNA in just *one* of your cells consists of about 3 billion bases on 46 chromosomes, and more than 99% of those bases are the same in all people. If you were to line up the DNA molecules from one cell, end to end, it would contain about 2 meters of DNA! Unfortunately, you can not see a single DNA molecule, even with a microscope. The width of a DNA molecule is approximately one billionth of a meter. This is much too small to see, even with the most powerful microscopes. The DNA you saw from the strawberry is a massive mess of many, many DNA molecules clumped together.

Sex Cell Formation

Each of us started life as a single cell containing 46 chromosomes. This single cell formed when a sperm from Dad joined an egg from Mom. It then divided again and again and again until it became a baby made up of millions of different cells. In each of the baby's cells are 46 chromosomes. Think about this for a minute.

How can a baby have 46 chromosomes in each of its cells when each parent has 46 chromosomes in each of his or her cells?

Work with paper models of chromosomes and read about sex cell formation to learn the answer to this question.

- Reading, *Meiosis*, page 451
- set of chromosome models
- two pairs of scissors
- three sheets of paper
- tape or glue
- stapler

Procedure   *(Work with a partner.)*

1. Get your set of chromosome models. Notice that Mother and Father each have four chromosomes. (Four chromosomes are used in this activity because it's too difficult to work with 46 chromosomes.)

2. Your goal: Use the chromosomes from the "mother" and the "father" to form a new "cell" that contains four chromosomes. This new cell must, of course, be able to grow into a "baby."

3. Mark your three sheets of paper as shown in **Figure 15.15**.

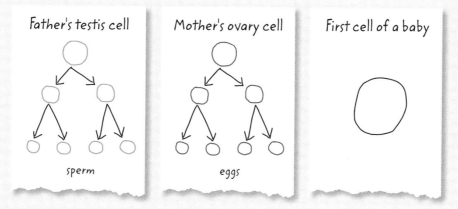

Father's testis cell

sperm

Mother's ovary cell

eggs

First cell of a baby

Figure 15.15

4. Prepare the chromosome models.
   a. Cut out Father's four chromosomes and put them in the testis cell.
   b. Cut out Mother's four chromosomes and put them in the ovary cell.
   c. Cut apart the four *replicated* chromosomes for the Father and the Mother. Place them to the side.
   d. Throw away the scraps of paper, but make sure you don't throw away any chromosomes.

5. Together with your partner, read *Meiosis*. (It's at the end of the activity.)

6. Select a *replicated* chromosome that matches (same size and pattern) a chromosome in each of the parent cells. Place it next to the matched chromosome in the cell.
   a. Continue to place the matched replicated chromosomes into the parent cells.
   b. You should now have a set of **two** *homologous (similar)* pairs of doubled chromosomes (same shape but different pattern) in the Mother's ovary cell and a set of **two** *homologous (similar)* pairs of doubled chromosomes (same shape but different pattern) in the Father's testis cell.

Includes SCi LINKS.
NSTA

Topic: Meiosis
Go to: www.scilinks.org
Code: MSLS3e447

7. Imagine that the testis and ovary cells are ready to divide to form sperm and egg cells. Using the information in the reading, *Meiosis,* and the meiosis diagram in **Figure 15.16** as a guide, move your chromosomes through Meiosis I and Meiosis II.

   a. Move the chromosomes from the testis cell into the two new cells *(Meiosis I)* and then into the four sperm cells *(Meiosis II).*

   b. Move the chromosomes from the ovary cell into the two new cells *(Meiosis I)* and then into the four egg cells *(Meiosis II).*

8. Make the first cell of a baby (called a **zygote**). It should contain 4 chromosomes.

   a. Take the chromosomes from one egg and one sperm and put them into the "baby cell."

   b. Check the chromosomes in the baby's cell. If it does not contain 4 chromosomes, redo the chromosomes in the egg and sperm.

   c. Do not tape or glue the chromosomes yet.

9. Check the chromosomes on your papers. Make changes if you need to.

   a. When you think you understand what happens to the chromosomes during meiosis, have your teacher watch as you move your chromosomes through the model.

   b. Tape or glue down the chromosomes in the baby's cell, sign your names to each of the papers, and staple the three pages together.

10. Answer the analysis questions and write your conclusion.

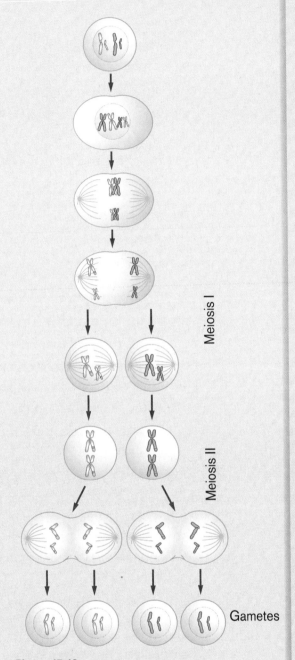

Figure 15.16

Meiosis is the type of cell division by which gamete cells (eggs and sperm) are produced.

Analysis Questions

1. Compare mitosis with meiosis.
   a. How are the two types of division alike?
   b. How are they different?

2. Imagine that the cell in **Figure 15.18** is from an ovary or a testis.
   a. How many chromosomes are in this cell?
   b. How many chromosome pairs?
   c. Sketch the new cells that will be produced by meiosis.
   d. How many chromosomes are in each of the new cells?
   e. How many matched pairs of chromosomes are in each new cell?

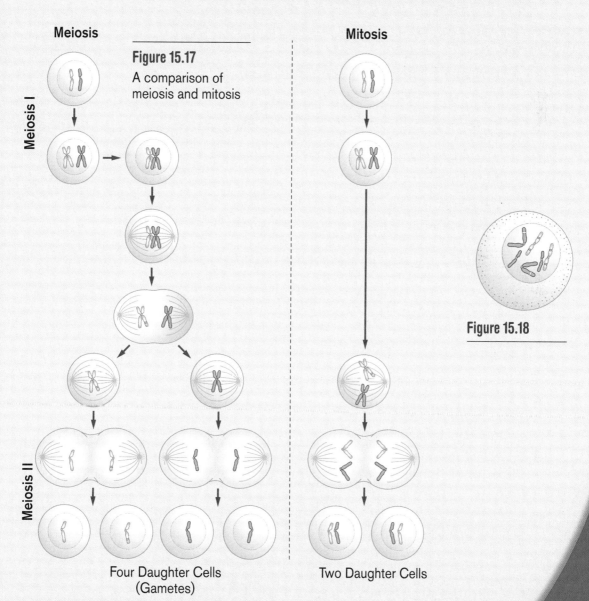

Meiosis

Meiosis I

Meiosis II

Figure 15.17

A comparison of meiosis and mitosis

Mitosis

Figure 15.18

Four Daughter Cells (Gametes)

Two Daughter Cells

3. A body cell from a frog has 26 chromosomes.

   a. How many chromosomes are in a frog egg?

   b. How many are in a frog sperm?

   c. How many are in a nerve cell from a male frog?

4. A sperm cell from a dog contains 39 chromosomes.

   a. How many chromosomes are in an egg from a dog?

   b. How many are in a cell from a dog's skeletal muscle?

5. Think about one of your skin cells.

   a. How many chromosomes does it contain?

   b. How many did you inherit from your mother?

   c. How many did you inherit from your father?

6. Imagine that the cell in **Figure 15.19** is about to divide.

   a. How many chromosomes are in this cell?

   b. How many chromosome pairs?

   c. Draw the cells that will form if this cell is a body cell.

   d. Draw the cells that will form if this cell comes from an ovary or testis.

Figure 15.19

Conclusion

Look back at the focusing question. With your partner, write one paragraph that clearly answers the question.

Meiosis

Humans have 23 pairs of chromosomes for a total of 46 chromosomes. Not just any combination of 46 will do. In order to develop and grow correctly, a person needs two of each of the 23 chromosomes. If a cell is missing just one chromosome, that cell is likely to die. If a cell has one too many, it too will probably die.

In order for offspring to end up with the right combination of chromosomes, a special kind of cell division occurs in the ovaries and testes. This type of division, called **meiosis** results in the formation of gametes (egg and sperm) that have half as many chromosomes as the original cell. If the original testis cell had 46 chromosomes, the sperm will have 23 chromosomes. Likewise, an ovary cell with 46 chromosomes will give rise to egg cells with 23 chromosomes.

Read the following stages of meiosis while looking at the diagram in **Figure 15.16**. See if you can follow along with your model chromosomes.

1. Chromosomes are carefully copied in the nucleus. The doubled chromosomes are held together at the center, forming the shape of an X.

2. **Meiosis I** Each doubled chromosome can be paired with another similar or **homologous** doubled chromosome. These pairs line up, two by two, in the center of the cell. Then, each pair of homologous chromosomes separate as each doubled chromosome move to opposite sides of the cell.

3. The cell divides and two cells are formed—each with only one member of the homologous pair of chromosomes.

4. **Meiosis II** Chromosomes are once again separated in the cell. This time the X-shaped chromosomes split and the copies separate in the cell. One copy (chromosome) is pulled to one side of the cell and the other copy (chromosome) is pulled to the other side of the cell.

5. Cells divide again, resulting in four cells with half the number of chromosomes as the original cell. For humans, each egg or sperm cell contains only 23 chromosomes.

Sex cell division must be very exact. One chromosome from each of the 23 pairs of chromosomes must end up in each gamete. When a sperm fertilizes an egg, the resulting cell is called a **zygote**. The zygote has 23 chromosomes from Dad and 23 chromosomes from Mom for a total of 46 chromosomes. The zygote then begins cell division (mitosis), growing into a fetus and then a newborn baby.

Male or Female?

When you studied meiosis, you may have wondered about the sex chromosomes. What happens to the X and Y chromosomes when eggs and sperm are formed? This is an important question to consider because these two chromosomes determine if a baby will be a boy or a girl.

Think about what happens during meiosis in a female. Find the circle that represents an ovary cell in **Figure 15.20** (left side). It contains 46 chromosomes—two X chromosomes and 44 others. When ovary cells divide, they produce egg cells that contain 23 chromosomes—one X chromosome and 22 others. Every egg cell contains one X chromosome.

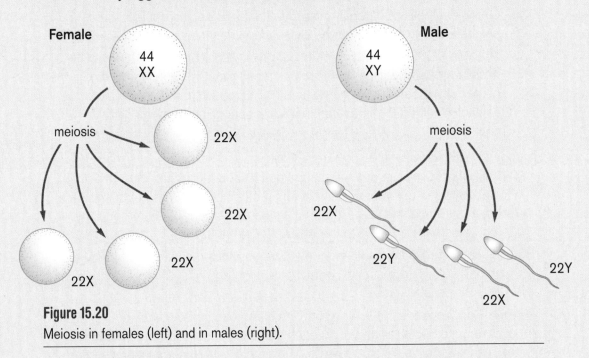

Figure 15.20

Meiosis in females (left) and in males (right).

Next, think about what happens during meiosis in a male. This time, look at the circle in **Figure 15.20** (right side) that represents a testis cell. It contains 46 chromosomes—an X, a Y, and 44 others. When testes cells divide, they produce sperm. Each sperm contains 22 chromosomes plus *either* an X *or* a Y chromosome.

If one of these sperm fertilizes an egg, a new baby will result. Depending on which sex chromosome is in the sperm, the baby will be either a boy or a girl.

What are the chances that a baby will be a girl? What are the chances it will be a boy?

In this activity, you will use a paper model to find the answer to this question. It is important that you record your data and look for patterns while you work.

- two sheets of paper
- Handout, *Quiz: Chromosomes and Cell Division*

Procedure

(Work with a partner.)

1. One of you should make a data table while the other makes the gametes.

   a. To make the data table, copy the table shown here onto your paper.

Number	Child's Chromosomes	Sex
1		
2 ↓ 19		
20		
Prediction: __ boys and __ girls		
Actual: __ boys and __ girls		

   b. To make the gametes, neatly fold and tear (or cut) a sheet of paper in fourths.

   - Use two of the pieces of paper to represent egg cells. Label each piece "Egg: X + 22 other chromosomes."
   - Use the other two pieces of paper to represent sperm. Label one "Sperm: X + 22 other chromosomes." Label the other "Sperm: Y + 22 other chromosomes."
   - Fold the four gametes and give your partner either the two sperm or the two eggs. You keep the other two.

2. With your partner, get ready to "have a family" of 20 children.

   a. Predict how many of the 20 children you think will be boys, and how many will be girls?

   b. Record your predictions on your data page.

3. Determine the sex of the first child in the "family."

   a. Without looking at the papers, pick one "egg" and one "sperm." Open the two pieces of paper and record the child's chromosome makeup on your data table. Example:

   $$44 + X + X.$$

   b. Decide if the child is a girl or a boy and record the child's sex on your data table.

4. Continue to collect data.

   a. Refold the papers that represent the gametes. Shuffle the two eggs. Shuffle the sperm.

   b. Again without looking, pick one egg and one sperm. Record the data for child number 2.

   c. Repeat this process until you have data for all 20 children.

5. Count the actual numbers of boys and girls in this family. Record these figures on your data page and on the class data table.

6. Answer the analysis questions and write your conclusion.

Analysis Questions _ _ _ _ _ _ _ _ _ _

1. From which parent does a boy inherit

   a. his X chromosome?

   b. his Y chromosome?

2. From which parent does a girl inherit

   a. her X chromosome?

   b. her other X chromosome?

3. In this *simulation,* what are the chances that each child will be a girl? a boy? Explain.

4. If you did this activity again and had another 20 children, how many boys do you predict you would have? How many girls?

5. In *real life,* what are the chances that a baby will be a girl? a boy? Explain.

6. A couple has three daughters. They are expecting another child. What are the chances that the baby will be a boy? Explain.

7. King John wants a son. After his sixth daughter was born, he told Queen Jane he was going to divorce her and marry someone else so he could have a son. Queen Jane says that a new wife won't solve his "problem." Who is right? Why?

8. Can a son inherit the information on his father's X chromosome? Explain. (You may want to use a diagram to illustrate your answer.)

9. Challenge: Can a grandson inherit the information on his grandfather's X chromosome? Explain.

Conclusion

You answered the focusing question when you answered question 5. So in your conclusion, instead of answering the focusing question, explain how using a model helped you understand how chromosomes determine the sex of a child.

Extension Activities

1. **Community Connection:** Find out how many babies were born in your state last year. You can get this information from your state health department or from an almanac. How many of the babies were girls? How many were boys? How do these figures compare with what you expected?

2. **Math Connection:** If you like solving problems, here are some more for you to try.
   a. What are the chances that a two-child family will have:

   a boy and a girl, in any order?

   first a girl and then a boy?

   two girls?

   b. In a three-child family, what are the chances that:

   there will be three girls?

   there will be two boys and one girl in any order?

   a fourth child will be a boy?

   c. What are the possible combinations of boys and girls that can be born to a four-child family? Which combination is most likely to occur?

Chapter 16

Genetics

Which traits do you think this mother and daughter share?

Whistling and Widow's Peaks

Can you whistle? Do you have a widow's peak? These are both examples of human characteristics and, like many human characteristics, they can vary from person to person. Even though we share many characteristics, we don't all look the same. We don't act the same, we don't have the same abilities, and we don't think alike. People are the same, and yet we are all different. When you are looking at the ways in which people differ, you are looking at human **variability**. Geneticists, people who study genetics, want to understand the causes of variability.

What causes variability?

List examples of human traits, especially variable ones, and then talk about the traits you listed. (When geneticists talk about the characteristics of an organism, they often call the characteristics **traits**.) Keep the focusing question in mind while you work.

- large piece of newsprint or chart paper
- three colored markers red, blue, black

Procedure *(Work in a group with two or three other students.)*

1. Get organized.
   a. Have one person get the materials.
   b. Decide who will be the recorder.
   c. All group members should help the others stay on task.
   d. Title the chart paper "Human Variability."

2. Think of ways in which people differ from each other.

   a. Recorder: When an example is mentioned, use the black marker to record it on the chart paper.

   b. Try to think of at least 40 examples to include on your list. (If you have trouble thinking of examples, start by listing the ways the people in your group differ from each other.)

3. Analyze the examples on your list.

   a. Look for traits on your list that could be **genetic**. That is, they depend on what you inherit from your parents. These traits are usually unchangeable. Have someone use the red marker to circle the traits you think could be genetic.

   b. Look for traits on your list that are most influenced by the **environment**. These are usually things you can change. They may result from decisions you make, the way you were raised, or the conditions where you live. Have someone use the blue marker to circle the traits you think are most influenced by the environment.

4. Hang your list where your teacher tells you.

5. If you finish before the other groups, start to talk over the answers to the analysis questions.

Analysis Questions — — — — — — — — — —

1. Define these two words:

   a. genetic

   b. environmental

2. Give one example of a trait you think is totally genetic. Explain.

3. Give one example of a trait you think is totally environmental. Explain.

4. List an example of a trait that all humans share and then describe ways the trait varies from one person to another. Try to think of an example that no one else will use. (For example: In general, humans have two eyes. However, human eyes vary with respect to their color and size, their distance vision, side vision, and ability to distinguish color, etc.)

5. Think about the variability in the human population. What are at least three advantages to all this variability?

Conclusion

This is the first of many times when you will consider the answer to this focusing question. Take a few minutes to gather your thoughts, and then write a few sentences to capture your thinking now, as you start to study genetics.

16-2

Is It Inherited?

By now you are probably beginning to realize there are thousands, probably even hundreds of thousands, of traits. Because these traits vary, each person is unique—no two people are exactly alike, not even identical twins. People are interested in the inheritance of these traits. They may be curious about why their hair is brown, red, or blond, or they may want to know the chances that they will get cancer, diabetes, or some other health problem that runs in the family. As a first step in understanding inheritance, you will want to start thinking about how both genes and the environment can affect the development of traits.

Thousands of Traits

A few of the many traits that vary from person to person are listed in the table. Notice that these traits include more than physical appearance. Personality traits and abilities also vary, as do body functions such as blood pressure and food allergies. It is often difficult to determine which traits are inherited and which are not.

Results of Research on Twins

Twin studies can help us understand whether or not particular traits are inherited. Identical twins are particularly helpful because they have identical genes. Knowing this, we can make the following three assumptions:

★ If identical twins are totally different for a trait, the trait cannot be genetic.

★ If identical twins are similar for a trait but not identical, the trait cannot be totally genetic. It must have an environmental component.

★ If all identical twins are always identical for the trait, the trait is genetic.

Human Traits		
Physical Appearance—Hair	**Abilities—Musical**	**Body Functions—Foods**
How many whorls are in your hair?	Can you carry a tune?	Are you allergic to peanuts?
Where are your whorls located?	Can you sight-read music?	Do you like most foods?
Do any whorls result in "cowlicks"?	Do you have perfect pitch?	Do you crave some foods?
Do you have a widow's peak? If so, is it centered on your forehead?	Can you play any tune you hear?	Can you digest milk?
What color is your hair?	Can you whistle?	Can you digest wheat?
How many hairs do you have per cm² of scalp?	Can you identify a tune after hearing it once?	Can you identify seasonings by taste?

Figure 16.1

Even though identical twins have identical genes, they are not identical people. They differ in many ways.

Observations of one, or even a few, sets of twins are not very useful. For example, if twins wear the same-colored shirt to school, does that mean choice of clothing colors is inherited? Or is it coincidence? If twins both have pimples, does that mean pimples are inherited? It's only after collecting data on hundreds of sets of twins that scientists have enough data to begin to look for patterns.

Totally Environmental

★ If identical twins are totally different for a trait, the trait cannot be genetic.

If a trait is not genetic, it must be environmental. The word *environment* refers to everything around you. As soon as a fertilized egg burrows into the wall of the uterus, the environment begins to play a part in shaping the new life. What nutrients are available to the developing embryo? How much stress does the mother experience? Is she taking medications? vitamins? Does she smoke? Think about the almost infinite number of differences that any two children experience from the moment they enter this world—the sounds, colors, smells, interactions with other people—everything is different. The way we are raised, what we learn in school, the friends we choose, the things we eat—all of these, as well as many other things, are part of the environment that shapes who we are. Environmental traits may seem to "run in families," but that's because family members share environments as well as genes. For example, people in the same family usually speak the same language. That's not because they inherited genes for speaking a particular language, but rather it's because they all learned the same language. Twin studies provide evidence to support this—when identical twins are raised in different homes by foster parents who speak different languages, the twins speak the language their foster parents speak. The twins do not speak the same language. Therefore, genes do not control which language a person speaks. Environmental traits can often be changed, sometimes easily (such as fingernail color) and sometimes with considerable effort (such as the language you speak).

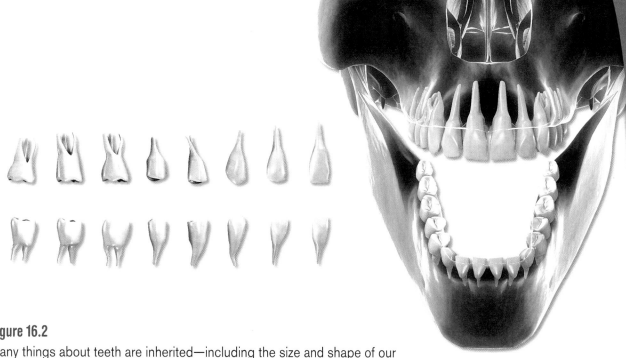

Figure 16.2

Many things about teeth are inherited—including the size and shape of our teeth and the thickness of enamel. Medications, diet, and dental care are examples of environmental factors that can affect the appearance of teeth.

Genes and Environment

★ If identical twins are similar for a trait but not identical, the trait cannot be totally genetic. It must have an environmental component.

Most human traits are controlled by many factors, some of which are genetic and some environmental. They are said to be **multifactorial**. Twin studies provide evidence that the tendency to dental caries (cavities) is multifactorial.

We probably inherit the thickness of the enamel on our teeth as well as the chemicals in our saliva. A person who inherits thin enamel and salivary chemicals that are good for bacterial growth is likely to have lots of cavities. However, nongenetic factors also influence tooth decay. For example, dental care is important for healthy teeth. People who routinely brush and floss their teeth generally have healthier teeth. Also, in most parts of the United States, fluoride is added to the drinking water to strengthen tooth enamel.

Many health problems, abilities, personality traits, and physical characteristics are multifactorial.

Genetic Traits

★ If all identical twins are always identical for the trait, the trait is genetic.

Genetic traits are ones you inherit from your parents. A genetic trait is determined by one or more **genes**. Each gene on DNA provides the information for one trait. Blood type is one example of a genetic trait. Blood type is determined the moment the sperm and egg join. That's because inside the fertilized egg are the 46 chromosomes that contain all the information needed to make a person. The information for blood type is located near the end of chromosome 9 (see **Figure 16.3**).

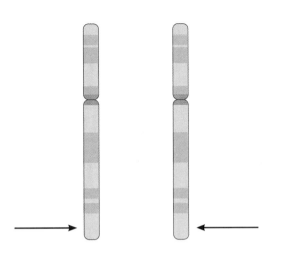

Figure 16.3
The arrows show the area on Chromosome 9 where the gene for blood type is located.

Research Continues

If we knew enough about the causes of traits, we could position every trait somewhere along a continuum, with traits that are totally genetic at one end of the continuum and traits that are totally environmental at the other end. The continuum might look something like this one.

Genetic ——————————— Environmental

blood	tooth enamel	dental	language
type	thickness	caries	spoken

Geneticists study genetic traits as well as multifactorial ones. In the rest of this chapter, you will be focusing on genetic traits and how they are inherited.

Analysis Questions

1. In your own words, explain what is meant by these terms:
   a. multifactorial trait
   b. genetic trait
2. Body weight is a multifactorial trait.
   a. Describe environmental factors that might affect body weight.
   b. How could genes affect body weight?
3. Define the word gene.
4. Draw a continuum and place the following traits where you think they belong:
   a. length of fingernails
   b. eye color
   c. height

Extension Activity

Going Further: Many medical problems have a genetic basis but are also influenced by the environment. Choose one, such as high blood pressure, cleft palate, or schizophrenia, and learn about it. How can the environment affect the condition? How can genes affect it?

Identical or Fraternal?

"Guess what? We're going to have twins!" Although people are usually surprised by the announcement that twins are expected, twins really aren't rare. In the United States, about 1 in 75 pregnancies results in twins. Twins can be either identical or fraternal depending on how they are formed.

Identical twins develop from the same egg and sperm. The fertilized egg starts to develop as it would for a single child, dividing repeatedly to form a tiny ball of cells. Then, for some unknown reason, this ball of cells may split into two groups of cells. Each group of cells then goes on to form a baby. When this happens, identical twins result. As a result, identical twins have identical genes.

Fraternal twins result when two eggs are fertilized. Each twin forms from its own egg and sperm. Because fraternal twins are formed from different eggs and sperm, they do not have the same chromosomes. As a result, fraternal twins are no more alike genetically than are ordinary brothers and sisters.

Meet A Geenoid

Genetics is the study of how traits are passed from parents to their off-spring. Ideally you would learn genetics by working with living organisms such as grass-hoppers, guinea pigs, or even penguins, and studying them for several genera-tions. However, studying real organisms is expensive and takes lots of time. Therefore, you will work with model organisms called "geenoids." In some ways, geenoids look similar to humans, and like humans, they have chromosomes.

Human chromosomes contain the information for making humans and geenoid chromosomes contain the information for making geenoids. Remember: A gene is a region of DNA that controls a specific hereditary char-acteristic. For instance, geenoids have one gene that controls ear shape and another that controls lip color.

What can you learn about genetics by using a model?

In this activity you will "read" the information on the chromosomes to create a "mother" and a "father" geenoid. While you work, keep the focusing question in mind.

Materials

- Handout, *Meet a Geenoid*
- colored markers (red, green, purple, blue, orange, black)
- ruler or compass
- two sheets of plain white paper
- 1 set of Mother's Chromosomes (models)
- 1 set of Father's Chromosomes (models)

Procedure

(Work with a partner.)

Part A: Creating Mother Geenoid

1. Get an envelope marked, "Mother's Chromosomes."
   a. Pair and arrange the chromosomes to make a karyotype.
   b. Record the mother's gene pairs on the handout.
   c. Put the chromosomes back in the envelope.
   d. Refer to the **Gene Key** on page 466 to fill in the column headed, "Mom's Appearance."
   e. Have your teacher check your work before you proceed.
   f. Draw Mother, using the information you recorded on your data table. **Do not add color to your drawing unless color is specified in the genes.**
   g. Go over the plain pencil lines (not the colored lines) with the black marker. (This is so students in the back of the room will be able to see your drawing.)
   h. In the lower right corner, label this drawing "Mother."
3. Answer analysis questions 1 through 3.

Part B: Creating Father Geenoid

1. Get an envelope marked "Father's Chromosomes" and follow the same procedure to create him that you did for Mother.
2. Hang your drawing of "Father" and "Mother" where your teacher tells you.
3. Answer the rest of the analysis questions and write your conclusion.

Analysis Questions _ _ _ _ _ _ _

1. Think about geenoid genetics.
   a. How many chromosomes does a mother geenoid have?
   b. How many genes does a mother geenoid have?
2. How many gene pair combinations are possible if a gene exists in:
   a. 2 forms?
   b. 3 forms?
   c. 4 forms?
3. How many drawings of the mother geenoid look exactly the same?

4. Remember, you are working with a *model*. List at least four ways you think this model is like real life.

5. List at least four ways this model is different from real life.

6. Think about one trait you could add to this model. Write down the forms of the gene, the trait it controls, and the appearance for the various gene combinations as they would appear on the Gene Key.

7. Think about how a "child" of these geenoid parents would look.

   a. Describe how you think the child would look for nose shape.

    b. Explain your decision.

Gene Key				
Gene Forms	**Trait**	**Appearance**		
A, a	head shape	AA=square	Aa=circle	aa=triangle
B, b	head "height"	BB=20cm	Bb=14cm	bb=8cm
C, c	eye shape	CC=quarter size	Cc=square	cc=dime size
D, d	eye color	DD=red	Dd=purple	dd=blue
E, e	ear shape	EE=circle	Ee=half circle	ee=square
F, f	eyebrows	FF=bushy	Ff=thin	ff=none
G, g	nose shape	GG=circle	Gg=pointed	gg=triangle
H, h	# of hairs	HH=8	Hh=5	hh=2
I, i	hair form	I I=curly	Ii=curly	ii=straight
J, j	length of hair	JJ=24cm	Jj=13cm	jj=2 cm
K, k	hair color	KK=blue	Kk=green	kk=orange
L, l	lip color	LL=blue	Ll=purple	ll=red
M, m	mouth	MM=frown	Mm=wide-open (O)	mm=smile
N, n	neck width	NN=4cm	Nn=1cm	nn=0

Conclusion

Write a paragraph that answers the focusing question. Use details to explain your answer.

Extension Activity

Animal Kingdom: Contact someone who breeds guinea pigs or rabbits. Find out which traits are inherited and how the breeder uses records to keep track of the genetic makeup of each animal. Ask him or her to explain the genetics of guinea pig coat color to you.

16-4

20,000 Genes

The 46 chromosomes in a fertilized egg contain all the information an individual inherits from his or her parents. This information directs the growth and development of the individual. Human genetics can be complicated, which is why you're using geenoids as a simple model to help you understand the basic concepts. As you read, keep asking yourself, how is geenoid genetics like human genetics? How is it different?

Genes Make Up Chromosomes

In geenoids, genes are on the chromosomes. In humans, too, genes make up chromosomes. Each gene is one small section of a chromosome. You can think of a chromosome as being like a chain, with each link on the chain being a gene. Each chromosome is made up of a particular combination of hundreds—even thousands—of genes. A particular gene is always located in the same place on the same chromosome. **Figure 16.4** shows the approximate location of a few of the genes that are on chromosome 9.

Chromosomes are paired so genes are, too. That means you have two copies of every gene. (One exception are male's sex chromosomes since male's have one "Y" chromosome and one "X" chromosome.) As you can see in **Figure 16.4**, the gene for producing blood type is located near the end of chromosome 9. Because you inherit two chromosome 9s, one from each of your parents, you have two copies of the blood type gene.

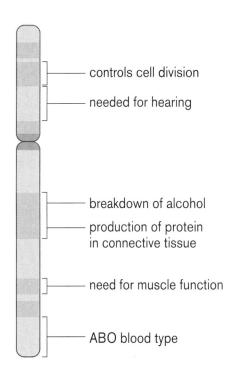

Figure 16.4
Hundreds of genes are on chromosome 9. The general locations of six of these genes are shown here.

A Gene Contains a Unit of Information

A geenoid has only 14 genes, but a human cell contains thousands of different genes—as many as 20,000 to 25,000. Almost every gene controls the production of a particular protein. Some of these proteins include the pigments that give color to hair, skin, and eyes. Others are used to make structures such as hemoglobin and fingernails. Most proteins do not make structures; most are enzymes. The enzymes

GENE → PROTEIN

Figure 16.5

Each gene controls the production of a particular protein.

do the work of the cell—whether it is to break down fat molecules, make cell membranes, digest a particular amino acid, or regulate the level of calcium in bones.

The same set of genes is in every cell in the body, but all the genes are not active in every cell. Each gene "works" at specific times in specific cells. The genes that produce hemoglobin do so only in red blood cells. The hemoglobin genes are "turned off" in nerve cells, bone cells, and all other cell types. Similarly, the genes that produce pigment are active in the cells that make up the iris, and the genes that produce enzymes to break down protein are active in the cells that line the stomach. If every gene "did its own thing," our bodies would be chaos. Cells would not specialize, tissues would not organize, and there wouldn't be organs or systems. There is order because genes are regulated.

Each Gene Can Exist in Many Forms

In the geenoid model, every gene exists in more than one form. The gene for lip color existed in two forms (L and l). Human genes also exist in many forms. For example, there are three forms of the blood type gene. Depending on the combination of forms a person inherits, he or she will have the A, B, AB, or O blood type. More than a hundred forms of some genes have been identified. Geneticists use a letter to represent a gene. Capital and lowercase letters represent the different forms of a gene.

Even though each gene can exist in many forms, a person can have only two forms of a gene—one gene form from the mother and one from the father.

Summary

The important concepts you should now understand about genes are summarized in these four statements.

1. A gene is a portion of DNA molecule.
2. Almost every gene controls the production of a particular protein.
3. Chromosomes are paired so genes are, too. (except male's sex chromosomes)
4. Each gene can exist in many forms, but a person can have only two forms of a gene.

Analysis Questions

1. Using your own words, define each of these terms.
   a. gene
   b. gene pair
   c. genetic trait
2. How many genes are in a:
   a. liver cell?
   b. nerve cell?
   c. fat cell?
3. Which of these six cell types do you think contain genes for blood type?

   skeletal muscle cells   fat cells
   connective tissue cells   nerve cells
   white blood cells   bone cells

4. Genes, cells, chromosomes, and nuclei are all extremely small structures.

   a. Make a sketch to show the relationship between these four structures.

   b. Which of these four structures is the smallest?

   c. Which is the largest?

5. Which human chromosome do you think is made up of the most genes? Why?

6. There are three forms of the blood type gene. Why can't a person have all three forms?

7. To some people, the colors red and green look the same. The gene for this form of color blindness is on the X chromosome.

   a. How many copies of the gene that can result in red-green color blindness does a female have in one of her cells?

   b. How many copies does a male have in one of his cells?

Genetic Similarities

Look at a human, a mouse, and a fruit fly, and you'll see all kinds of differences. But if you compare their genes, you'll find that these three kinds of organisms actually have many similarities. About 70% of a fruit fly's genes are similar to a human's genes. This figure jumps to 80% when mice genes are compared with human genes, and there is only a 1% difference between humans and chimpanzees.

At first, these similarities may seem surprising. But remember, genes are like a set of directions that control the structure and function of a cell. All organisms are made of cells, and cells are similar in many ways—for example, they all have a nucleus that is surrounded by a membrane and located in cytoplasm. The "directions" for all of these structures and for the control of the chemical reactions are in the genes. These genes are similar from one organism to the next.

Even though two humans may look very different, they share 99.9% of their genes.

The Human Genome Project

What if you discovered a great big book that held all the secrets of life, but the pages were blank. You and your friends found a way to reveal symbols on the pages, so you could tell the book was full of information. You couldn't quite put all of the symbols together yet, but you were beginning to figure out that the symbols, when put together, formed certain words. In fact, you were beginning to understand a few paragraphs here and there in the book. You've even started to decipher whole pages.

That is where scientists are with the *Human Genome Project*. Over the past decade, scientists have created tools and techniques that have helped them sequence the human genome. A **genome** is an organism's complete set of DNA. Remember, a DNA molecule is made of four chemical bases, adenine (A), thymine (T), guanine (G), and cytosine (C). Scientists have figured out the order (sequence) of the bases that make up the entire human genome (and some other organisms like the chimpanzee, dog, rat, and chicken).

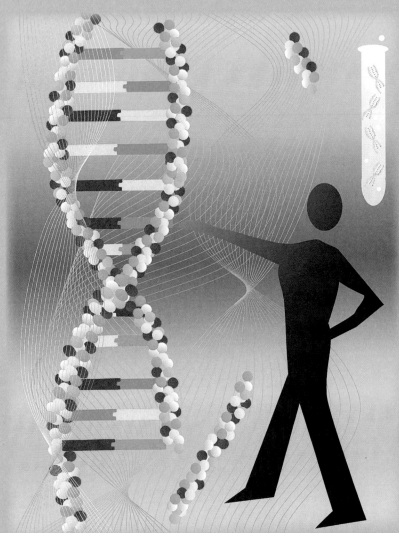

Scientists used to think that we had over 100,000 genes. Now, they believe that we have around 20,000 to 25,000 genes in our genome. And, they are even mapping the genes to certain chromosomes and figuring out where the gene is located in that chromosome. One way that scientists can find genes is by looking at a family with a certain trait or disease. They can compare the DNA sequence from that family to others and find out where there are differences in the DNA.

One day, scientists hope to have the entire human genome mapped, which means they would know exactly what and where every gene is on each chromosome. This can be used to diagnose or develop treatments for genetic diseases. Other people think that this information should not be shared. Would you like to have your genome information available? Would you have tests done to see if you had genes for certain traits or disorders? This kind of information can be powerful.

Topic: Human Genome Project
Go to: www.scilinks.org
Code: MSLS3e471

16-5

Patterns and Predictions

When parents are expecting a child, they always wonder what the child will look like. Sometimes it's possible to predict the traits the child will inherit. If you know the parents' gene forms (different forms of a gene are called **alleles**), you can figure out the possible combinations a child could inherit. That's because, for each trait, a child receives one allele of each gene from his or her mother and one from his or her father.

Four Combinations

When predicting the traits of a child, a table called a **Punnett square** can be used. A Punnett square shows all the possible combinations of alleles from the parents. You will always be able to identify these combinations if you understand how they are formed.

In **Figure 16.6**, the Punnett square illustrates the possible genetic combinations that can occur during fertilization of an egg by a sperm. If both parents have homologous pairs of chromosomes with the **Bb** alleles, then as a result of meiosis, half of the sperm cells will contain

one of dad's gene allele **B** and half will have the other allele **b**. Half of mom's eggs will contain one gene allele **B** and half will have the other allele **b**.

The inner squares in the Punnett square show the possible genetic combinations in the zygotes resulting from fertilization. Remember the sperm and eggs can have either the **B** allele or the **b** allele. The sperm with the **B** allele could join with either egg. That makes two possible combinations. (These are represented by the top two inner boxes containing BB and Bb.) The sperm with the **b** allele could also join with either egg. That makes two more combinations. (The bottom two inner boxes containing Bb and bb.) Altogether the genes from the parents can combine in four different ways.

Probability of Each Combination

The four combinations in the example are BB, Bb, bB, bb. Each combination is equally likely to occur. Once you know this, you can make

Figure 16.6

The Punnett square illustrates the possible genetic combinations that can occur during fertilization of an egg by sperm.

Mother Bb
Eggs (produced by meiosis)

		B	b	
Father Bb Sperm (produced by meiosis)	B	BB	Bb	zygotes resulting from fertilization
	b	bB	bb	

predictions about the traits you would expect to find in a *group* of children and about what an *individual* child is likely to inherit.

Consider the *group* of children first. You would expect 1/4 of them to inherit the BB combination, 1/4 the Bb gene pair, 1/4 bB, and 1/4 bb. But because Bb and bB mean the same thing, it's better to combine the chances and report them as being 2 out of 4, or 1/2. It is always clearer to combine the combinations that are the same before you write the fractions. In summary, the probabilities for this example are show below.

Probability	Gene Pair
1 in 4 (25% chance)	BB
2 in 4 (50% chance)	Bb
1 in 4 (25% chance)	bb

This means that in a group of children whose parents have the same gene combinations, on the average, one-fourth will have the BB combination, one-half the Bb combination, and one-fourth will be bb.

Now consider an *individual* child. The child will inherit one of the four possible combinations. The probabilities for an individual child are the same as for the group of children, except this time we say that the child has a one-fourth chance of inheriting the BB combination, a one-half chance of inheriting the Bb combination, and one-fourth chance of inheriting bb.

Dominant and Recessive Patterns

When you worked with the geenoid model, the letter I represented hair allele. A geenoid with the ii gene pair has straight hair, but both the II and the Ii gene pairs result in curly hair. The I allele of the gene is said to be **dominant**. If one I is present, curly hair results; it doesn't matter what the other gene allele is. By comparison, the i is said to be **recessive**. You only know it's there when it isn't paired with a dominant gene. (Scientists use capital letters to indicate dominant alleles and lowercase letters to indicate recessive alleles.)

In humans, the allele of the "hairline gene" for a widow's peak is dominant (W) and the one for a straight hairline is recessive (w). Both traits are shown in **Figure 16.7**.

Of the three alleles of the gene for blood type, A and B are both dominant. That's because people who have either the AA or the AO gene pair are type A. Likewise, the BB and BO gene pairs both result in type B blood, and the AB gene pair results in type AB blood. The O gene form is recessive. The OO gene pair is the only one that results in type O blood.

Topic: Probability and Genetics
Go to: www.scilinks.org
Code: MSLS3e473

Figure 16.7

The man on the left has a straight hairline. The woman has a widow's peak. What gene pairs for hairline could each of these people have?

Children Are Not Exactly Like Either Parent

Even though a child may look a lot like his mother or may seem to have her father's personality, each person has a unique combination of genes. In fact, a child can never be exactly like either parent because the child inherits only half their genes from any one parent. This is true, not only for humans, but for all organisms that reproduce sexually.

For some traits, children may be very different from their parents. For instance, parents with widow's peaks (Ww x Ww) can have a child with a straight hairline (ww). This may seem confusing at first, but after you practice solving several problems, it will make more sense.

Analysis Questions

1. Assume the geenoid parents have the following gene pairs for mouth shape:

   Mm × Mm

   a. Use a Punnett square to show the four gene combinations the offspring could inherit.

   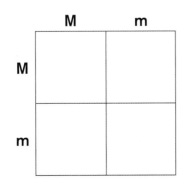

   b. Which combinations result in the same mouth shape?

   c. Write the probabilities of a child inheriting each of the combinations.

2. Assume the geenoid parents have the following gene pairs for lip color:

   Ll × ll

   a. Use a Punnett square to show the four gene combinations their offspring could inherit?

   b. What is the probability for each of the possible gene pairs?

   c. What lip colors could the children inherit and what is the probability of each?

3. Assume the geenoid parents have the following gene pairs for curliness of hair:

   Ii × Ii

   a. Use a Punnett square to show the four gene combinations their offspring could inherit?

   b. What is the probability for each of the possible gene pairs?

   c. What is the probability that a child will have straight hair?

   d. What is the probability that a child will inherit curly hair?

4. Make a chart to show which gene pairs result in which blood types.

5. Read *Mendel's Findings* on page 475.

   Use a Punnet square for the first experiment and then again for the second experiment to show how Mendel arrived at the conclusion that about one-fourth of the peas in the second experiment will be green. Record the correct gene pair for each pea, using Y for yellow colored peas and y for green colored peas.

6. Challenge: Coat color is an inherited trait in shorthorn cattle. Cattle with the RR gene pair are reddish brown in color; WW results in white, and RW in roan (reddish brown with white hairs).

   a. If a reddish brown cow is crossed with a white bull, what are the possible gene combinations the offspring can inherit?

   b. What color would you expect the offspring to be?

Mendel's Findings

The first explanation for the inheritance of traits wasn't based on humans but on pea plants. In the 1860s an Austrian monk named Gregor Mendel was working in the monastery gardens. He was curious about some of the traits he saw in the peas, in the flowers, and in the plants themselves, so he chose seven traits to study. One of these traits was the color of the seed. Some of the peas were yellow and some were green. When he crossed plants that always produced yellow peas with plants that always had green peas, all the new plants had yellow seeds.

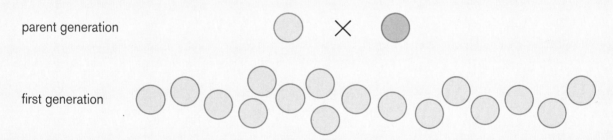

parent generation

first generation

This was puzzling. Where did the "green" trait go? Mendel took two pea plants from the new batch and crossed them. Some of the peas in this new generation were green!

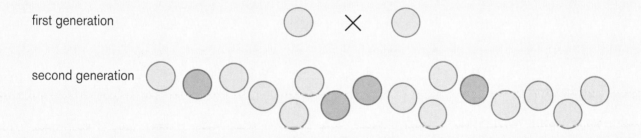

first generation

second generation

Mendel knew that mathematics can help a person recognize patterns, so he counted the peas and analyzed his data. Of the 8,023 peas he harvested and counted, 6,022 were yellow; the other 2,001 were green. This pattern was similar to the pattern he saw for some of the other traits: About three-fourths were of one type and the other one-fourth were the other.

Mendel made up the terms **dominant** and **recessive** to describe the results he observed. In this case, Y (for yellow peas) was dominant and y (for green peas) was recessive.

Geenoid Generations

Like humans and many other organisms, geenoids reproduce sexually. That is, the females produce eggs and the males produce sperm. Do you remember what happens to the number of chromosomes when gametes are produced?

When an egg and sperm merge, a young geenoid is created. You know what the mother and father geenoid look like.

Will geenoid children look like their parents? Explain.

In this activity you will use the chromosomes from a mother and a father geenoid to create an egg and a sperm. When you put the egg and sperm together, you will produce their "'child." By "reading" the information on the child's chromosomes, you can draw the young geenoid.

Materials

- colored markers (red, green, purple, blue, orange, black)
- ruler or compass
- one sheet of plain white paper
- one set of Mother's Chromosomes (models)
- one set of Father's Chromosomes (models)
- Handout, *Geenoid Generations*
- Gene Key, page 466
- Handout, *Geenoid Class Data*

Procedure   (Work with a partner. Share the tasks so you work efficiently.)

Part A: Egg, Sperm, and Fertilized Egg

1. Get your materials.
   a. Make sure both sets of chromosomes are complete and that they are in the correct envelopes.

    b. In the first column of the *Geenoid Generations* handout, record the information about the mother geenoid's genes.

    c. In the second column, record the information about the father geenoid's genes.

2. This mother and father are going to have a child.

    a. Have one partner use Mother Geenoid's chromosomes to produce an egg.

       • Set up Mother's chromosomes in pairs.

       • Randomly choose one chromosome from each pair. Try not to pick specific ones.

       • Under the column marked "Mom's Gene Pair," circle the genes that ended up in the egg.

    b. Have the other partner use Father Geenoid's chromosomes to produce a sperm.

       • Set up Father's chromosomes in pairs.

       • Randomly choose one chromosome from each pair. Try not to pick specific ones.

       • Under the column marked "Dad's Gene Pair," circle the genes that ended up in the sperm.

3. Put the egg and sperm together to create the fertilized egg. Record the child's genes.

4. Have your teacher check your work before you proceed.

5. Return the chromosomes to the correct envelopes.

Part B: Geenoid Junior

1. Look at the Gene Key on page 466 to fill in the last column of the handout.

2. Draw the child according to the instructions "in the genes."

    a. **Do not add color to your drawing unless color is specified by the genes.**

    b. Go over the plain pencil lines with black marker so the students who are sitting at the back of the room will be able to see your young geenoid when you hang it up.

3. Name the child. Write its name on the top of the paper. In the lower right corner, write whether the child geenoid is a boy or a girl.

4. Hang the picture of the child geenoid where your teacher tells you.

5. Answer analysis questions 1 and 2.

Part C: Class Data

1. Fill in the first column of the handout, *Geenoid Class Data*.
2. Record the total number of young geenoids that were created by your class. Then calculate the number of young geenoids you predict will have each of the traits.
3. Collect the data for the trait your teacher assigns to you and your partner.
4. Fill in the actual counts on your handout during the class discussion.
5. Answer analysis questions 3 through 6.

Analysis Questions _____

1. In one cell, how many chromosomes would:
   a. a mother geenoid have?
   b. a father geenoid have?
   c. a child geenoid?
2. How does the number of gene pairs in a young geenoid compare with the number of gene pairs each parent has?
3. How can two purple-lipped parents have a red-lipped child? Explain your answer.
4. Remember, you are working with a *model*. List at least three ways you think this model is similar to inheritance in humans.
5. List at least three ways you think this model differs from inheritance in humans.
6. What can you learn by looking at all the young geenoids together that you cannot learn by looking at only your geenoid?

Conclusion

In conclusion, write a paragraph that answers the focusing question.

Extension Activity

Going Further: Volunteer to summarize the data for the young geenoids produced by all the life science classes.

A Third Generation

Here's your chance to find out if you are beginning to think like a geneticist. Your teacher will show you the drawings of two geenoids—a male and a female. Your task is to draw what their first child is most likely to look like. You will need to predict, as accurately as possible, which form of each trait a child is most likely to inherit.

Procedure

1. Plan how you want to proceed and decide what materials you will need.

2. Organize your work so your teacher will be able to understand your thinking.

3. When you are confident about your predictions, draw the child.

4. Answer the analysis questions.

Analysis Questions

1. What problems did you encounter when you did this task?

2. How confident are you that your drawing is correct? Explain.

UNIT VIII
Ecosystems and Ecology

Chapter 17

The Living Environment

Is everything shown in the photo alive? How can you tell the difference between living and nonliving things?

It's Alive

Living things are different sizes, shapes, and colors. Another word for living thing is organism. Imagine a worm, a tree, a dog, or even the mold growing on a piece of bread. Each is a different kind of organism. How many kinds of organisms can you think of? You probably pass dozens while on your way to school. In fact, millions of different kinds of organisms live on Earth.

How are living things different from each other?

How are living things similar to each other?

Keep these questions in mind while you, on your own and as a class, list organisms. You will think about how the organisms are similar and different. Then you will group similar organisms into new lists.

Materials

- notebook paper

Procedure

1. Think about different kinds of living things you see every day. On your own, write down at least 10 different organisms. Be as specific as you can. For example, instead of writing *bird,* write *sparrow.*

2. Be prepared to share your list of organisms with the class. Your teacher will call on you and your classmates to record your ideas on the board.

3. Study the list your teacher has recorded. Think about how the organisms are similar and different. Write down your ideas on your paper.

4. With your classmates, decide how to divide the list of organisms into two groups.

5. Watch as your teacher records the two groups in a T-chart. A T-chart is a type of chart that arranges information into two categories. You write the categories at the top of the chart as headings.

6. Copy the T-chart on your paper.

7. Think of other ways the organisms can be divided into categories. Share your ideas with the class.

8. Copy onto your paper any other T-charts your teacher writes on the board.

Analysis Questions _ _ _ _ _ _ _ _ _ _

1. What were some of the characteristics that made organisms similar to each other? List at least three characteristics.

2. What were some of the characteristics that made organisms different from each other? List at least three characteristics.

Conclusion

Write complete sentences to answer the focusing questions.

Sorting Animals

You have started thinking about the different kinds of organisms that live on Earth. You might have listed different kinds of plants and animals. Plants and animals have different characteristics. Both plants and animals have cells but plants don't have to eat food. You will learn more about plants in the next chapter. Right now, focus your attention on the different kinds of animals. If you had to explain to someone what an animal is, what would you say? All animals share some characteristics, but they are also very different. There are so many animals on Earth it would be impossible to know about all of them. One way to learn about animals is to group similar animals together and compare them.

What characteristics are most useful for sorting animals?

Figure 17.1

What is common between all the photos? Are there any differences in the organisms in the photos?

To answer this question, design a way to sort animals into groups. Then compare your sorting system with your classmates' systems. Sorting organisms into groups based on their characteristics is called **classification**. At the end of this activity, you will have constructed your own classification system for animals.

- notebook paper
- Handout, *20 Animal Cards, Set A and B*
- envelope

Procedure
(Work with a partner.)

Part A: Designing a Sorting System

1. Get 20 animal cards from your teacher.
2. Sort the 20 animals into groups. Put similar animals in the same group. You decide how many groups you need.
3. On a clean piece of paper, number (or name) the groups and then list characteristics of the animals that fit into that group. (In other words, what "clues" did you use to put an animal in one of the groups?)
4. Answer analysis questions 1 through 3.
5. Participate in a class discussion. Be prepared to share the characteristics you used to divide your animals into groups.

Part B: A Second Sorting

1. Get ready to sort another team's set of animals using *your* system. Do you think your system will work with a different set of animals? Follow these steps and change the names of your groups if necessary.
   a. Name your groups. Use the names as column headings.
   b. Describe the characteristics of each group under the headings.
   c. List the names of your 20 animals under the appropriate group headings.

2. When you are satisfied with your system for sorting animals, trade your set of animals with another team. Make sure they have a different set of animals.

   a. Use your system to sort the new set of 20 animals.

   b. Add the names of the other team's animals to the appropriate groups.

   c. Adjust your system if you end up with animals that don't fit into any groups.

3. Meet with the team from the previous step and compare your sorting methods.

   a. Have the other team explain what characteristics they used to sort the animals into groups and you explain your method to them.

   b. Talk about the advantages of each method.

4. Put your animal cards into an envelope to save for the next lesson. Write your names on the envelope.

5. Answer analysis questions 4 and 5 and write your conclusion.

Analysis Questions _ _ _ _ _ _ _ _

1. Which of the animals fit into more than one of your groups?

2. What problems did you have when you put the animals into groups?

3. Think of at least two reasons why someone might want to put animals into groups.

4. Think about how your method for sorting animals compared with the other team's method.

   a. How were they alike?

   b. How did they differ?

5. Write down at least one new idea you learned from the other team and change your groups if necessary.

Conclusion

Write complete sentences to answer the focusing questions.

Extension Activity

Going Further: Start with one group of animals and come up with a way to sort it. For instance take a group such as "pets" or "birds" and divide it into subgroups. Keep track of the characteristics you use. Once you have created the subgroups, subdivide it again.

Classification of Living Things

You just created your own classification system by sorting 20 kinds of animals into groups. Actually, if you are thinking like a scientist, it is more accurate to talk about "species" of animals rather than "kinds." Organisms that are so similar that they can mate and produce fertile offspring make up a species. For example, dogs and cats are two different species. However, poodles and cocker spaniels are two breeds of dogs. Poodles and cocker spaniels can mate and produce live offspring. They are from the same species; they are both dogs. Scientists have named over 1.75 million species on Earth. They estimate that a total of 13 to 14 million species exist when unnamed species are included. Now you will learn how scientists make sense of all these organisms.

What is an animal?

How do scientists group organisms?

Materials

- Reading, *Animal Groups*, page 489
- Reading, *Major Animal Groups*, pages 490–495
- Reading, *The Other Organisms*, page 496

- notebook paper
- envelope with 20 animal cards
- Handout, *10 Non-Animal Cards*

Procedure

Part A: Animal Groups

1. Read *Animal Groups* and the information on major animal groups on the following pages.
2. Answer analysis questions 1 through 6.

Animal Groups

Scientists put organisms in groups to help them make sense out of all the species. It would be impossible for them to share their findings if they didn't agree on a method for grouping the animals. For example, imagine that someone told you they found a "conehead" and a "coachwhip." You'd probably say, "Huh?" However, if they tell you that a conehead is an insect and a coachwhip is a snake, you immediately know something about the two species.

By putting organisms in groups, it makes it easier to share information. When scientists put organisms in groups, they look for similarities. They look for similarities in things such as physical appearance and structure.

All animals have several things in common despite how different they can look. All animals are multicellular organisms. They are made up of many cells that are organized into tissues and organs. Animals must get their food by eating other organisms such as plants and other animals. Animals also require oxygen to keep their cells alive. The cells use the oxygen to turn food into energy and to make more cells.

Animals share some basic similarities, but they can also be very different. Think about a sponge and a fox. Both are animals but there is one big difference; foxes have backbones and sponges do not. Animals without backbones are called **invertebrates**. Some examples are insects, jellyfishes, clams, and worms. Animals with backbones are called **vertebrates**. Vertebrates include fish, humans, and birds. The next three boxes describe some of the major animal groups that you might be familiar with. Not all the animal groups have been included because there are too many to describe.

Includes *sci*LINKS. NSTA

Topic: Classification
Go to: www.scilinks.org
Code: MSLS3e489

Major Animal Groups: Invertebrates

Figure 17.2

Animals include many different kinds of organisms. All of the animals shown here are invertebrates—they do not have backbones. *The names in parentheses such as Porifera and Cnidaria are the names scientists use to identify each group of animals.

Tube Sponge

Boring Sponge

Soft Coral

Sponges (Porifera)* (about 5,000 species) Sponges have the simplest body type and live in water. Their cells function independent of each other; they are not organized into tissues or organs. Sponges do not have mouths. Instead, they have tiny pores in their walls that filter food into their body.

Winged Comb Jelly Fish

Jelly Fish

Sea Rod

Orange Elephant Ear Sponge

Jellyfish, Corals, and Comb Jellies (Cnidaria and Ctenophora) (about 9,000 species) These animals are mostly marine (they live in salt water). They move freely and most are shaped like either a tube or a bowl. They do not have body organs. Most jellyfish and corals have mouths surrounded by stinging tentacles.

Segmented Worms (Annelida) (at least 15,000 species)

These animals have segmented bodies with a head and distinct body organs. They live in soil, marine sediment, and water. Examples include earth worms, leeches, and marine worms.

Clamworm

Earthworm

Leech

Squid

Deep-water Clams

Mollusks (Mollusca) (at least 50,000 species)

They have soft bodies with a head and foot region. Many produce a hard exoskeleton or shell. They are found almost everywhere on Earth from cold mountain tops to deep in the sea. Mollusks are a varied group that includes snails, octopuses, squid, clams, scallops, and oysters.

Tree Snail

Sea Star, Sand Dollars, and Sea Urchins (Echinodermata) (about 7,000 species)

All echinoderms live in the ocean. They have an internal skeleton and their bodies are they have 5-part symmetry, with rays or arms in fives or multiples of five. They also have a system of water-filled canals that are used for feeding and moving.

Sea Star

Sea Urchin

Sand Dollars

Major Animal Groups: Arthropods

Figure 17.3

Arthropods are the most common animal on Earth. They have hard exoskeletons. They also have segmented bodies that have at least one pair of appendages such as legs, antennae, or tails. All of the following animals are arthropods. *The names in parentheses such as Arachnida and Crustacea are the names scientists use to identify each group of animals.

Spiders, Mites, and Scorpions (Arachnida) (at least 65,000) These arthropods have four pairs of legs and their bodies are divided into two sections. They do not have antennae or wings. All are terrestrial; they live on the land not in the water.

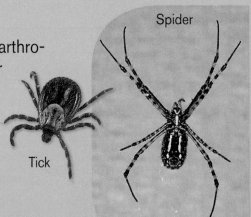

Spider

Tick

Scorpion

Shrimp

Crab

Lobsters, Crabs, and Crawfish (Crustacea) (about 40,000) These arthropods have segmented bodies usually divided into three sections. All have exoskeletons and two pair of antennae. Most live in fresh water or marine habitats, but a few such as hermit crabs and pill bugs live on land. Other examples of crustaceans include brine shrimp, barnacles, and water fleas.

Lobster

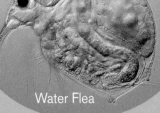

Water Flea

Millipede

Centipede

Millipedes and Centipedes (Myriapoda)

(about 13,000 species) These arthropods have ten to over 200 pairs of appendages and one pair of antennae. All live on land and most live in soil, leaf litter, or under stones and wood.

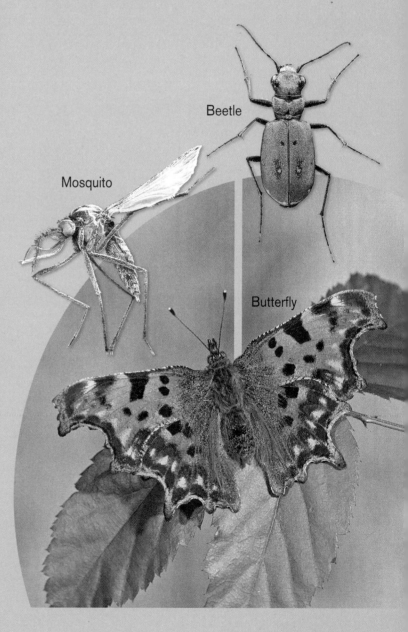

Beetle

Mosquito

Butterfly

Insects (Insecta) (over 1 million) These

arthropods have three pairs of legs and their bodies are divided into 3 sections. Most have one or two pairs of wings. Silverfish and bristletails are examples of wingless insects. They live almost everywhere on Earth, but are uncommon in the ocean. Examples include flies, grasshoppers, dragon flies, beetles, moths, ants, and praying mantis to name just a few.

Major Animal Groups: Vertebrates

Figure 17.4

Vertebrates are easily seen because of their size. For this reason, they are also more familiar to us than the other groups of animals. All vertebrates have a backbone with nerves located inside. They also have a relatively well-developed brain, two eyes, and a circulatory system with a heart. Vertebrates can be ecotothermic or endothermic. Ectothermic animals are also called cold-blooded. This means their body temperature changes with the outside temperature. They regulate their temperature by changing their behavior such as moving into a sunny area or a shaded area. Endothermic animals are also called warm-blooded. This means they maintain their body temperature using internal means.

Angelfish

Seahorse

Fish (28,000 species) Fish are cold-blooded. Fish live in the water and breathe with gills. Most have two sets of paired fins and are covered in scales. Fish have skeletons made of bone or cartilage. Sharks, skates, and rays have skeletons made of cartilage. Most fish have bony skeleton such as trout, bass, catfish, and salmon. Some fish including hagfish and lampreys, do not have jaws.

Amphibians (about 6,000 species) Amphibians are cold-blooded. They have a bony skeleton and thin, moist skin. They usually live in or near water, at least part of the time. Most have four legs and lay eggs. They include frogs, toads, and salamanders.

Bullfrog

Red-eyed Tree Frog

Reptiles (about 8,000 species) Reptiles are cold-blooded. They have a bony skeleton and skin covered in scales. Most have four legs and lay eggs. Some are covered by a shell. They include turtles, crocodiles, snakes, and lizards.

Kingsnake

Crocodile

White Pelican

Penguin

Wood Thrush

Birds (about 9,000 species) Birds are warm-blooded. Birds lay eggs. They have bony skeletons, feathers, and no teeth. The limbs in the front of their body are wings instead of legs. Most but not all can fly. Examples include sparrows, ostriches, and ducks.

Long-tongued Bat

Pine Squirrel

Humpback Whale

Mammals (about 5,000 species) Mammals are warm-blooded and produce milk to feed their young. They have bony skeletons, four limbs, hair, and teeth. Examples include dolphins, humans, buffalo, and mice.

Part B. Classification of Organisms

1. Read *The Other Organisms*.
2. Look through your envelope of 20 animals and get 10 *non-animal cards* from your teacher.
   a. Sort the organisms into groups based on what you learned from the reading.
   b. Get a clean piece of paper and write down the names of the groups as column headings.
   c. List the names of your organisms under the appropriate group headings.
3. Answer analysis questions 7 through 9.

The Other Organisms

Only some of the organisms that live on Earth were in the groups you just learned about. Many of the organisms that live on Earth are not animals. You have seen plants and heard about bacteria. Fungi and mushrooms also are not animals. Scientists group these other organisms based on their similarities just like animals.

Plants are one large group of organisms on Earth. They are found everywhere on Earth. So far scientists have found and named over 200,000 species of plants. Plants share several things in common. All plants make their own food. You will learn how they do this in the next chapter. Most plants get the energy to make food from the Sun. Some examples of plants are ferns, trees, flowers, and aquatic plants such as kelp.

Fungi are another group of organisms. Fungi can be single-celled or multicellular, made of many cells. Fungi get the nutrients they need from other organisms. They live on plants, other organisms, or decaying material and absorb nutrients from them. Some species are an important part of decomposition. Some fungi cause diseases in plants and animals such as root rot in plants or athletes foot in humans. Yeast, molds, and mushrooms are other examples of fungi.

Bacteria are one of the most numerous groups of organisms. They are found everywhere on Earth. Bacteria are a type of single-celled organism.

Finally, there are tiny organisms that are different from other fungi and bacteria. Most are single-celled, but a few are multicellular. Some of these organisms are called protists. They eat by surrounding the food with their cell wall or by filter feeding. Euglena and Amoebas are examples of protists.

Scientists group organisms in other ways than using the characteristics you have been learning about. They look at how related organisms are to one another. For example, you know that you are more closely related to your grandmother than your friend's grandmother. Scientists have come up with several different ways to classify organisms using how related they are. They are always debating which way best shows the relationships between organisms.

Scientists change their ideas about classification when they learn new information. **Figure 17.5** shows two examples of the classification systems used by scientists. The five kingdom system has been used for more than 40 years. Recently scientists learned new information about microbes. Microorganisms are tiny organisms that can only be seen with a microscope, such as bacteria. They have discovered differences in their characteristics. Now many scientists think that the microorganisms called Monera should be divided into two groups. Scientists have also learned that protists, fungi, plants, and animals share more in common with each other than with bacteria. As a result, many scientists now place all the organisms on Earth into three groups or domains.

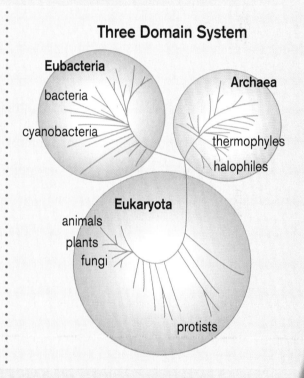

Five Kingdom System

Plants   Fungi   Animals

Protists

Monera

Three Domain System

Eubacteria

bacteria

cyanobacteria

Archaea

thermophyles

halophiles

Eukaryota

animals

plants

fungi

protists

Figure 17.5

Five Kingdoms or Three Domains? Both systems show how related organisms are using lines. The five kingdoms include Monera, Protists, Fungi, Plants, and Animals. Monera are organisms such as bacteria and other microorganisms. The three domains include Eubacteria, Archaea, and Eukaryota. Eubacteria are bacteria that you are familiar with such as the bacteria that grow on food. Archaea are microorganisms that grow in extreme environments such as where it is hot or very salty. Eukaryota includes protists, fungi, plants, and animals.

Analysis Questions

1. What is the largest group of animals?
2. What are the five groups of vertebrates?
3. What characteristics do vertebrates have in common?
4. List two ways that sponges and all vertebrates are alike.
5. Explain why all small, crawling things with an outer skeleton are not insects.
6. There are over a million species of animals on Earth. Even though very few of these are mammals, most people usually think of mammals when they hear the word "animal." Why do you think this is so?
7. How are the systems in **Figure 17.5** different than the other ways you learned to group organisms?
8. Compare your system of sorting animals with how scientists group animals. Use the chart you created and complete the following.
   a. What is similar?
   b. What is different?
   c. Describe at least one way you could make your system more similar to a classification system used by a scientist.
9. Explain why some scientists changed what classification system they use.
10. Write a sentence or two about something new you learned or something that surprised you.

Conclusion

Write a paragraph that answers each focusing question.

Extension Activities

1. **Math Connection:** Make a graph to illustrate the numbers of species in the various animal groups. You could make a bar graph, a pie graph, or a pictorial graph (plan how many species each symbol will represent).
2. **Going Further:** "Adopt" one of the animal groups and do a research project to learn more about it.

Who Survives?

Sometimes scientists study individual organisms, but they are more likely to study populations. A **population** is a group of organisms, all of the same species that live in the same area. For example, someone might study the fur colors of the population of mice that live on your school grounds and compare them with the fur colors of a population of the same species of mice that live on the school grounds of a school in another part of the state. If they noticed any differences, they would try to find out why. Could there be something about the environment that is different?

How can you model a population of animals?

What happens to the characteristics of a population over time?

Plan to answer the second question after you model a population of "platyzoans" for 10 "days." You will be a predator who devours two platyzoans every "day." Observe the platyzoans and record population data every five "days."

- Handout, *Who Survives?*
- platyzoan population envelope containing 48 platyzoans

Procedure *(Work in a group with 4 other students.)*

Read the entire procedure and make sure you understand what you will need to do.

1. Read the information about platyzoans on the next page.
2. Assign roles. Four students should be predators and the fifth will be the monitor.

    a. The four "predators" will "hunt" platyzoans.

    b. The monitor will distribute the platyzoans across their habitat and watch for problems.

Includes *sci*LINKS® NSTA

Topic: Natural Selection
Go to: www.scilinks.org
Code: MSLS3e499

3. Have the monitor get your platyzoans. As a team, count the number of each type and record it on the handout.

4. Get in position around the habitat mat your teacher gives your group (see Figure 17.6).

5. Prepare to model a population of platyzoans as follows.

   a. Predators: Decide which two will feed first and which two will feed second. Close your eyes.

   b. Monitors: Spread the 48 platyzoans evenly across the habitat map.

Figure 17.6

Platyzoans

Platyzoans are gentle, docile herbivores—animals that eat plants. Their name comes from the Greek, *platy* meaning flat, and *zoon* meaning animal. Platyzoans live in a variety of habitats. You will study platyzoans in one of two habitats: a tropical forest or a desert landscape. Each habitat supports a population of 48 platyzoans.

The predators that eat platyzoans are very lazy. They sleep most of the time, awakening only long enough to hunt. They hunt at dusk. Each hunting foray is brief. The predators awaken, swoop down and grab two platyzoans for a meal, then fall back asleep. Half the predators feed at a time.

Platyzoans reproduce every five days. Each platyzoan that survives five days in a habitat produces five offspring of the same color.

6. Follow these steps for a "day":

    a. The teacher will call "Eat!" Then two predators should open their eyes and swoop down and pick up the **first two** platyzoans they see. They should then set the two platyzoans on their laps and close their eyes immediately.

    b. Repeat with the other two predators. Once they finish, your team has completed one day in the life of a platyzoan.

    c. Repeat this four more times.

7. Follow these steps to model reproduction for the platyzoans:

    a. Record the colors of the eight surviving platyzoans on the handout.

    b. Give the platyzoans that the predators "ate" to the monitor.

    c. Figure out how many platyzoans will be "born." Each surviving platyzoan will have five offspring of the same color as the adult. (For example, for every red survivor, add five more red platyzoans to the population.)

    d. Fill in the next two columns of your handout.

    e. Work as a team to count the platyzoan offspring that are to be added to the population. (You can reuse the platyoans that are eaten. If you run out of a color, the monitor will get the number needed from a supply tub.)

8. Start a new 5-day period by giving the monitor the surviving platyzoans and the offspring.

    Repeat Step 6. When the predators have closed their eyes and the monitor has spread out the platyzoans, the teacher will announce feeding times as before.

9. Count and record the colors of the remaining platyzoans.

    a. Assume each survivor has five offspring like in Step 7.

    b. Fill in the last two columns of the handout.

10. Return the correct number of platyzoans to the platyzoan population envelope. (There should be 12 each of four different-colored platyzoans.) Return the extra platyzoans to your teacher.

11. Look at your data while you answer analysis questions 1 through 5.

12. Participate in a class discussion of your results.

13. Answer analysis questions 6 and 7.

Analysis Questions

1. Think of the characteristics of platyzoans.

    a. What do they all have in common?

    b. How do they differ from each other?

2. How did the characteristics of the platyzoans at the end of 10 days compare with the characteristics of the starting population?

3. What color platyzoan was more likely to survive in your habitat? What color was least likely to survive? Write a sentence to explain each answer.

4. List at least two ways this activity models what happens to real populations in a real habitat.

5. List at least two ways that a real situation would differ from this model.

6. What color platyzoans were more likely to survive in:

    a. the tropical forests?

    b. the desert landscape?

7. What do you predict would happen to a population of these platyzoans living on red sandstone? Explain your answer.

Conclusion

Write a paragraph that answers the focusing questions. Use evidence from your data to support your answer.

Extension Activity

History of Science: Read about the famous naturalist Charles Darwin. He spent much of his life thinking about, observing, and describing how different species of plants and animals evolved the characteristics that make them unique. He completed some of his experiments in his own backyard, but also used data he collected on a long ocean voyage.

Characteristics That Survive

You have learned about the unique characteristics of organisms. You used these characteristics to classify organisms into groups. Then you learned that scientists also use characteristics to classify organisms, but the characteristics can't always be seen. By modeling a population, you observed that differences in characteristics can make some individuals more likely to survive than others. Now you will work on your own to show what you have learned. You will investigate organisms that live on the Hawaiian Islands. First you will classify organisms using their physical characteristics. Then you will predict which crickets are more likely to survive on a Hawaiian island.

Materials

- Handout, *Animal Key*
- notebook paper
- Reading, *Does Chirping Help or Hurt a Cricket?*, page 505

Procedure

1. Get the *Animal Key* handout from your teacher.
2. Study **Figure 17.7** and carefully read the caption.
3. Use the handout to identify each of the animals pictured.
   a. For Animal A, write down the number and letter combination you choose as you go through the key. (for example: 1A, 2B, 8A, 9A)
   b. Then write down your final identification by copying down the scientific name.
   c. Repeat this process for Animals B, C, and D.
4. Read the information about crickets on page 505.

5. Describe what you learned about the cricket population on Kauai.

   a. What are some of the characteristics of the crickets?

   b. What are some things that affect whether the crickets survive and reproduce? Explain how each thing would help the cricket survive or reproduce.

6. Predict what will happen to the cricket population over time. To do this, write a paragraph that describes the characteristics of the population. Then describe any changes you think might occur.

Figure 17.7

Four kinds of animals found in Hawaii are shown here. Notice the drawings to the right of the photos. It can be difficult to see all the physical characteristics of animals in photos. The drawings provide more details.

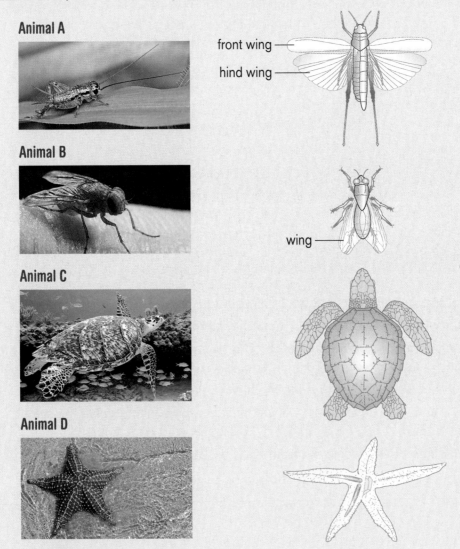

Animal A

front wing

hind wing

Animal B

wing

Animal C

Animal D

Does chirping help or hurt a cricket?

Have you ever been bothered by the constant chirping of a cricket? You were hearing a male cricket rubbing its wings together. Male crickets have a section of their wing that is toothed; it has a rough surface. When they rub this part of the wing, they make a chirping sound. Female crickets have smooth wings and cannot make chirping sounds. Male crickets make chirping sounds to attract female crickets, mate, and have offspring.

Crickets have many predators such as birds, lizards, and small rodents. But on the Hawaiian island of Kauai, they have another predator you might not suspect—maggots. Maggots are an immature form of flies that hatches from fly eggs. One species of fly on Kauai lays its eggs on cricket's backs (see Figure 17.8). Maggots hatch from the eggs, burrow into the cricket, and feed on it. The fly finds crickets by following their chirping sounds.

Scientists studied the cricket population on Kauai and learned that some male crickets were silent. These male crickets had wings more like a female cricket. They were unable to make the chirping sound with their wings that attracts females. Scientists wondered how this characteristic would affect the ability of the males to survive and reproduce. What do you think?

Figure 17.8
Maggots inside a cricket.

Chapter 18

Energy in Living Things

By eating foods such as corn, you get the energy you need to survive. Where do you think plants get the energy they need to survive?

Plant and Animal Needs

You know that all living things are made of cells. You also know there are millions of different kinds of living things. Some are single-celled organisms and some are multicellular organisms. Even though organisms can look very different, all have some basic structures in their cells that are the same. All organisms also need some of the same things to function, to live.

The living things around us have basic needs to live. All living things need energy to power processes that occur in their cells. They also need matter to build their cells and other parts of their bodies, if they have them. Living things meet these needs in different ways.

Where do plants and animals get energy?

Where do plants and animals get the matter they need to grow?

Work on your own to tell what you already know about what living things need to live.

- notebook paper

Procedure

1. Think about what you already know about rabbits and answer the following questions. Copy the questions and your answers on a piece of notebook paper.
   a. What do rabbits need to stay alive?
   b. Are these things living, nonliving, or both?
   c. Where do rabbits get what they need?

2. Think about what you already know about pine trees and answer the following questions. Copy the questions and your answers on a piece of notebook paper.

   a. What do pine trees need to stay alive?

   b. Are these things living, nonliving, or both?

   c. Where do pine trees get what they need?

Analysis Questions ― ― ― ― ― ― ― ― ― ―

1. What are some things that both plants and animals need?

2. Think about how the needs of plants and animals are different. List at least one difference.

3. How do plants use energy from the Sun?

4. Can animals survive without plants?

Conclusion

Write complete sentences to answer the focusing questions.

Material Exchange

All organisms continually take in and release materials with their surroundings. For example, you might have already thought about how plants and animals need air and water to live. Animals also must eat food and excrete waste. Food, water, and waste are either used by cells or are produced by cells. Processes occur in cells that use or produce these materials. By studying the materials produced by plants, you can learn about two processes that occur in plant cells.

What gases do plants exchange with their surroundings?

How does light influence the gases exchanged?

Find the answers to these questions by carrying out an investigation with an aquatic plant. You will predict what will happen, carry out a procedure, and collect data. Then you will analyze your results and compare them to your predictions. Your teacher will also provide clues to help you identify the gases released by the plant.

Materials

- 1 fresh sprig of *Elodea*
- 1 250-mL beaker
- 100-mL graduated cylinder
- water
- lamp with 100-watt spotlight or fluorescent bulb
- Handout, *Elodea Investigation*
- Reading, *Gas Exchange in Plants*, page 510
- Reading, *Designing Scientific Investigations*, page 512

- timer or watch
- ruler
- notebook paper

Procedure *(Work in a group of three or four students.)*

Part A: Elodea Investigation

1. Watch as your teacher gently blows into a cup with "blue" water. Do you notice a change?

Gas Exchange in Plants

Plants can live on land and in the water (aquatic). All plants take in and release gases. Carbon dioxide and oxygen are examples of gases.

They can be present in the air or they can dissolve in water. Bubbles form on the outside of aquatic plants when they release gases. One way to measure the amount of gas released by an aquatic plant is to count the number bubbles.

Figure 18.1

Elodea is a kind of aquatic plant. It is found in ponds, lakes, and streams. It is often sold as an aquarium plant.

2. Think about what might have caused the water to change. Write your ideas down on a piece of notebook paper.

3. Participate in a class discussion about the change in the water. Write down any new things you learn from your classmates and teacher.

4. Read about gas exchange in plants in the box above.

5. Get the *Elodea Investigation* handout from your teacher. Then write down the following question on your handout:

   "How does light influence the gases exchanged by an aquatic plant?"

   This is your testable question. Questions are an important part of scientific investigations. Scientists start with one question when they begin an experiment, but often add new questions as they gather information.

6. Study **Figure 18.2** and read about designing scientific investigations starting on page 512. Think about the different variables for an experiment with Elodea.

7. List the variables on a piece of paper.

Figure 18.2

Set up for an experiment with Elodea.

8. Complete the following tasks to begin designing your investigation.

    a. Write down different ways you could vary the amount of light to answer your question. Remember that light is the variable you are testing.

    b. Write down how you could control the other variables you have listed.

9. Make a prediction about how light will affect gas exchange in Elodea. Write your prediction on your handout.

10. Discuss with your group how you could design an investigation to answer your testable question. Use the information in the reading, *Designing Scientific Investigations,* as a guide. Then participate in a class discussion and share your ideas.

11. Fill out the remaining sections of your handout using what you learned from the class discussion. Use a blank sheet of paper for your data table.

12. Have your teacher approve your handout and data table.

13. Conduct your investigation using the procedure on your handout.

14. Return the materials as directed by your teacher once you have finished recording your data.

15. Calculate the average number of bubbles released in one minute.

16. Make a graph showing number of bubbles (on the y-axis) against time (on the x-axis).

17. Compare your results to the predictions you made. Be prepared to share your group's results with the class.

18. Participate in a class discussion about the results from each group.

19. Write one sentence for each group that summarizes their results.

20. Answer analysis questions 1 through 4.

Part B: Identify the Gas Released by Elodea

1. Study the results of the experiment with "blue" water and Elodea that your teacher will show you. Then answer the following questions.

    a. What color is the water from the setup that was in the dark?

    b. What color is the water from the setup that was in the light?

    c. Did the results match your prediction? Write a sentence or two to explain why or why not.

2. Answer analysis questions 5 and 6.

Designing Scientific Investigations

An important part of science is asking questions. Scientists answer questions by making observations and conducting experiments. For example, scientists might want to know if fertilizer makes plants grow taller. First they have to think about all the variables that might affect plant growth. Variables are conditions or factors that can affect the results of an experiment. One variable might be the type of plant. Other variables could be the amount of water, the amount of light, the amount of nutrients, or the type of soil.

To test questions, scientists must focus their experiments on the variable that will answer their question. In this case, the variable is fertilizer or nutrients. This is the variable being tested. The other variables (type of plant, light, water, and soil) must be kept constant—controlled. Scientists control the other variables so they will know for sure that a plant grew taller because of fertilizer, and not the other variables.

One way to setup up this experiment would be to grow plants of the same type. The plants will be grown in four identical pots with the same type of soil and under the same light and water conditions. One pot could be left alone. This pot is the control for the experiment. Then 1 gram of fertilizer could be added to the second pot, 2 grams to the third pot, and 3 grams to the fourth pot. All the conditions are the same in each of the pots except for the addition of fertilizer. This allows scientists to focus on the question of whether fertilizer makes plants grow taller. If the plants in one pot grew taller, the scientists would know that fertilizer caused the difference in results.

Scientists make sure they control variables and collect the data they need by following a procedure. A step-by-step procedure for the above scenario is shown. Scientists also must record their data in an organized way. To do this, they create a data table. By using a data table, scientists record their data in a consistent way. An example is shown here.

1. Obtain four pots of the same size.
2. Fill all four pots with the same amount and type of soil. The soil should be moist.
3. Plant four of the same kind and size of plants in each pot.
4. Leave one pot alone. Label it control.
5. Add 1 gram, 2 grams, and 3 grams of fertilizer to the remaining pots. Label each pot.
6. Place all four plants near each other so conditions such as temperature and light are the same. Water each plant with the same amount of water once a week.
7. Observe the pots daily for one month.
8. Write down the height of each plant in a data table. Also record any other observations you make.

Data Table for Pot with 1 Gram of Fertilizer

Date of observation	Height of plant	Other observations
March 3, 2009	5 cm	none
March 4, 2009	5 cm	The plant has more leaves than before.
March 5, 2009	5.5 cm	none

Analysis Questions

1. Describe how you controlled the variables in your investigation. Why is that important?

2. How did light affect the number of bubbles released by Elodea? Why do you think the number of bubbles was affected?

3. Write down one thing that surprised you during your investigation. Or, write down one thing you would do differently if you did the investigation again.

4. Predict what will happen in the following experiment after 24 hours.

   Your teacher has two setups with "blue" water and Elodea sprigs. One is placed in the dark. The other is placed in front of a light.

   a. What color will the water be in the setup placed in the dark? Write one or two sentences explaining your answer.

   b. What color will the water be in the setup placed in the light? Write one or two sentences explaining your answer.

5. Think about what gas might have been present in each situation described below. Then write one sentence explaining your answer. Share the ideas you have now. Later you will learn more information to help you answer these questions.

   a. The gas that your teacher blew into the "blue" water.

   b. The gas in the bubbles when Elodea was placed near a light.

   c. The gas that was more abundant in the water with Elodea kept in the dark.

   d. The gas that was more abundant in the water with Elodea kept in the light.

6. You saw the "blue" water change color as your teacher blew into it. You saw the same thing happen to the "blue" water with Elodea when it was placed in the dark overnight. Remember that materials that are released from organisms are produced by cells.

   a. Do you think these color changes could be caused by the same process? Why or why not?

    b. Could the same process occur in human cells and plant cells? Why or why not?

Conclusion

Write complete sentences to answer the focusing questions.

Making Food

In the last activity, you observed the products of two processes that occur in plants. One of these processes is called **photosynthesis**. This process allows plants to make their own food in the form of sugars. Plants need matter and energy to make their food (sugars). You might already know that plants get their energy from the Sun to make their food. However, it might seem strange to talk about making food from matter. What is matter? Everything around us including the water, air, and food we need are made up of matter. Living things are made up of matter that is mainly composed of carbon molecules. In this chapter you will learn how plants take in carbon molecules and make food from them.

Where do plants get the matter and energy they need to make food?

What happens in plant cells so plants can produce their own food?

You can begin answering the first question by learning about carbon molecules in different forms. Then you will discover why carbon is important for the process of photosynthesis. You will also learn what happens in plant cells when photosynthesis occurs.

Figure 18.3

Most plants, such as this tree, use energy from the Sun to make their food.

Materials

- notebook paper

Procedure *(Work on your own and with a partner.)*

Part A: Carbon in Different Forms

1. Think of things (living or nonliving) that have carbon in them. List as many as you can think of, but try to come up with at least three things.
2. Watch as your teacher brings out a piece of dry ice, charcoal, a piece of wood, corn, and sugar.
3. Listen as your teacher describes the items and answer the following questions.
   a. Write one or two sentences describing what dry ice is.
   b. Explain what the items have in common.
4. Study **Figures 18.4** and **18.5** and read the captions in the box. Then participate in a class discussion.

Ways to Represent Atoms and Substances Made of Carbon

Atom or molecule	Symbol	Model
carbon	C	
oxygen	O	
carbon dioxide	CO_2	

Figure 18.4

Different symbols and models are used to represent elements and compounds. Letters can be used as symbols for elements. Circles of different colors also can be used to illustrate atoms of elements. Carbon and oxygen are elements. Carbon dioxide is a compound made of carbon and oxygen. Compounds contain two or more elements that are combined.

5. Follow your teacher's instructions to model atoms in different states of matter.

6. Answer analysis questions 1 through 3.

Part B: Photosynthesis

1. With a partner, read the following paragraphs and study the figures on the next pages.

All living things need energy for growth and other processes. There are different forms of energy (light, chemical, electrical, and mechanical). Light energy from the Sun is one form of energy. Organisms store energy in sugars, starch, fat, and other carbon-based molecules. This form of energy is known as chemical energy. Food and gasoline are also examples of chemical energy. Energy can be transformed from one form to another. For example, cars get energy from gasoline. Burning the gasoline in the engine converts the chemical energy into mechanical energy, and the car moves. Plants and animals also convert energy.

Unlike animals, plants do not eat food. Since they don't eat food, where do plants get their energy? Plants get their energy from the Sun. But plants cannot use the light energy to grow. Plants must convert the light energy into chemical energy (food). They produce food through photosynthesis. **Photosynthesis** is process by which cells use light energy to make sugars (food) from carbon dioxide and water.

Includes sci*LINKS*.
NSTA

Topic: Photosynthesis
Go to: www.scilinks.org
Code: MSLS3e517

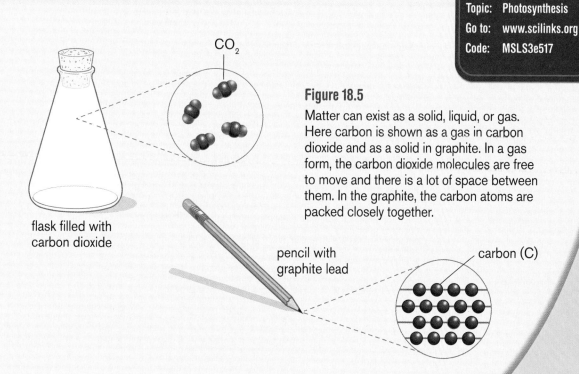

CO$_2$

flask filled with carbon dioxide

pencil with graphite lead

carbon (C)

Figure 18.5

Matter can exist as a solid, liquid, or gas. Here carbon is shown as a gas in carbon dioxide and as a solid in graphite. In a gas form, the carbon dioxide molecules are free to move and there is a lot of space between them. In the graphite, the carbon atoms are packed closely together.

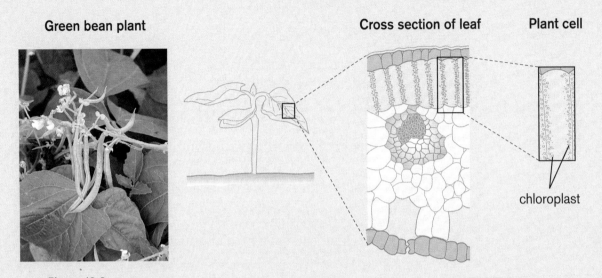

Green bean plant   **Cross section of leaf**   **Plant cell**

chloroplast

Figure 18.6

All plants, such as the bean plants shown here, have cells in their leaves. If you could cut a leaf in half, you could see a cross section of a leaf and all the cells inside. Some of the cells have numerous chloroplasts. Photosynthesis takes place inside chloroplasts.

Photosynthesis begins when light strikes a plant's leaves. Some of the cells in a leaf contain structures called **chloroplasts**. Photosynthesis takes place in chloroplasts (see **Figure 18.7**). You saw these small green structures in some plant cells when you looked at them through a microscope. Chloroplasts contain a green pigment called chlorophyll. **Chlorophyll** absorbs light and starts a series of chemical changes. These chemical changes involve water and carbon dioxide.

Photosynthesis takes place in two stages. The first stage of photosynthesis involves light. Light energy splits water molecules (H_2O) into oxygen and hydrogen. Oxygen gas (O_2) is released. In the second stage of photosynthesis, carbon atoms are used to build sugars (food or chemical energy). Chemical changes cause carbon to combine with oxygen and hydrogen. This process produces sugars. The carbon atoms come from carbon dioxide in the air. Water supplies the hydrogen and oxygen. See **Figure 18.7**.

Photosynthesis is a complex process. Many chemical changes take place that combine atoms in different ways. The most important thing for you to know is what these atoms start out as and where they end up. One way to show this is by using a chemical equation. The following equation summarizes all the changes that occur during photosynthesis.

$$\text{light} + \text{carbon dioxide} + \text{water} \longmapsto \text{sugar} + \text{oxygen}$$

energy CO_2 H_2O $C_6H_{12}O_6$ O_2

Representation of Photosynthesis Occurring in a Chloroplast

light

water
(H_2O)

oxygen
(O_2)

carbon dioxide
(CO_2)

sugar
($C_6H_{12}O_6$)

chloroplast

Figure 18.7

Chloroplasts are found inside plant cells. When light enters a chloroplast, the process of photosynthesis begins. Chemical changes cause the release of oxygen from water. Then carbon atoms from carbon dioxide combine and rearrange to form sugars.

The sugars produced during photosynthesis provide food for plants. Plants can use the food in several ways. The sugars can be broken down to provide energy for the plant. The energy is used by leaf cells or is carried to other cells in the plant. Sugars also can be stored for use later. Plants store sugar by converting it into starch. Starch is a form of stored energy for plants. Think about a potato plant. As the plant grows, much of its energy is stored as starch in the potato underground. Finally, sugars can be broken down and then used to form other molecules needed by plants. This is the matter plants use to form all the molecules they need to grow bigger and repair damaged cells. See **Figure 18.8**.

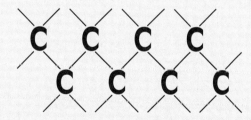

Figure 18.8

Plants can breakdown sugars to make other carbon molecules. Carbon joins together in long chains. These chains form the many compounds found in our bodies and other living organisms. Other elements, such as oxygen, hydrogen, and nitrogen, are also found on these chains. But the chains are mainly composed of carbon.

2. Answer analysis questions 4 through 9.

Analysis Questions

1. Imagine a glass jar filled with carbon dioxide.

   a. Draw the jar and the carbon dioxide molecules inside. Use dots to represent the carbon dioxide molecules.

   b. Write a caption that explains your drawing, including how you decided to space the dots.

2. Imagine what the carbon molecules in a log might look like. All living things are made of many carbon molecules connected in chains.

   a. Draw a log and the carbon molecules inside. Use dots to represent the carbon molecules.

   b. Write a caption that explains your drawing, including how you decided to space the dots.

3. Does carbon have mass? In other words, if you could put carbon atoms on a scale could you measure the mass? Explain your answer.

Parts of a Leaf

Leaves have several layers of cells. The cells covering the upper and lower surface of the leaf provide protection. Scattered throughout the leaf surface are tiny openings called stomates. Each stomate is surrounded by two guard cells. The guard cells regulate the opening and closing of the stomates. It is through these openings that gases enter and exit the leaf. For example, plants exchange oxygen and carbon dioxide through these openings. Plants can also close the stomates to prevent water loss from the leaf.

Unique cells are also found in the inner part of the leaf. Some of the inner cells contain chloroplasts for photosynthesis. Other cells move materials throughout the plant and are called vascular bundles. The vascular bundles found inside the leaf are made of xylem and phloem cells. Xylem cells transport water and nutrients in the plant. This is how

4. Answer the following questions about photosynthesis.

   a. Where does photosynthesis take place?

   b. What three things must be present for photosynthesis to occur?

   c. What two things are produced as a result of photosynthesis?

5. List at least three ways your life would be different if your skin cells contained chloroplasts.

6. Explain why you think this statement is true or false:

   "Blocking light from a plant is like starving a person."

7. Explain why the following statement is *not* true:

   "Plants get their food from the soil."

8. Imagine that a plant is placed in a tightly sealed bottle. The plant is in soil and was given water, but it can't get any more air than is already trapped in the bottle. Can the plant make food (sugars)? Why or why not?

9. Where does the matter come from that makes up a piece of wood or a green bean plant?

chloroplasts in leaf cells get the water needed for photosynthesis. Phloem cells transport the sugars produced during photosynthesis to all parts of the plant. Vascular bundles are found in roots, stems, and leaves of plants.

Figure 18.9

Cross section of a leaf showing cells inside. The cells inside leaves perform different functions. Guard cells help regulate flow of gases in and out of the stomates. The cells that make up the vascular bundles help transport water and food to all parts of the plant.

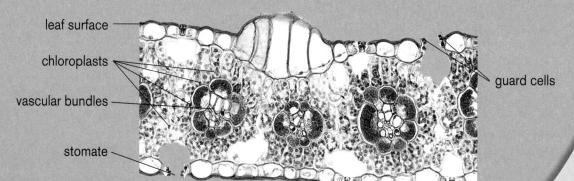

leaf surface

chloroplasts

vascular bundles

stomate

guard cells

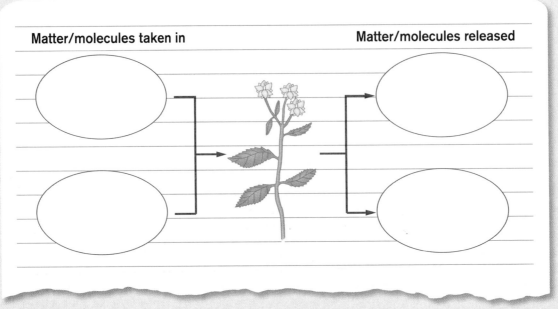

Matter/molecules taken in   Matter/molecules released

Figure 18.10

10. Copy the drawing from **Figure 18.10** on to your paper. Use dots to represent the matter taken in and released by plants during photosynthesis. (See **Figure 18.5** on page 517 for ideas on how to represent different forms of matter with dots.) Label each circle with the kind of matter used and produced by the plant.

Conclusion

Write complete sentences when you answer the focusing questions.

Extension Activity

1. **Going Further:** Look at an Elodea leaf under a microscope. Look carefully at the cells. Locate the cell walls and chloroplasts.
2. **Going Further:** How are plants important in your own life? Write about a day in your life. Mention the things you do, what you see, and what you eat. Then, circle everything in your story that comes from plants.

Making Energy for Cells

You have been learning about processes that occur in plants. Now you know that plants release oxygen during one of these processes—photosynthesis. But you also saw that Elodea release carbon dioxide. All plants release both oxygen and carbon dioxide. This might surprise you. In this activity, you will learn about the process in plants that causes them to release carbon dioxide. This process is called **cellular respiration**. It converts food into energy that cells can use.

How do plants use the food they produce?

How do animals use the food they eat?

Find the answers to these questions by reading about cellular respiration. As you read, think about why cellular respiration is important for all living things.

Many processes that are important for life happen inside cells. Photosynthesis is one of these important processes. But it only occurs in plants because it takes place in chloroplasts. Chloroplasts are an example of an organelle—tiny parts or structures inside cells. **Figure 18.11** shows you a few of the organelles that are common to cells in all organisms. Mitochondria are found in the cells of most organisms and provide cells with energy. The energy comes from the process of cellular respiration.

Cellular respiration takes place in the cytoplasm and mitochondria of cells. The process begins with the breakdown of sugars. Hydrogen and carbon molecules are left over. Carbon molecules enter the mitochondria (**Figure 18.11**).

Figure 18.11

Cells have organelles that act similar to the parts of a car. Each organelle performs a function. The nucleus and mitochondria are examples of organelles. The cells of most organisms have a cell membrane, cytoplasm, nucleus, and mitochondria. In addition to these structures, plant cells also have cell walls. Some plant cells also have organelles called chloroplasts.

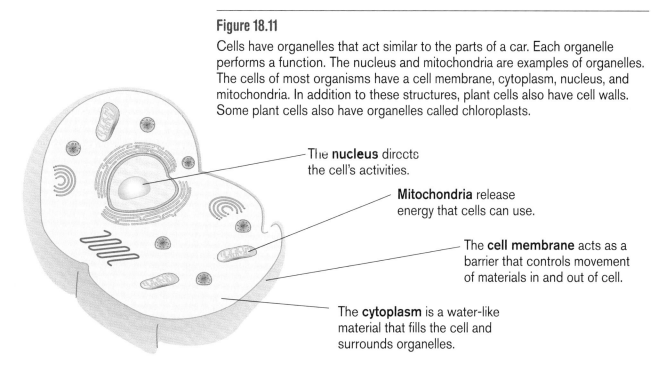

The **nucleus** directs the cell's activities.

Mitochondria release energy that cells can use.

The **cell membrane** acts as a barrier that controls movement of materials in and out of cell.

The **cytoplasm** is a water-like material that fills the cell and surrounds organelles.

Chemical changes cause the release of carbon dioxide from the cell. More chemical changes take place, leading to oxygen combining with hydrogen to form water. This process releases energy that can be used by the cell (see **Figure 18.12**).

Like photosynthesis, cellular respiration is a complex process. Many steps and chemical changes take place. But the end result of cellular respiration is the same for all organisms. Energy from food is converted into energy for cells. The following equation summarizes cellular respiration.

sugar + oxygen ↦ carbon dioxide + water + energy
$$C_6H_{12}O_6 \quad O_2 \quad\quad CO_2 \quad\quad H_2O$$

The energy produced by cellular respiration can be used in many ways. It might be used to transport molecules between cells. Energy is also used by cells to make different molecules the organism needs. In animals, the energy might be used to make muscle cells contract or send signals in the nervous system.

Cellular respiration is constantly taking place inside plant and animal cells. This is because cells need a constant supply of energy. As a result, cellular respiration occurs in plants even during the day. It also occurs constantly inside your cells.

Your cells need a supply of energy even when you are sitting still. Sugar, in the form of glucose, travels from your blood into your cells. Then cellular respiration takes place to convert the sugar into energy for your cells. When you are active, such as playing sports or exercising, your cells need more energy. So the rate of cellular respiration increases to supply more energy. This requires more oxygen and sugar. The amount of carbon dioxide produced by cells also increases. You start breathing faster to get more oxygen to your cells and get rid of excess carbon dioxide. In animals, carbon dioxide and oxygen are exchanged by breathing. And the blood transports the gases to and from cells. In plants, the gases are exchanged through openings (stomates) in the leaves.

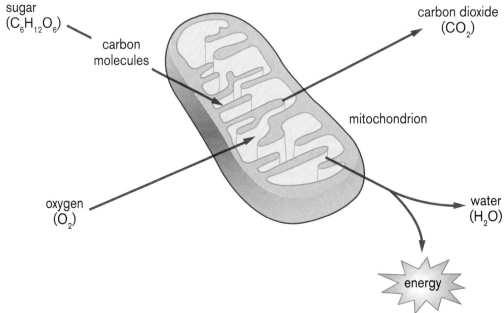

Representation of Cellular Respiration Occurring Inside a Cell

Figure 18.12

Cellular respiration takes place inside cells. First, *sugars* are broken down into *carbon molecules*. The carbon molecules enter the mitochondria. Chemical changes occur and *carbon dioxide* is released. *Oxygen*, then, combines with *hydrogen*. This process releases *water* and *energy*. Why do you think mitochondria are sometimes called the "powerhouses" of the cell?

Analysis Questions

1. Answer the following questions about cellular respiration.

   a. Where does part of cellular respiration take place?

   b. What two things must be present for cellular respiration to occur?

   c. What three things are produced as a result of cellular respiration?

2. List at least three ways that the energy produced by cellular respiration is used.

3. Do you think it is important that plant cells have mitochondria? Explain why or why not.

4. Think back to your investigation with Elodea. What evidence do you have that cellular respiration occurs in plants?

5. Why is cellular respiration important for all living things?

6. Do plants carry out cellular respiration during all parts of the day? Write a sentence or two explaining your answer.

Conclusion

Write complete sentences to answer the focusing questions.

Following Carbon and Energy

Photosynthesis converts energy from the Sun into food energy. At the same time, carbon atoms from the air are rearranged into sugars that make up the food energy. Cellular respiration breaks down stored food energy (sugars). This produces energy that cells can use. Do you think photosynthesis is necessary for animals to get the energy they need for their cells? In other words, would you have the energy you need to survive without plants?

What is the relationship between photosynthesis and cellular respiration?

You will answer this question by reading how different organisms get the energy they need to live. Then you will follow the path of energy and carbon molecules. You will do this by using models to act out photosynthesis and cellular respiration.

Materials

- Reading, *Photosynthesis and Cellular Respiration*, page 528
- one photosynthesis and cellular respiration kit
- notebook paper

Procedure *(Work with a partner.)*

1. Read *Photosynthesis and Cellular Respiration* to learn how photosynthesis and cellular respiration are connected.

2. Get a photosynthesis and cellular respiration kit from your teacher. You will use the materials in the kit to model these processes.

Figure 18.13
What connection is there between photosynthesis and a hamburger?

3. Spend some time looking over the materials in the kit. The materials are representations of different things involved in photosynthesis and respiration. Locate the following items in your kit:

plant cell	carbon dioxide
animal cell	oxygen
light energy	hydrogen
energy for cells	carbon
sugar	

4. Discuss in your team how you could use the materials to model photosynthesis and cellular respiration. Notice you can use carbon, hydrogen, and oxygen atoms to create compounds such as sugars or carbon dioxide. You can use the equations below to help you.

photosynthesis

$$\text{light} + \text{carbon dioxide} + \text{water} \longmapsto \text{sugar} + \text{oxygen}$$
$$\text{energy} \qquad 6CO_2 \qquad 6H_2O \qquad C_6H_{12}O_6 \qquad 6O_2$$

cellular respiration

$$\text{sugar} + \text{oxygen} \longmapsto \text{carbon dioxide} \longmapsto \text{water} + \text{energy}$$
$$C_6H_{12}O_6 \quad 6O_2 \qquad 6CO_2 \qquad 6H_2O$$

5. Decide how you will divide the roles when you model the processes. For example, one of you could use the materials to model photosynthesis and the other could model cellular respiration.

6. Practice modeling both cellular processes. Then change roles so you both understand each process.

7. Model photosynthesis and cellular respiration for your teacher after you have each had time to practice.

Analysis Questions _ _ _ _ _ _ _ _ _

1. How does photosynthesis provide energy for both plants and animals?

2. What is the original source of the energy in the food that plants make?

3. Think about the energy you get from meat after you digest it. What is the original source of energy in the meat we eat? Explain your answer.

4. How do plants and animals use the carbon molecules that are in food?

Conclusion

Write complete sentences to answer the focusing questions.

Photosynthesis and Cellular Respiration

Plants get the energy they need by making their own food. They make sugars through photosynthesis. Some of the sugars they produce are immediately used for energy. But first the sugars must be broken down through cellular respiration. Fortunately for humans, plants usually produce more sugar than they need for energy. The excess sugar is stored in the stems, roots, seeds, and fruits of plants. Humans and other animals can obtain this energy by eating plants. Study **Figure 18.14**.

Animals get the energy they need by eating plants directly or indirectly. We get energy directly from plants by eating the plant itself, such as lettuce. We also eat fruits, seeds, and roots produced by plants. Think of all the fruits and vegetables you see in the grocery store such as apples, green beans, carrots, and potatoes. When we eat plants, we break down, through digestion, the carbon molecules that make up the plant. Some of these molecules are sugars. Just like plants, we get energy from the sugars through cellular respiration. Where did the energy in the plants we ate originally come from? It came from the Sun.

We get energy from plants indirectly when we eat other animals. For example, we are eating other animals when we eat beef. But the cows didn't make their own food. Cows get their food by eating grass and other plants.

Includes sci**LINKS**®
NSTA

Topic: Energy Cycle
Go to: www.scilinks.org
Code: MSLS3e528

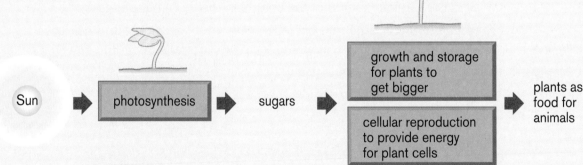

Figure 18.14

Energy passes from the Sun, to plants, and then, to animals. This is because plants make enough food (sugars) for their own energy needs and also store excess energy. The stored energy can be obtained from plants when animals eat them.

When we eat other animals, our bodies break down the meat through digestion. The meat is broken down into carbon molecules. These molecules can be used in cellular respiration to produce energy. Where do you think the energy from the meat originally came from?

Food supplies both energy and carbon molecules for organisms. In plants, food is in the form of sugar. In animals, food can come from plants or animals. The carbon molecules from food are used to repair cells and build new cells. Building new cells allows organisms to grow. So without the carbon molecules to build new cells, a pine seedling couldn't grow into a tree that is 50 feet tall. A green bean plant couldn't produce green beans. You also couldn't grow taller or build bigger muscles without carbon molecules.

Carbon molecules cycle between plants and animals (see **Figure 18.15**). The carbon molecules start as carbon dioxide in the air. Plants use energy from the Sun and carbon dioxide to form sugars through the process of photosynthesis. Then the carbon (sugars) is broken down through cellular respiration. This process releases carbon dioxide into the air. Remember that both animals and plants break down sugars through cellular respiration. So both plants and animals release carbon dioxide into the air. But plants use more carbon dioxide than they produce.

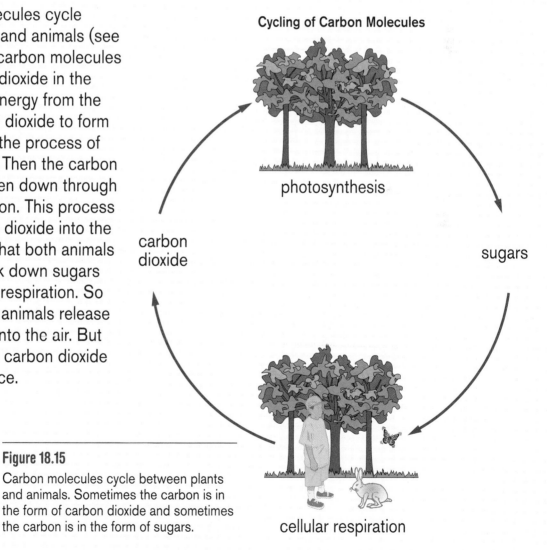

Cycling of Carbon Molecules

photosynthesis

sugars

carbon dioxide

cellular respiration

Figure 18.15

Carbon molecules cycle between plants and animals. Sometimes the carbon is in the form of carbon dioxide and sometimes the carbon is in the form of sugars.

Important Processes for Life

You started this chapter by thinking about the matter and energy that plants and animals need to live. You learned how photosynthesis and cellular respiration are important for all living things. Now it's time to reflect on the things you've been learning. First, you will review your answers from activity 18-1, *Plant and Animal Needs*. Do you think you will want to change some of your answers? Then you will use diagrams to show what you now know about photosynthesis and cellular respiration.

Materials

- answers from activity 18-1, *Plant and Animal Needs*
- large sheet of paper

Procedure *(Work with a partner and on your own.)*

1. Get your answers from activity 18-1, *Plant and Animal Needs,* from your teacher.
2. Find a partner and go over your answers together. Make changes to your answers based on what you have learned.
3. Get a large sheet of paper from your teacher.
4. Working alone, study the following list of terms.

photosynthesis	carbon dioxide
cellular respiration	water
chloroplast	oxygen
mitochondrion	sugars
light energy	energy for cells

5. Using the large sheet of paper, create two drawings. One drawing should illustrate photosynthesis. The other drawing should illustrate cellular respiration.

   a. Use all the terms as labels on your drawings. Some terms may be used more than once. You might also want to add words that aren't listed here.

   b. Write the formula for each process next to the appropriate drawing.

   c. Write one sentence describing where the energy comes from in photosynthesis and where it ends up.

   d. Write one sentence describing where the carbon comes from in photosynthesis and where it ends up.

   e. Write one sentence describing where the energy for cellular respiration comes from and how it is used.

Chapter 19

Ecosystems

Ecosystems are made up of living and nonliving things. What living and nonliving things do you see in this photo?

Your Environment

All organisms require certain things in order to live. In the last chapter, you learned that plants need light and carbon molecules to make food. You also learned that all organisms need food as a source of energy and to supply materials to build and repair their bodies. Food is one of the most important things that affects whether organisms will survive. However, organisms require more than just food to survive.

Most organisms need specific conditions to live. Do you think a polar bear could survive in a tropical rainforest? All organisms depend on their surroundings or environment to survive. An ice sheet and a rainforest are two very different environments. An **environment** includes all living and nonliving things in a particular place. It also includes conditions such as light, temperature, and moisture.

What things do you notice in the environments that surround you?

Begin answering this question by observing your surroundings. Pay attention to the organisms that you see every day. Is it warm or cool where you live? Do you get a lot of rain or snow or is it dry? You will compare your observations with the rest of the class.

Materials

- notebook paper

Procedure

1. Make observations of the environment you live in. You can do this on your way to school or when you are out for an errand with your family or friends.

2. Write down the living and nonliving things you observed. You will probably notice living things more than the nonliving. Make sure you list the nonliving things, too.

3. Share your observations with the class.

4. Think about the environmental conditions where you live. What is the temperature like? How often to you get rain? Be prepared to share with the class the conditions that you think affect the organisms that live around you.

Analysis Questions _ _ _ _ _ _ _ _ _ _

1. Write a one sentence definition for environment.

2. Do the conditions in your environment change during the year? Explain why or why not.

3. Think about an environment that is different than yours. Write a short paragraph describing the living and nonliving things that you might find there.

Conclusion

Write complete sentences to answer the focusing question.

How Many Coyotes?

Do you think the environment can affect the size of a population? Remember that a population is a group of individuals of the same species living in the same place. For example, a population of trout is all the trout found in a particular stream. A population of water lilies would include all the water lilies in a specific pond. In this activity, you will study a coyote population.

How does the environment affect the number of coyotes in an area?

Since you do not have time to go out and study a coyote population, you will do an activity that models the changes in the size of a coyote population over several generations. Keep track of your data and analyze it in order to write a quality answer to the focusing question.

- Handout, *How Many Coyotes?*
- Reading, *A Survivor*, page 539

Procedure

Part A:

Read the entire procedure and make sure you understand what you will be doing.

1. Imagine you are a coyote. You must select a "den" somewhere on the edge of a field. Coyote "food" will be scattered around the field.
   a. Write your name on your handout and use it to mark your den. (If a pencil is not heavy enough to hold the paper in place, use a rock or book to weigh it down.)
   b. Once you identify your den, stand by it.
2. Be aware that, even though you are all coyotes, you are not all exactly equal. Your teacher will assign these roles.

a. One young male coyote caught his foot in a trap. As a result, he has a broken leg. This coyote must hunt by hopping on one leg.

b. One coyote is a female with two pups. She must gather twice as much food as everyone else.

3. *Walk* through the field to gather food when the teacher says to start. Follow the *Guidelines for Searching for Food* listed below.

a. When you find something that looks appetizing, pick it up and return to your den with it. (Coyotes actually would not return to their dens; they'd eat on the spot.)

b. After "eating" an item, return to the field to find something else to eat.

c. Take only one thing at a time to eat.

4. Fill out your handout for Part A when all the food has been gathered (one round is over). Use your handout to calculate the number of kilograms of food you collected. (The numbers on the markers represent kilograms of food.)

a. Record the number of kilograms of small mammals you collected by adding up the numbers on the orange markers.

b. Total and record the number of kilograms for each of the other food types.

c. For every type (color) of food you're missing, penalize yourself 5 kilograms.

Guidelines for Searching for Food

You need to know these things about coyote behavior:

- It is natural to snatch food from a crippled coyote; wild animals are seldom generous.

- Coyotes do not steal from each other's dens.

- Coyotes seldom fight. If they did, they could be injured and unable to gather sufficient food; causing them to starve.

You need to know these things about the food choices:

- There are five different kinds of food represented by five different colors.

- As a coyote, you must eat from each of the five "food groups." Like people, coyotes need a variety of foods in order to stay healthy.

Figure 19.1
Coyotes often have more difficulty finding food during the winter.

5. Decide whether you survived or not and whether you can have a pup.

    a. A coyote needs 20 kilograms of food to survive for ten days. If you collected fewer than 20 kilograms of food, you died. "Dead coyotes" must go to the "graveyard."

    b. Any coyote with at least 15 extra kilograms of "food" can have a pup. Any one coyote cannot have more than three pups. (At that point, let a "coyote" from the graveyard return to the field and choose a new den.)

6. Complete a total of four rounds. Repeat steps 3–5 for each round.

7. Fill in the handout for Part A and answer analysis questions 1 through 3.

Remember: The object is not to see who "wins." There is no winner. The object is to see how many coyotes are able to survive in this area.

Part B:

1. Model a coyote population for another four rounds. This time the environment will change.

    a. In rounds 1 and 2, a drought will occur and affect the amount of food.

    b. In rounds 3 and 4, rain will return and will bring more animals and plants to the field.

2. Complete a total four rounds and record the amount of food gathered after each round on your handout for Part B.

3. Complete the handout for Part B.

4. Create graphs showing the number of coyotes that survived each round. Label the y-axis with the number of coyotes. Label the x-axis with Round # because each round represents a passage of time.

    a. Create one graph using the data from Part A.

    b. Create one graph using the data from Part B.

5. Read *A Survivor*. Think about the following question as you read: How could conditions in the environment affect a coyote's diet?

6. Answer analysis questions 4 through 6 and write your conclusion.

Analysis Questions _ _ _ _ _ _ _ _ _

1. Since a coyote needs 20 kilograms of food to survive for 10 days, how many coyotes should have been able to live in the field? (Use your answer to question 1 on the handout to help you answer this question.)

2. The amount of food never changed, so why did the number of surviving coyotes change after each round?

3. Why do you think there was a 5 kilogram penalty for every food type you were missing?

4. Think about how changes in the environment affected the coyote population. Did the population increase or decrease? Why?

   a. What happened when there was a drought?

   b. What happened when rain returned?

5. List at least three ways this models what happens to a real coyote population in its environment.

6. List at least four ways that a real situation with coyote population would differ from this model.

Conclusion

Write a paragraph that answers the focusing question.

Extension Activity

Going Further: Do some research on wolves. Look for ways that they are like coyotes and ways that they differ. While you work, think about this question: Why are coyotes surviving despite the growing human population and yet wolves are endangered?

Figure 19.2

What similarities do you see between this wolf and the coyotes pictured in Figure 19.1? What differences do you see?

A Survivor

Coyotes look a lot like dogs, especially shepherd-collie mixes. Like medium-sized dogs, coyotes are 40–60 centimeters (16 to 24 inches) high at the shoulder and they weigh 11–13.5 kg (25 to 30 lbs). Their fur color varies depending on where they live. Coyotes that are native to woodlands tend to have dark fur, allowing them to blend in with shadows. Those that live in deserts have a lighter colored coat, which makes them hard to see against the sand.

Figure 19.3

Coyotes are known for their mournful, evening howls. This is a true sound of the wild.

Coyotes are very flexible about the environments they can live in. They live in a wide variety of environments including forests, grasslands, deserts, and swamps. Coyotes also have adjusted to living near humans. They often live in agricultural areas, as well as in suburbs and urban areas.

A Coyote Diet

Coyote pups learn to hunt by stalking insects. At first they seem to think it's a game and that grasshoppers, crickets, and other insects are toys. It isn't long, however, until they realize that the "game" is serious and means survival. When they venture out on their own, they must constantly be on the lookout for food and danger.

Many scientific studies have looked at what coyotes eat. Most of these were done by analyzing the contents of dead coyotes' stomachs; but some involved examining coyote droppings. All of the studies have shown that coyotes eat whatever is available. Exactly what each eats depends on the season of the year and where the coyote lives. Like all wild animals, the coyote must struggle for its survival. Its preferred food might not be available when hunger strikes. The results of a typical study look like those shown on page 540.

As you can see, small mammals make up most of a coyote's diet. However, this varies considerably from one year to another. Some coyotes are mainly vegetarians, relying almost totally on plants for food. Others are strictly carnivores, eating only meat. Occasionally, a coyote will discover that livestock are easy prey.

Figure 19.4

Coyotes eat insects such as grasshoppers to supplement their diet. Their diet usually consists mostly of mammals.

Analysis of a Coyote's Diet

Rodents and other small mammals   **63–70%**

This includes mice, cottontail rabbits, gophers, ground squirrels, meadow voles, skunks, porcupines, beavers, hares, marmots, and woodrats.

Deer, livestock, and other large mammals   **5–16%**

In addition to deer, coyotes also eat large mammals such as antelope, sheep, elk, and cattle. Coyotes usually do not kill these animals. They eat the carcasses of animals that have died from other causes. When they do take live prey, they often end up killing the ones that are easiest to catch (the sickly, old, wounded, or very young).

Insects and other invertebrates   **7–21%**

This could include moths, butterflies, caterpillars, crickets, beetles, spiders, dragonflies, and anything else that may catch a coyote's attention.

Plants and plant products   **4–8%**

This varies considerably depending on the location and time of year. Coyotes eat apples, corn, watermelon, berries, grasses, and other vegetation.

Birds, reptiles, and amphibians   **3–4%**

This includes ground feeding birds such as juncos, robins, road runner, and sparrows. It also includes frogs, toads, snakes, lizards, tadpoles, and salamanders.

Miscellaneous items   **2–6%**

This includes anything that's easy to find. When starved, a coyote will eat just about anything, even garbage such as string, rags, peels, and paper wrappers.

Populations and Ecosystems

So far you have been learning about the living parts of the environment—populations. You also have thought about the nonliving parts of the environment. You did this when you described conditions in your area such as temperature and moisture. The study of the living and nonliving parts of the environment is called **ecology**. The organisms are the living parts of the environment. Temperature, water, and light are parts of the nonliving environment. Together, all the populations living together, and the nonliving environment in a particular area make up an **ecosystem**.

Ecosystems

Ecosystems can be large regions or a small area. A freshwater pond is an ecosystem; so is a mountain forest, a marsh, prairie grassland, or even a cornfield. A fallen log can be an ecosystem for small organisms such as insects, bacteria, and fungi. Every ecosystem has a unique set of living and nonliving things. The living and nonliving things interact and are dependent on one another.

The Nonliving Parts of an Ecosystem

We often overlook the nonliving or **abiotic** parts of an ecosystem. If you are like most people, you probably pay the most attention to the living parts of an ecosystem. The abiotic parts of an ecosystem are very important. Abiotic factors such as water, temperature, light, nutrients, and soil type determine what can live in a given area. Abiotic factors can vary considerably from one ecosystem to another. For example, water is an abiotic factor. An area that receives a lot of rain might be a rain forest. An area that receives only a few inches of rain might be a desert.

> The words **biotic** and **abiotic** come from the Greek language. *Bio* means "life" and *a* means "not" or "without." Putting these together you can see why **abiotic** means "not living."

The Living Parts of an Ecosystem

Living organisms make up the living or **biotic** part of ecosystems. Ecosystems consist of many populations of different organisms. They interact with one another in different ways. These interactions often involve food and energy. In other words, organisms are always eating or being eaten. Scientists place organisms into categories based on how they obtain their food.

Includes sci**LINKS**®
NSTA

Topic: Abiotic and Biotic Factors
Go to: www.scilinks.org
Code: MSLS3e541

Organisms get their food in different ways. Organisms that produce their own food are called **producers**. Recall that plants make their own food through photosynthesis. All plants are producers. They do not eat food.

Organisms that cannot produce their own food are called **consumers**. They must eat (consume) in order to get the matter and energy they need to survive. All animals are consumers. No animal can live without eating. Scientists often group consumers according to the kinds of food they eat. There are four kinds of consumers—those that eat plants, those that eat animals, those that eat both plants and animals, and those that feed off dead organisms.

Consumers that eat living plants are called **herbivores**. There are thousands of species of herbivores. They include everything from tiny leaf-eating insects to large mammals. Two examples of herbivores are shown in **Figure 19.5**.

Consumers that eat both plants and animals are called **omnivores**. Pigs, bears, raccoons, and rats are all omnivores. Some omnivores such as sponges, clams, and oysters, get their food by filtering the water they live in. They eat anything that they happen to catch. Since they take in microscopic plants and animals, they, too, are omnivores. **Figure 19.7** shows two examples of omnivores.

Figure 19.7
Bears and sponges are omnivores. Their diets include both plants and animals.

Figure 19.5
Elk and caterpillars are herbivores. What do you think they eat?

Consumers that feed on other animals are called **carnivores**. A trout is an example of a carnivore, since it lives off smaller fish and other animals that live in the water. All spiders and snakes are carnivores. Mountain lions, wolves, and bobcats are all examples of large mammals that are carnivores. Birds that are carnivores include hawks and owls. See **Figure 19.6** for other examples of carnivores.

Animals that eat dead animals or plants are called **scavengers**. Some scavengers such as vultures feed only on dead and decaying animal carcasses. One type of vulture is shown in **Figure 19.8**. Most scavengers are much smaller.

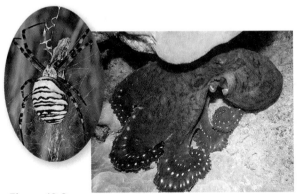

Figure 19.6
Carnivores such as spiders and octopuses eat other animals. What kind of animals do you think they eat?

summer
year round

Figure 19.8
Turkey vultures can locate dead carcasses or carrion by smell as well as sight. The range map shows where turkey vultures live.

Slugs, sow bugs, tadpoles, grubs, and some beetles are examples of scavengers. Their small size allows them to crawl through the tiny spaces in rotting plant and animal matter. See **Figures 19.9** and **19.10**. Some animals only scavenge when they happen to come upon an animal carcass. Coyotes, lions, hawks, and sharks are all examples of animals that will take advantage of a "free meal" if they find it.

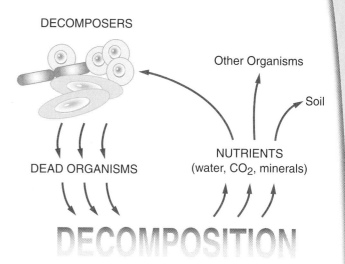

DECOMPOSERS

Other Organisms

Soil

DEAD ORGANISMS

NUTRIENTS
(water, CO_2, minerals)

DECOMPOSITION

Figure 19.11

Decomposers such as bacteria, yeast, and fungus break down dead organisms into water, minerals, and carbon dioxide (CO_2).

Figure 19.9

A single rotting apple may contain as many as 100,000 roundworms. Without these worms, it would take longer for apples to decompose.

Figure 19.10

Dung beetles lay their eggs in dung. When the eggs hatch, the developing larvae feed on it. Beetles are important in recycling waste.

Organisms that break down dead plant and animal material are called **decomposers**. They live off things such as fallen leaves, dead insects, and animal carcasses. You seldom see decomposers because they are so small. Most of them are microscopic. Bacteria, yeast, and fungus are all examples of decomposers. They convert dead plant and animal material and waste from organisms into materials that are usable. Without them, we would be living on stacks of dead leaves, old trees, animal bodies, and insect skeletons. The decomposers break down the matter in the dead organisms. When decomposers are done, all that's left is water, minerals, carbon dioxide, and other nutrients. These things are released back into the ecosystem where they can be used again by other organisms. This process is called **decomposition** (See **Figure 19.11**).

Every ecosystem has its own assortment of decomposers. The decomposers in a desert are different from the ones that live in a pond; which are different from those that live in a grassland. Even the decomposers living next to one another are different. You could search a decomposing log and find one set of decomposers, and then you could inspect the soil under the log and find another set. See **Figures 19.12** and **19.13**.

Figure 19.12

Orange cup or orange peel fungus (Aleuria aurantia) gets its name from its appearance. This fungus grows 1 to 10 centimeters wide. It grows on dead and decaying material.

Figure 19.13

Liberty cap mushrooms (Psilocybe semilanceata) grow in meadows and fields that are well fertilized by cow or sheep dung.

Carnivore, herbivore, omnivore... all of these words come from Latin. If you understand the Latin bases, it is easier to remember their definitions.

carn = flesh   *onmi* = all
herb = grass   *vorous* = to devour or eat greedily

By referring to this list, you can see that
carnivorous means "to eat flesh greedily."

There's a Limit

You will never see 50 squirrels living in someone's backyard...unless, of course, it is a very large yard. Nor will you ever hear about 5 adult coyotes living on an acre of land. Why? There is a limit to the number of organisms that a given amount of land can support. That is, there are limits to how much a population can grow.

Population size changes over time. A population increases when new individuals are added to it. This occurs when organisms reproduce. The new offspring add to population. Populations decline when individuals leave. This occurs when organisms die. Population size also changes when organisms move into or out of an area. This is only possible for organisms that can move. For example, a population of aphids might leave a plant if they have eaten most of the leaves and very little food is left. Or, in spring, birds might return to a forest after wintering somewhere else.

Limiting Factors

Something will always limit population size. These "things" are called **limiting factors**. Limiting factors can be biotic or abiotic.

Food is one of the major biotic limiting factors. There has to be enough food to feed the animals year round, not just during the summer months. If there isn't, some animals may die. Or some animals might need to move from season to season. For instance, during the summer months, deer and elk may live high in the mountains. However, when winter arrives and deep snow covers the grasses and other edible plants, they move down to the lower valleys and plains to graze.

There are many abiotic limiting factors. *Water* is a limiting factor for many organisms. Without a regular water supply, organisms die. *Temperature* also is an important factor. For instance, **Figure 19.14** shows how organisms such as fish are most abundant in a specific range of temperatures. Fish can be very abundant in their optimal zone. They are less abundant when temperatures are very low or very high. *Light* is important for organisms too. Imagine a tomato plant. When it is growing in bright sunlight, it can produce many tomatoes. If a tomato plant is in the shade most of the day, the plant doesn't grow as large and may never produce tomatoes.

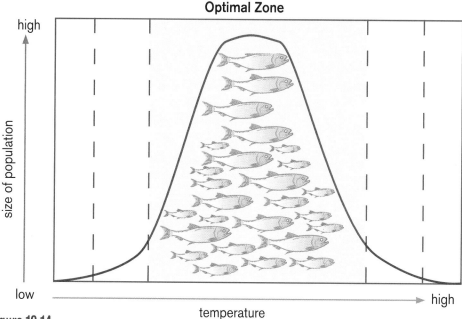

Figure 19.14

There are optimal conditions where an organism can survive. Environmental conditions might include temperature, moisture, light, salinity, or other factors. If the conditions are extreme, fewer organisms will survive.

Shelter is another limiting factor. Animals need shelter for protection from enemies while they sleep and for raising their young. Shelter means different things to different animals. A spider might be satisfied with a crack in a rock. But some owls might not be satisfied unless they can find a hole at the right height in a specific kind of dead tree. A pile of brush can provide cover for a chipmunk, but an alligator couldn't survive any place but in water.

Some limiting factors are based on how closely the organisms are living. The *amount of space* an organism needs is related to the animal's size and its need for food. A grizzly bear needs much more space than a shrew does. Not only is a grizzly bear bigger, but it is also a solitary animal and requires a lot of space for its daily activities. If there are too many animals in an environment, disease is more likely to spread. This could increase the number of deaths in a population.

Interactions between populations can affect their sizes. That is, *organisms* themselves can be limiting factors. For example, foxes are limiting factors for mice. Because foxes eat mice, they *limit* how big a mouse population can get. *Competition* between populations can also be a limiting factor. Lions and cheetahs are a good example. Because both animals eat the same things, they compete for food. A population of lions will be smaller in size when they are living near cheetahs.

Analysis Questions

1. Decide if the following characteristics are abiotic factors. Answer by writing *yes* or *no* for each characteristic.

    a. amount of rainfall

    b. number of days the temperature goes below freezing

    c. heights of the tallest plants

    d. number of decomposers in the soil

    e. type of soil

    f. type of consumers

    g. average wind speed

2. Several abiotic conditions are described here. For each one choose the value you think best matches the ecosystem you live in. Briefly explain your answer.

   a. coldest winter temperature
      below –10°F   11°F to 35°F
      –10°F to 10°F   above 36°F

   b. elevation
      0 to 3000 feet   6001 to 9000 feet
      3001 to 6000 feet   above 9000 feet

   c. annual rainfall
      0 to 10 inches   21 to 30 inches
      11 to 20 inches   above 30 inches

3. Give three examples of each of the following types of animals. Think of examples other than those that are given in the text.

   a. carnivore   c. herbivore

   b. scavenger   d. omnivore

4. Based on your diet, which consumer group do you belong to? Explain your answer.

5. If all the producers suddenly died, what do you think would happen to the

   a. consumers?

   b. decomposers?

   c. abiotic factors?

6. When decomposers are finished eating, what products are left?

7. Mountain goats live atop high mountain ridges in Alaska, a few other states, and Canada. They are herbivores that graze on low shrubs, daisies, strawberry blossoms, grasses, ferns, mosses, and other plants.

Imagine that it has been a particularly long and harsh winter. Months of blizzards and bitter winds have made it difficult for the goats to find food. One starving mountain goat comes across a mouse that recently died. Do you think the goat will eat the mouse? Explain your answer.

8. Get the handout, *An Ecosystem: Parts and Interactions,* from your teacher. Think about the local ecosystem when you fill in the spaces.

   a. Using a pencil, list at least eight consumers that live in the local ecosystem in the section marked "consumers."

   b. List at least eight local producers in the correct section of the drawing.

   c. List some examples of decomposers.

   d. In the center, describe the abiotic conditions of your ecosystem. Be specific.

9. Use the following words and phrases to fill in the equation.
   (births, deaths, organisms that move into an area, organisms that leave an area)
   Population size = _____ + _____
   – _____ – _____

10. List at least five factors that can limit the size of a population. Include both biotic and abiotic factors.

11. What do you think limits the size of coyote populations? (List at least three things.)

12. Give an example of how weather might limit the size of a population.

Extension Activity

Going Further: Look through magazines to find a picture that includes lots of information about a particular ecosystem. Try to find one that shows examples of producers, consumers, decomposers, and abiotic factors.

Food Webs and Energy Flow

All the populations of organisms interact in different ways to form a web of life. Scientists learn how organisms interact by studying their relationships. One of the most common relationships is a feeding relationship. When you show feeding relationships in a diagram, you have made a food chain or a web. You can show the transfer of energy in these relationships using an **energy pyramid**.

How does a food web show the flow of energy through an ecosystem?

How do energy pyramids show the energy transferred in an ecosystem?

With other students in your class, create food webs and energy pyramids. While you work, keep the focusing questions in mind.

Materials

- Reading, *Energy in Ecosystems*, page 548
- Handout, *Food Web Pieces*
- notebook paper

Procedure

(Work with a partner.)

Read the entire procedure to make sure you understand what you will be doing.

1. Read *Energy in Ecosystems* below.
2. Get the food web pieces from your teacher.
3. Look through the pieces and decide how to categorize each organism. Write down whether each organism is a producer or consumer. If it is a consumer, identify which kind of consumer it is.
4. Decide how to arrange the pieces into a food web. Draw the food web you created on a piece of paper. Make sure you use arrows to show how the energy flows from one organism to another.
5. Draw an energy pyramid and write the names of the organisms from your food web into the correct levels.
6. Participate in a class discussion about the food webs and energy pyramids you created.

Energy in Ecosystems

Energy is one of the most important resources in an ecosystem. The source of energy for almost all ecosystems is the Sun. Plants convert the Sun's energy into food. This energy is then transferred from one organism to another through feeding relationships.

Food Chains and Food Webs

To learn about feeding relationships, you have to know how organisms get their food. There are two common feeding relationships. One type of feeding relationship is herbivory. **Herbivores** eat living plants. Rabbits and cows are examples of herbivores. The other feeding relationship is predator-prey. A **predator** is an animal that captures other animals for food. The animal the predator eats is called **prey**. A hawk eating a rabbit or a frog eating an insect are examples of predator-prey relationships.

You can represent feeding relationships using a food chain. A **food chain** shows how organisms are linked by what they eat. Study the food chain in **Figure 19.15**. The arrows in a food chain always point in the direction of energy flow. Grass seeds provide energy for the mouse. The mouse provides energy for the skunk. The skunk provides energy for the owl. Each organism can be thought of as a link in the chain.

However, you might realize that owls don't always eat skunks. In fact, they're more likely to eat mice or small birds. Likewise, skunks don't live

Figure 19.15

This food chain shows the feeding relationships between grass seeds, a mouse, a skunk, and an owl. Mice eat grass seeds. Skunks eat mice. Owls eat skunks. Can anything be added to this chain? Think about where the grass gets its energy.

on mice alone. They also eat things such as frogs, small rabbits, and eggs. When you show all of these feeding relationships in one diagram it is called a food web. Food webs contain several food chains. They can be very complex because they show all the possible feeding relationships in an ecosystem.

A **food web** is a series of interrelated food chains. Food webs give a more accurate picture of the feeding relationships in an ecosystem because organisms usually eat many different things. See **Figure 19.16** for an example of a food web.

Figure 19.16

Food webs show that organisms *interact* with many other organisms through feeding relationships. What happens to these organisms when they die?

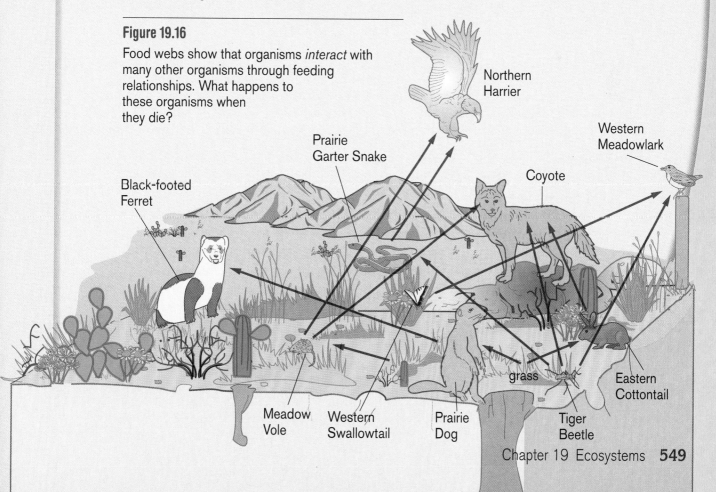

Energy Pyramids

Energy pyramids show the amount of energy transferred from one group of organisms to another. Energy pyramids are divided into levels. Producers make up the bottom level of the pyramid. They are the foundation of ecosystems. Most living organisms rely on plants for food because they either eat plants themselves or eat animals that eat plants. The next level of an energy pyramid is made up of herbivores. They eat plants. Thus, organisms in the second level eat organisms in the first level. The top level is made up of carnivores and omnivores.

Includes SCILINKS NSTA

Topic: Energy Pyramid
Go to: www.scilinks.org
Code: MSLS3e550

Energy Pyramid

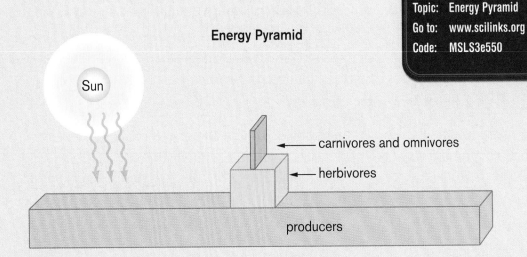

Figure 19.17

Energy pyramids show the amount of energy stored in producers and consumers. Herbivores get about one-tenth of the energy that producers have. Carnivores and omnivores get about one-tenth of the energy that herbivores have. The sun provides energy for producers.

The blocks in an energy pyramid represent the amounts of energy stored at each level. The energy is stored in the matter that makes up the organisms. The largest amount of energy is stored in producers. Although many plants are small, there are a lot of them so as a whole, they produce and store a lot of energy. Notice that the amount of energy stored in herbivores is less. And the amount of energy stored carnivores and omnivores is even smaller. One way to think about it is that it takes a lot of plants to support a small population of rabbits. If there more rabbits than plants, the rabbits would eat all the plants and there would be no food left for future generations of rabbits. The number of rabbits is limited by the amount of food. In the same way, prey such as rabbits, snakes, and mice are found in greater numbers than owls.

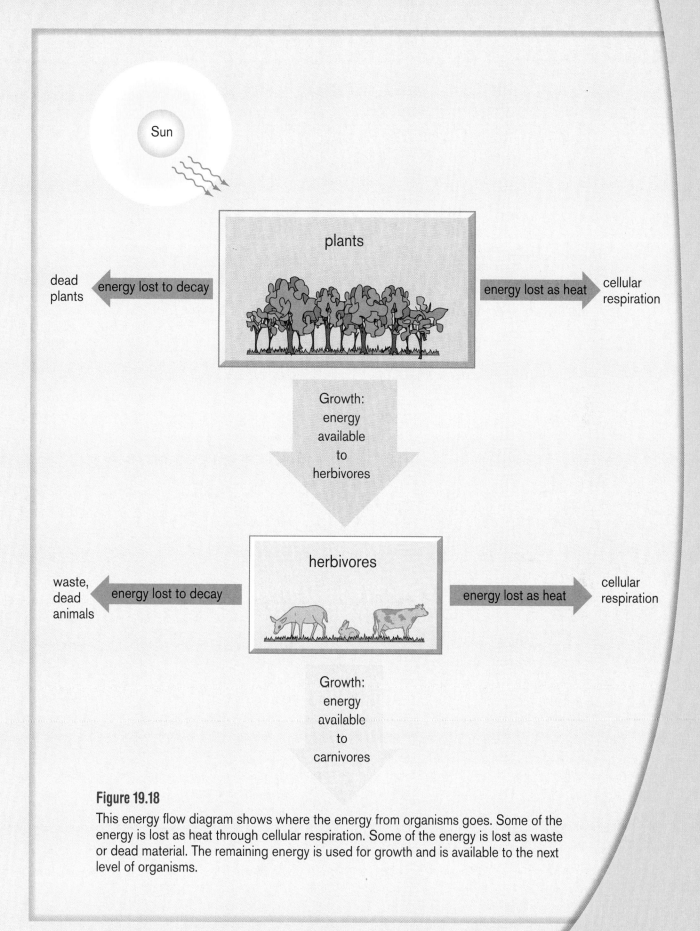

Figure 19.18

This energy flow diagram shows where the energy from organisms goes. Some of the energy is lost as heat through cellular respiration. Some of the energy is lost as waste or dead material. The remaining energy is used for growth and is available to the next level of organisms.

Energy is lost as it moves through the levels from producers to consumers. This happens because organisms use much of their energy in life processes. One of these life processes is cellular respiration. As the energy produced in cellular respiration is used, heat is released. Some energy is lost by organisms through their waste and their decaying bodies when they die. The remaining energy is used for growth of the organism. This is the material that will be available as food. **Figure 19.18** on page 551 shows where the energy from a plant and animal goes.

Analysis Questions – – – – – – – – – –

1. What is the source of energy in ecosystems?
2. What is the difference between a food chain and a food web?
3. Why do you think scientists say that "energy flows through an ecosystem?"
4. Why does each level of an energy pyramid have less energy than the level below it?
5. Which organisms would be more abundant in an ecosystem—caterpillars or birds? Write a sentence or two explaining your answer.
6. Explain what scientists mean when they say "energy is lost as it flows through an ecosystem."

Conclusion

Write a paragraph that answers the focusing questions.

Ecosystem Analysis

You started this chapter by thinking about living and nonliving things. Now you've learned how these things interact in ecosystems. You've also learned how energy flows through ecosystems. Now it's time to show what you've learned by creating your own ecosystem in a bottle. Then you will identify all the parts of your ecosystem and diagram the interactions in a food web. By doing this you will show that you have learned the answers to the following questions about ecosystems.

How does energy flow in ecosystems?

What is the relationship between plants and animals in ecosystems?

What factors limit the size of populations in ecosystems?

Follow the procedure carefully and keep in mind all that you have been learning. The purpose of this activity is for you use what you know about ecosystems to create a small ecosystem of your own.

- clear 2-liter bottle
- living and nonliving materials collected from outside
- clear packing tape
- notebook paper

Procedure

1. Look around your backyard and neighborhood for materials you can use in your ecosystem. Keep the following guidelines and questions in mind.

   a. You can only collect seeds or small plants that can easily fit inside the bottle.

   b. You can only collect small animals such as insects, small fish, and worms. Do not collect birds, mammals, reptiles, or amphibians!

    c. The bottle will be sealed with only a few small holes so everything the organisms need must be inside it.

    d. What kind of environment will you mimic? It can be an aquatic (water) environment or a terrestrial (land) environment.

    e. Where will you keep the bottle in the classroom? Its position will determine the conditions in the bottle.

2. Make a list of the materials and organisms you plan to use and where you plan to get them from.

3. Have your teacher approve your list before you begin collecting materials.

4. Bring a bottle to class and have your teacher cut it in half.

5. Create your ecosystem in the bottle.

    a. Poke small holes in the bottle.

    b. Add materials to the bottle.

    c. Use tape to seal the bottle.

    d. Record your name and the date on the bottle.

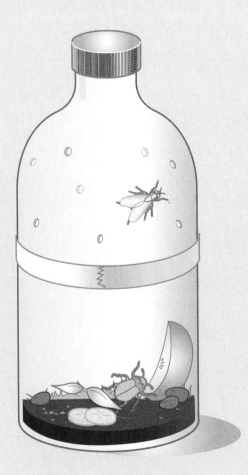

6. Complete the following tasks based on the ecosystem you created.

   a. List the abiotic parts of your ecosystem.

   b. List the biotic parts of your ecosystem.

   c. Describe any limiting factors that might affect the populations in your ecosystem. Explain how the factors might increase or decrease populations.

   d. Draw a food web.

   e. Create an energy pyramid.

7. Predict what will happen over the next two weeks in your ecosystem.

8. Create a data table to record observations of your ecosystem. Make sure you make observations on a regular basis.

   Record your observations in your data table. Note where the organisms are in your ecosystems and the behavior of any animals you included. You should record any changes you see.

9. At the end of the two weeks, your teacher will give you instructions on how to dispose of your ecosystem and bottle.

10. Write a paragraph or two describing what happened in your ecosystem.

   a. Write a few sentences that compare your prediction (from Step 7) to the results you described. Was your prediction similar to or different from the end result?

   b. Answer the analysis questions.

Analysis Questions _ _ _ _ _ _ _ _ _

1. How did the plants and animals in your ecosystem get the energy they need? Write a sentence for each organism.

2. How did the choices you made affect what your ecosystem looked like after two weeks? For example, did some organisms survive and others die? Write a paragraph describing what happened as a result of what you selected for your ecosystem.

3. What worked well in your ecosystem and what would you do differently? Describe at least one thing that worked well and one thing you would change.

Chapter 20

People and Ecosystems

What resources do you use every day?
Where do they come from?

Taking Stock

You have been learning about the different organisms that live on Earth. Processes occur in these organisms that are necessary to support all life. Think about the importance of photosynthesis to all living things. You also learned that organisms are connected to each other through food webs. They also depend on the environment for their survival. Have you ever thought about how you are connected to ecosystems?

When you woke up this morning, what was the first thing you did? Maybe you sat down to eat breakfast, or you brushed your teeth. To get ready for your day, you had to get dressed and gather the things you need for school. You use a lot of things in one day. Have you ever thought about all the things you use in a day and where those things come from? Like animals, humans have basic needs such as food, water, and shelter. But we also rely on things we make such as computers and cars.

What resources do we rely on to live?

Work with your classmates and list things you use every day. Then think about where the resources came from that make up the things you use.

Figure 20.1
Lakes and rivers often provide the water that people use in cities. Do you know where the water you use every day comes from?

Materials

- notebook paper

Procedure

1. Work with a partner and list all the things you can think of that you use in a day. Write down at least three things in each of the categories shown below.

   a. Food

   b. Shelter

   c. Clothing

   d. Transportation

   e. Communication and entertainment

2. Make a list of all the resources that are necessary to make the things you use. For example, plants supply food and rocks provide building materials. Write down as many resources as you can think of for each category above.

3. Now think about energy as a resource. Write down the different sources of energy that are necessary for the things you use every day. For example, natural gas might be used to heat the water where you live.

4. Share your list with the class. Listen carefully and add any resources that weren't on your list.

Analysis Questions _ _ _ _ _ _ _ _ _ _

1. How do ecosystems provide the resources we use? To answer this question, complete the following tasks.

   a. Make a T-chart on your paper with the headings "biotic" and "abiotic."

   b. Look at your list of resources. Do they come from the biotic (living), or the abiotic (nonliving) parts of an ecosystem?

   c. List the resources under the appropriate heading.

2. Where do you think the resources you use come from? You can write down a description such as a forest or a location such as a city or country. Write down any ideas you have. You can change your answers later as you learn more.

3. Do you think where you live might affect the resources you use? For example, do you think a person living in Africa uses the same resources as a person living in the United States? Explain your answer.

Conclusion

Write complete sentences to answer the focusing question.

Extension Activity

Going Further: Search for an ecological footprint calculator on the Internet and get an estimate of your ecological footprint. An ecological footprint estimates the amount of land area required to produce all the resources you use. Write a paragraph describing what influences the size of your ecological footprint. Then list five ways you could reduce your ecological footprint.

Nature Provides

Ecosystems provide the resources we use. There are different types of ecosystems. Each type is unique and has different conditions and organisms. Forests and grasslands are two examples of major ecosystems of the world. Major ecosystems are also called **biomes**. Each major ecosystem has a distinctive environment and vegetation. The conditions and organisms found in a major ecosystem are determined by its climate. **Climate** is the average weather over a long period of time. For example, forests grow in moist climates. The organisms that live in a forest ecosystem are different than the organisms that live in a drier grassland ecosystem.

When studying ecosystems, it helps to have an idea of where they are located. Perhaps you already know where marine (ocean) ecosystems are located, but you might want to know where to find rain forest or tundra. Would you know where to find these ecosystems?

What resources are found in different ecosystems?

Color a map of the major ecosystems found in North and Central America to begin to answer this question. Then you will work in teams to describe the climate and organisms found in each type of ecosystem.

Materials

- Handout, *Ecosystems of North and Central America*
- crayons or colored pencils (gray, green, yellow, orange, and blue)
- Handout, *Major Ecosystems*
- notebook paper
- Reading, *Major Ecosystems*, page 563

Procedure

1. Get the *Ecosystems of North and Central America* handout from your teacher. Notice the dashed lines showing the boundaries between countries. The solid lines are the boundaries for major ecosystems.

2. Neatly label these places on your handout:
   a. Canada
   b. Caribbean Islands
   c. Central America
   d. Mexico
   e. United States, including labels for Alaska and Hawaii

3. Put a star on the handout to indicate where you live.

4. Look at the northern part of Canada and Alaska. This area is tundra. Color the tundra and the appropriate box in the key a light gray.

5. Locate the forest ecosystems. Color these areas and the appropriate box in the key a light green. Forests are located throughout the following areas:
   a. most of Canada
   b. the eastern United States
   c. the west coast of the United States
   d. western parts of both Mexico and Central America

6. Locate the desert ecosystems. Color these areas and the appropriate box in the key a light yellow. Desert ecosystems are located throughout the following areas:
   a. the western United States
   b. northern Mexico

Figure 20.2

Do any of these ecosystems look similar to where you live?

7. Locate the grassland ecosystems. Color these areas and the appropriate box in the key orange. Grasslands are located throughout the following areas:

   a. the central United States

   b. a small area in southern Canada

8. Locate the tropical rain forest ecosystems. Color these areas and the appropriate box in the key a dark green. Tropical rain forests are located throughout the following areas:

   a. Hawaii

   b. Caribbean Islands

   c. the eastern edges of Mexico and Central America

9. Locate the fresh water ecosystems. Color these areas and the appropriate box in the key a dark blue. Freshwater ecosystems are located in the following areas:

   a. the Great Lakes

   b. the three lakes shown in Canada

   c. rivers, small lakes, ponds, and wetlands are also freshwater ecosystems but they are too small to show on the handout

10. Locate the marine (ocean) ecosystems. Color these areas and the appropriate box in the key a light blue.

11. Get into a team of 3 students and listen as your teacher assigns you an ecosystem.

12. Read about your ecosystem in the reading, *Major Ecosystems*.

13. Fill in the *Major Ecosystems* handout using the information from the reading. Be prepared to share what you learned with the rest of the class.

14. Participate in a class discussion about all the types of ecosystems. Write down information about each ecosystem type as you listen to the other teams.

15. Answer the analysis questions.

Major Ecosystems

Tundra

Tundra is the coldest of all the ecosystems. In the winter, the average temperature is below freezing. In the summer, the temperature can range from 37°F (3°C) at night up to 60°F (16°C) during the day. Winters are very long and summers are very short. There are two main types of tundra: Arctic tundra and alpine tundra. Most of the tundra on Earth is found in the Arctic. The Arctic includes the North Pole and extends into parts of North America, Europe, and Siberia. Alpine tundra is found at the tops of very high mountains. Some tundra is found on islands near Antarctica. But most of Antarctica is too cold for plants to grow because it is covered with ice. The plants that grow in both types of tundra include grasses, low shrubs, lichens, and mosses.

Arctic tundra has a harsh climate. In addition to being very cold, tundra is often very desert-like. It receives very little precipitation. Arctic tundra has a layer of permanently frozen soil called permafrost. In the summer, the top surface of the soil thaws. This allows plants to grow and makes the soil very soggy. However, the soil deeper down is still frozen. Trees rarely grow in tundra because a short growing season and frozen soils make it nearly impossible for them to survive. The animals found in Arctic tundra include lemmings, voles, caribou, arctic hares, arctic fox, and wolves. In the summer, warmer temperatures and wet conditions make it an ideal place for insects such as mosquitoes and blackflies. The insects provide food for migratory birds such as terns and shorebirds.

Figure 20.3

These pictures show Arctic tundra (left) along the border of Alaska and the Yukon Territory in Canada and alpine tundra (right) in the Rocky Mountains of Colorado.

Alpine tundra is found at high altitudes anywhere on Earth. For example, the tops of the Rocky Mountains, the Himalayas, and the Andes all have alpine tundra ecosystems. If you look at a world map, you will find that the Himalayas and the Andes are close to the equator. Trees do not grow in alpine tundra. In fact, the edge of the tundra is often called the tree line because trees aren't found beyond a certain point on mountain tops. Several factors prevent the growth of trees. In many alpine areas, the temperatures stay too cold for too much of the year for trees to grow. Temperatures are cold enough for deep snow to persist until late summer at some locations. Sometimes the slopes are so steep and unstable that it is difficult for trees to grow. Strong winds can reduce tree growth because trees can blow over and freeze.

Figure 20.4
This arctic fox (Alopex lagopus) has captured a lemming. The diet of most arctic foxes consists mainly of lemmings, but also arctic hares, insects, berries, and carrion.

Because alpine tundra is found in different parts of the world, the animals living there are very different. Common insects are butterflies, beetles, and grasshoppers. Some of the animals include ptarmigan, grouse, marmots, pikas, llamas, mountain goats, and yaks.

Forest

Forest ecosystems are found in parts of Earth where the climate is very different between seasons. They usually have warm summers and cool winters. There must be enough precipitation to support the growth of trees. The temperature can range from below freezing in the winter to 90°F (32°C) in the summer. The average amount of precipitation varies between regions. It can range from 24 to 60 inches (61 to 150 cm). Regions with similar temperatures but less rainfall usually have grassland ecosystems.

Figure 20.5
These pictures show (top) deciduous trees in the Great Smokey Mountains and (bottom) coniferous trees in Michigan.

Forests are made up of two types of trees: deciduous and coniferous (evergreen). Deciduous trees lose their leaves in the winter. These trees adjust to cold temperatures by going into a dormant state when the leaves fall off. Oaks, maples, cottonwood, willow, and hickory are types of deciduous trees. Coniferous trees keep their leaves, which are usually shaped like needles, all year long. Coniferous trees grow well in areas where it is very cold or dry. Some examples of conifers are spruce, fir, and pine trees. Forests can be deciduous, coniferous, or a combination of both.

Forests contain many other organisms in addition to trees. Plants include shrubs, herbs, ferns, mosses, and lichens. Forests are also home for many animals. Some of those animals include foxes, bears, woodpeckers, skunks, deer, squirrels, mice, sparrows, beetles, centipedes, and snakes. This list just names a few. Can you think of more animals that live in forests?

Figure 20.6

Forests are great places for many birds to build their nests. This photo shows a yellow warbler nest.

Desert

Desert ecosystems are the driest ecosystems on Earth. Deserts are found where there is 20 inches (50 cm) or less of precipitation per year. Desert soils are sandy and rocky and usually have very little organic material. Organic material is material that comes from living organisms. Fewer organisms live in the desert so there is less decayed plant and animal material in the soil. Most deserts are hot and dry and found closer to the equator. These deserts usually have an average temperature around 68°F to 77°F (20°C to 25°C). But temperatures can get as high as

Figure 20.7

The Sonoran Desert is located in southwest Arizona.

Figure 20.8

The fringe-toed lizard burrows into the sand to escape the heat during the hottest part of the day.

120°F (49°C) during the day and drop below freezing at night. Deserts are also found in places like Antarctica and Greenland. In these cold deserts, temperatures range from 28°F to 39°F (-2°C to 4°C) in the winter to 70°F to 79°F (21°C to 26°C) in the summer.

The organisms living in deserts are very specialized. Plants such as cacti store water in their stems to survive. Other plants such as grasses and bushes have very small leaves to conserve water. Some plants have very deep root systems that reach water beneath the soil surface. Most animals cannot store enough water and survive the temperature extremes in deserts. The most common animals are insects such as spiders, beetles, grasshoppers, ants, and scorpions. Reptiles such as snakes, lizards, and tortoise are also common. The mammals that can survive are usually smaller. For example, squirrels, rabbits, badgers, and mice are found in deserts. Birds, such as some species of owls, wrens, and thrashers, also survive there.

Grassland

Grasslands grow in drier areas. These ecosystems have just enough precipitation to support grasses but there is too much rain for a desert ecosystem. Common plants in grasslands are grasses, flowers, herbs, and some shrubs. A few trees grow in grasslands, but they are mostly absent. Since precipitation is inconsistent, grasslands often experience droughts and fires. These events keep large numbers of trees from growing. Grasses survive fires. The dry steams and leaves of grasses burn, but the roots underneath the ground are unharmed. Grasses also die back to their roots when conditions are very cold or dry. When the soil becomes moist again after a fire, winter, or drought, new shoots of grass grow from the roots.

Figure 20.9

This photo shows native plants in an Iowa grassland.

Figure 20.10

Today most fires in grasslands are put out to save houses and other structures. Sometimes fires, such as the one pictured here, are intentionally set in controlled areas. This helps maintain the grassland in its natural state.

Grassland ecosystems cover large areas and are found on every continent except Antarctica. Grasslands are given different names such as prairies, steppes, pampas, and savannas. Grasslands get 20 to 50 inches of rain (50 to 1270 cm) per year. The average temperatures for grasslands are -4°F to 86°F (-20°C to 30°C). The temperatures vary depending on where the grasslands are located. Grasslands in tropical areas are warm all the time and have dry and wet seasons. The savannahs in Africa are an example. Other grasslands are located where temperatures vary from summer to winter. For example, the prairies in North America have freezing temperatures in the winter.

The animals living in grasslands are very diverse and vary with location. All types of insects are common including flies, ants, spiders, beetles, and grasshoppers. In African savannahs, you would also find termites. Snakes, mice, gophers, ground squirrels, badgers, and jack rabbits are all found in grasslands. Birds such as owls, hawks, sparrows, and quail are also common. You might be most familiar with the larger grazing animals that live in grasslands. In prairies you might see bison. In the African savannah, zebras, rhinoceros, gazelles, and elephants are common among many others.

Figure 20.11

Pronghorn antelope like the one pictured here are found in grasslands throughout western North America. This photo was taken in Montana.

Figure 20.12
The rain forest pictured here is located in Costa Rica.

Tropical Rain Forest

Tropical rain forest ecosystems are located near the equator. The average temperature ranges from 68°F to 77°F (20°C to 25°C) and rarely gets higher than 93°F (34°C). It is a hot, moist environment. It rains throughout the year. Total annual rainfall is at least 80 inches (203 cm). Rain forests are found in Central America, the Amazon River basin, parts of Africa and Madagascar, as well as parts of India and Southeast Asia.

Rain forests are known for having distinct layers of trees and other plants. Tall trees form a canopy or ceiling that shades plants on the forest floor. Vines grow into the canopy to reach sunlight. Beneath the tall trees you find smaller trees, ferns, mosses, orchids, vines, and palms. Hundreds of different species of plants can live within only one acre of rain forest. Many of the medicines we use come from rain forest plants.

In addition to plants, thousands of species of animals live in rain forests. Scientists estimate that more than half of plant and animal species on Earth are found in rain forests. Insects are the most abundant animal species. Some of the insects include ants, butterflies, mosquitoes, wasps, and beetles. Interesting birds such as parrots, toucans, peacocks, and cockatoos live in rain forests. Tree frogs, snakes, geckos, and iguanas are also residents. Mammals that live in rain forests include bats, monkeys, ocelots, wild boar, anteaters, and sloths.

Figure 20.13
Red-eyed tree frogs like the one pictured here are found in Costa Rica and other countries in Central America.

Marine

Marine ecosystems cover a large portion of the Earth's surface. About 70 to 75% of the Earth is covered with water and most of the water is found in the ocean. Marine ecosystems are different than ecosystems on land. Temperatures change very slowly in water and are less extreme than on land. Also, water is always available so precipitation is less of a factor. Light is one of the most important factors in water. Even in the clearest water, light can only penetrate 328 feet (100 meters). Most plants and algae only grow where there is light. Other animals are dependent on plants and algae for food.

Figure 20.14
Coral reef off the Cayman Island in the Caribbean.

There are three main types of marine ecosystems: oceans, coral reefs, and estuaries. Oceans cover the largest area and are home to many different animals such as marine worms, clams, crabs, shrimp, fish, shorebirds, whales, dolphins, and sharks. Common plants include algae, seaweed, and kelp. Coral reefs are located in warm, shallow waters near the edge of continents. In addition to coral, organisms such as algae, sea urchins, sea stars, octopus, and many types of fish live in coral reefs. Estuaries are located where freshwater streams mix with ocean water. Estuaries have plants such as marsh grasses, mangrove trees, and seaweeds. You will also find algae, marine worms, crabs, oysters, snails, and fish. Birds are very abundant and include ducks, herons, and shorebirds.

Figure 20.15
Aerial view of an estuary.

Figure 20.16

This satellite image shows the Gulf of Mexico off the coasts of Texas and Louisiana. Notice the green swirls near the coast. The green area is actually large populations of marine organisms, most of which are algae. Scientists call these organisms phytoplankton. Also notice the tan-colored swirls coming off the edge of the land. That is sediment flowing out of rivers such as the Mississippi River.

One of the most abundant and important organisms living in all marine ecosystems is algae. Marine algae supply a large portion of the oxygen in the Earth's atmosphere. The algae undergo photosynthesis just like land plants. So when the algae give off oxygen, they also take in carbon dioxide.

Freshwater

Freshwater ecosystems are found almost everywhere on Earth. They include ponds, lakes, streams, rivers, and wetlands. Temperature has more of an effect in freshwater ecosystems than marine ecosystems. In cold climates, ponds and lakes may have a layer of ice in the winter. Water temperatures in streams can also be very different. Mountain streams and streams in northern latitudes are supplied by melting snow. This water is much colder than the water in steams from other locations.

A variety of organisms live in freshwater ecosystems. Algae and tiny microorganisms such as daphnia and protozoa are found in all freshwater ecosystems. Grasses, trees, bushes, and other plants grow along the edges of ponds, lakes, streams, and rivers. Aquatic plants such as lilies grow in the water. There are many different types of wetlands such as swamps, bogs, and marshes. Trees and bushes such as cypress, gum, and tamarack grow in some. But in others, you will find only find grasses, sedges, rushes, cattail, and other non-woody plants.

Freshwater ecosystems support a variety of animals. Insects such as dragonflies, mosquitoes, beetles, and caddisflies are very common. Invertebrates such as snails, worms, and crayfish are also present. Other animals include frogs, toads, turtles, snakes, fish, muskrats, river otters, and raccoons. Many different birds are attracted to freshwater ecosystems such as ducks, herons, swallows, eagles, and red winged blackbirds.

Figure 20.17

Some freshwater ends up in streams as snow melts. In other places, rainwater collects in lakes, ponds, wetlands, and rivers. Birds, such as the great blue heron pictured here, are often found in wetlands.

Analysis Questions

1. Which ecosystem on the map:
   a. do you live in?
   b. is largest?
   c. is smallest?

2. Which of the major ecosystems do you think has the:
   a. most producers? Why?
   b. biggest consumers? Why?
   c. most decomposers? Why?
   d. harshest abiotic conditions? Why?

3. If you could live in another ecosystem for six months, which one would you choose? Why?

4. Choose one of the major ecosystems and name at least five resources the ecosystem provides.

5. Choose one of the ecosystems and write a paragraph describing how you think humans might affect the ecosystem. Think about how humans might influence the environment or population.

Conclusion

Answer the focusing question using complete sentences.

Harvesting Earth's Bounty

The resources we use are often referred to as natural resources. **Natural resources** are materials that come from the Earth. Examples of natural resources include water, minerals, and fuels such as oil and natural gas. Air and water are also resources.

Why are some resources renewable and others nonrenewable?

How do processes in ecosystems sustain us?

Find the answers to these questions by reading about the natural resources found on Earth. As you read, think about how ecosystems support human life. We need natural resources from various ecosystems to survive. But, in addition, what about the processes that occur in ecosystems? How do processes, such as photosynthesis, help to sustain us?

- Handout, *Natural Resources*
- notebook paper
- Reading, *Natural Resources* (below)

Procedure

1. Get the *Natural Resources* handout from your teacher.
2. Fill in the handout as you read *Natural Resources*.
3. Participate in a class discussion about what you learned.

Natural Resources

Natural resources can be divided into two groups: renewable and nonrenewable. **Renewable resources** are resources that can be renewed or replace themselves. **Nonrenewable resources** can only be depleted, and will eventually run out.

Renewable Resources

Renewable resources are usually available in large amounts and replace themselves within a human lifetime. They are replaced through natural processes. For example, trees are a renewable resource because new trees are always growing as some die. All living things are renewable resources such as bacteria, animals, plants, and crops. Other examples of renewable resources include air, soil, and water.

Living organisms are important renewable resources. Plants supply many resources. The fruits and vegetables we eat are all plants. Forests supply people with wood for fuel, paper, and building materials. Animals also provide us with food. Some of the animals are gathered from the wild such as fish or game animals (deer, elk, ducks, etc.). Other animals such as cows, pigs, and chickens are domesticated but they still rely on nature because the food we feed them comes from nature. You might not think that insects are important to you, but they are. Many insects are crucial to the survival of our crops. Some insects are pollinators and others are predators for pests like aphids. Bacteria and fungi are important because many are decomposers that break down our waste. Can you think of other ways we rely on organisms?

Figure 20.18

Trees are a resource that provides many products such as paper and lumber (top). Bees and other insects are an important resource that pollinates plants. Without pollination, plants wouldn't produce fruits, seeds, or vegetables (middle). Lots of land in the midwestern United States is used for growing food crops (bottom). What kind of ecosystems would you find these natural resources in?

Figure 20.19

Soil is a valuable resource for growing food. It is a home for many decomposing organisms. These organisms break down dead plants and animals and recycle nutrients.

The air you breathe is constantly being recycled by plants. Remember that during photosynthesis plants take in carbon dioxide and give off oxygen. Carbon dioxide is added to the air by plants and animals during cellular respiration. Oxygen and carbon dioxide naturally occur in the air, but sometimes pollutants end up in the air. Examples of pollutants are particles and gases given off by things such as cars and manufacturing plants. Scientists have learned that plants not only add oxygen to the air, they also can help clean the air. Many plants can absorb pollutants in the air and improve the quality of the air we breathe. This is important because air pollution can increase our risk of disease.

Soil is a mixture of broken down rock and decaying plant and animal material. Decomposers are an important part of soil. They break down the matter in dead plants and animals. The nutrients in the soil come from the dead plants and animals. This is important because plants get nutrients they need, such as nitrogen, phosphorus, potassium, calcium, magnesium, and sulfur, through the soil. As the plants grow, they use up these nutrients. Without decomposers to break down dead organisms, soil would eventually run out of nutrients. In fact, farmers often add fertilizer to soil to provide additional nutrients for their crops. Sometimes nutrient rich soil is carried away by erosion. Erosion is the wearing away of soil by water or wind. If too much soil is eroded, the top layer of soil is removed. Then only the lower layers that contain mostly rock particles are left. There are fewer nutrients in the lower layers of soil and it is more difficult for plants to grow. As a result, it is important to protect the soil even though it is a renewable resource.

We use water as a resource in many ways. Each person needs to drink water daily for their body to function properly. However, we also use water indirectly. We use water for bathing, cleaning our households, and for recreation. Farmers often water their crops and water is also necessary for raising livestock. Manufacturers also need water to produce steel, paper, and other products. Water is also a crucial part of every ecosystem because all organisms need at least some water.

Plants in freshwater ecosystems maintain water quality by breaking down waste and pollutants. Wetlands are especially good at purifying water. Wetlands are areas that are flooded for a period of time such as marshes and river floodplains. Plants and soils in wetlands absorb excess nutrients and pollutants. Many wetland plants can also remove toxic substances from industrial waste. To learn how water is recycled naturally in ecosystems, read *The Water Cycle* on page 578.

Although water is a renewable resource, we often alter our water supplies. For example, we build dams to store water and supply energy. Altering rivers helps supply needed resources for towns and cities. However, it also affects the way water flows in rivers and affects fish such as salmon. When human populations increase, we add more waste to rivers. Rivers can also become polluted from runoff from cities or factories. In the United States, waste water is treated before it is released back to waterways. But many countries do not have waste treatments systems and people have difficulty getting clean water.

Nonrenewable Resources

Nonrenewable resources are those that cannot be replaced during a human lifetime. Once the supply of a nonrenewable resource has been used, we cannot get any more of it. Nonrenewable resources include minerals and fossil fuels.

Minerals provide the raw materials for many of the things we use every day. Aluminum, tin, iron, gold, zinc, and silver are all minerals. Geological processes formed these minerals over millions of years. Since minerals are in limited supply, it is a good idea to recycle them. Minerals like aluminum and tin can be melted down and used over and over again in products.

Fossil fuels are formed over millions of years from the remains of plants and animals. Because they are made up of organisms, fossil fuels are sometimes called carbon-based fuels. This is because all organisms contain large amounts of carbon. Examples of fossil fuels include coal, oil, and natural gas. Coal formed where plants accumulated in large amounts, such as in wetlands. Across time, the plant material was buried. After millions of years, chemical changes transformed the plant material into coal. Oil and natural gas formed in coastal areas. Marine organisms were buried under sediment. Over time, high temperatures and chemical changes transformed the marine organisms into oil and natural gas.

Includes sci LINKS.
NSTA

Topic: Renewable and Nonrenewable Resources
Go to: www.scilinks.org
Code: MSLS3e575

Figure 20.20

Coal burning power plants such as the one pictured here provide electricity for many cities. The electricity is produced by a steam turbine that spins a generator. The material you see that looks like smoke is water vapor produced by the steam turbine. Power plants like these produce pollution, but the pollutants are invisible to the naked eye.

Energy Resources

Energy resources can be renewable or nonrenewable. Fossil fuels are nonrenewable sources of energy. Nuclear energy also is nonrenewable. This is because nuclear energy requires a specific type of uranium that is very rare. Examples of renewable energy include hydropower, solar, geothermal, and wind energy.

Right now, we get most of our energy from fossil fuels. Most of the power plants that supply energy for our homes burn fossil fuels to produce electricity. Cars and other vehicles also burn oil that has been converted into fuels such as gasoline and diesel. Because the supply of fossil fuels is limited, it is a good idea to practice energy conservation. For example, turning down the thermostat reduces the amount of energy needed to heat a home. Can you think of other ways to conserve energy? Burning fossil fuels can also potentially cause problems. It releases pollutants. Some of the pollutants include carbon dioxide, carbon monoxide, sulfur dioxide, and nitrogen oxide. These gases can also cause acid rain and smog.

Nuclear energy has advantages and disadvantages. Nuclear power plants will be able to provide power after the supply of fossil fuels is gone. They also don't produce air pollution. However, nuclear power plants produce dangerous waste. This waste gives off radiation which could give people cancer if they were exposed to it.

Renewable energy sources account for only a small amount of the energy that we use. Hydropower creates energy by using water from a river to spin turbines that generate electricity. Solar energy is converted into electricity through solar cells. Geothermal energy is obtained from hot water below the Earth's surface. Steam is used to produce electricity. Wind generates energy by turning windmills. Renewable energy sources are advantageous because they do not produce as much pollution as fossil fuels and nuclear power. The disadvantage of these energy sources is that they are not always available. The Sun doesn't always shine and the wind doesn't always blow. Drought can also reduce the amount of water available for hydropower. Sometimes renewable energy is expensive to develop as well. For example, the materials used in solar cells can be expensive.

Figure 20.21

The United States consumes or uses all the different types of energy sources. This figure shows how much of each type of energy is used.

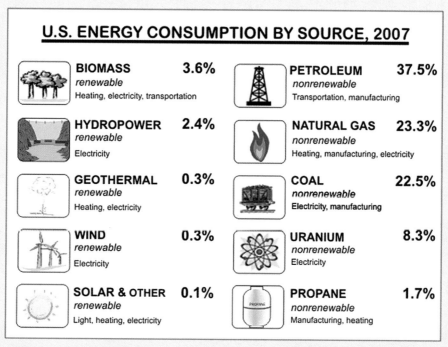

U.S. ENERGY CONSUMPTION BY SOURCE, 2007

BIOMASS 3.6%
renewable
Heating, electricity, transportation

HYDROPOWER 2.4%
renewable
Electricity

GEOTHERMAL 0.3%
renewable
Heating, electricity

WIND 0.3%
renewable
Electricity

SOLAR & OTHER 0.1%
renewable
Light, heating, electricity

PETROLEUM 37.5%
nonrenewable
Transportation, manufacturing

NATURAL GAS 23.3%
nonrenewable
Heating, manufacturing, electricity

COAL 22.5%
nonrenewable
Electricity, manufacturing

URANIUM 8.3%
nonrenewable
Electricity

PROPANE 1.7%
nonrenewable
Manufacturing, heating

Source: Energy Information Administration

The Water Cycle

Did you know that the water we use today is the same water the dinosaurs used more than 100 million years ago? The same water has been evaporating into the air and returning as rainfall since Earth first formed and cooled. Water is not used up and recreated. Instead, it is constantly recycled.

You can imagine water falling to the ground in the form of rain. Then it evaporates and returns to the atmosphere as water vapor. Time and again, it falls and evaporates, falls and evaporates. This continual movement of water is called the **water cycle**. However, water does not always fall as rain. There are other forms of precipitation. Also, water doesn't always evaporate immediately. It often collects on different parts of the Earth.

Earth's Water

The water on Earth is continually moving and exists in different forms. It exists as water vapor, liquid water, and ice. The water on Earth is found in different places. A small amount of the water on Earth is found in the atmosphere. Some of the water collects in rivers and streams. This water eventually ends up in lakes or oceans. Water also collects on ice caps and glaciers. Some water even collects underground. The movement of water from one part of Earth to another occurs through processes such as precipitation and evaporation.

Precipitation

Precipitation is water, in any form, that falls to the Earth's surface. It can be rain, snow, sleet, or hail. All forms of precipitation start as water vapor in the atmosphere. When the water vapor cools, it forms droplets that eventually become heavy enough to fall to the Earth's surface.

Most of the precipitation falls into the oceans and seas, but some falls on the ground. It may run off the surface of the ground into streams and rivers, or it may soak into the ground. The water that collects in the spaces between the grains and sand and particles of dirt in the ground is called groundwater. Water sometimes remains underground for thousands of years. Many communities tap into this groundwater for their water supply.

Evaporation

Water on the surface of the ground absorbs heat from the Sun. If the water is warm enough, it may change from a liquid to a gas and become water vapor. Water vapor returns to the atmosphere. The change from liquid to gas is called **evaporation**. The time that a drop of water spends on Earth varies considerably. A drop that lands on a piece of hot metal will be there for only a few seconds. Whereas, a drop of water that lands on an ice cap might remain there for thousands of years.

Water also travels up and out of the small openings in plants' leaves called stomata. This process is similar to the way that you release water vapor when you breathe. The process of plants releasing water vapor into the atmosphere is called **transpiration**. Scientists have found that a field of corn releases enough water to cover the field with 30 to 40 centimeters of water. When you think about all the plants releasing water into the atmosphere, you can see that plants are an important part of the water cycle.

Analysis Questions

1. Write down at least four examples of natural resources in each category below.

   a. renewable resource

   b. nonrenewable resource

2. Write down at least four examples of energy resources in each category below.

   a. renewable energy

   b. nonrenewable energy

3. Write a one or two sentence definition for renewable resources.

4. Write a one or two sentence definition for nonrenewable resources.

5. List the advantages and disadvantages of both types of energy resources.

   a. renewable energy

   b. nonrenewable energy

6. Choose *one* of the resources below and describe how it is recycled through natural processes.

   a. soil

   b. water

7. List at least two examples of processes in an ecosystem. Think of processes that help support human populations. Try to come up with examples that are different than your answer to question 6.

8. How do human activities affect ecosystems and the resources we use? Write a short paragraph for your answer.

Conclusion

Write two short paragraphs, one in answer to each of the focusing questions.

Extension Activity

Going Further: Do research on what type of energy is used to provide electricity for your home and school. You may need to contact your utility company. Or you could research different ways to conserve energy.

Burning Carbon

We rely on fossil fuels to supply the energy to heat our homes and power our cars. Fossil fuels are made mostly of carbon. Have you ever thought about what happens when we burn that carbon? Maybe you have heard that burning fossil fuels causes global climate change. How can these two things be related? To learn how they are related you first need to know a little about what happens to the Sun's energy as it enters the Earth's atmosphere. You also need to know how carbon dioxide in the atmosphere affects climate.

What affects the temperature on Earth?

How can climate change affect the survival of organisms?

Answer the first question by working with models of the Earth's atmosphere. Then read about what keeps the Earth warm, and climate change to answer the next question. As you read, think about how humans affect the ecosystems on Earth and the organisms that live in them.

Materials

- Handout, *Modeling Earth's Atmosphere*
- large clear container
- lid or plastic wrap
- small beaker
- dropper
- spoon
- baking soda
- vinegar
- Reading, *What Makes Earth Warm?*, page 585
- soil
- light colored material (white sand or gravel)
- thermometer
- clear tape
- ruler
- stopwatch
- lamp
- notebook paper
- graph paper

Procedure

Part A: Modeling the Earth's Atmosphere   *(Work in a group with three or four other students.)*

1. Read the entire procedure for Part A so you understand what you are to do.
2. Create a model of Earth and its atmosphere according to your teacher's instructions. You will create one of the models described below in *Modeling Earth and Its Atmosphere*.

Modeling Earth and Its Atmosphere

- Model the Earth and its atmosphere using a large clear container.
- Model the different surfaces on Earth using water, soil, and light-colored sand.
- Model clouds by covering the large container with a lid or plastic wrap.
- Model carbon dioxide in the atmosphere by adding a heaping spoonful of baking soda to a small beaker. Place the beaker in the large container and then add 6 to 8 drops of vinegar to the baking soda in the beaker. Adding vinegar to baking soda causes a reaction that releases carbon dioxide. You can see the release of carbon dioxide as fizzing.

Read below for instructions on the different ways of modeling Earth and its atmosphere.

Model 1: Light Surface/no carbon dioxide

Fill the bottom of a large container with 3 cm of light-colored sand.

Model 2: Light Surface/carbon dioxide

Fill the bottom of a large container with 3 cm of light-colored sand. Add a heaping spoonful of baking soda to a small beaker and place the beaker on top of the sand. Wait to add 6 to 8 drops of vinegar to the beaker.

5. Get the *Modeling Earth's Atmosphere* handout from your teacher. You will record the data you collect on this data table.
6. Predict which model will be the warmest and which model will be the coolest when placed under a lamp. Record your predictions on the handout.
7. Place a thermometer in your model by taping it to the inside of the large container. Make sure you tape it with the number facing out so you can read it. Also, be careful not to let the end of the thermometer touch any surfaces.

Model 3: Dark Surface/no carbon dioxide

Fill the bottom of a large container with 3 cm of soil.

Model 4: Dark Surface/carbon dioxide

Fill the bottom of a large container with 3 cm of soil. Add a heaping spoonful of baking soda to a small beaker and place the beaker on top of the soil. Wait to add 6 to 8 drops of vinegar to the beaker.

Model 5: Water/no carbon dioxide

Fill the bottom of a large container with 3 cm of room temperature water.

Model 6: Water/carbon dioxide

Fill the bottom of a large container with 3 cm of room temperature water. Add a heaping spoonful of baking soda to a small beaker and place the beaker in the container. Wait to add 6 to 8 drops of vinegar to the beaker.

8. Place a lamp about 15 to 25 centimeters away from your model. Make sure the thermometer is on the opposite side of the lamp. Do not turn on the lamp.

9. Record the beginning temperature in your model on the data table.

10. Have your teacher check your model.

11. Turn on the lamp and record the temperature at 2 minute intervals for 20 minutes or until the temperature levels off. If your model includes carbon dioxide, add the vinegar to the baking soda.

12. Prepare a graph of your results. Time should be on the x-axis and temperature on the y-axis.

13. Participate in a class discussion of the results.

    a. Record the maximum temperature of each of the models on your handout.

    b. Compare your prediction to the results. Explain why you think your prediction was correct or incorrect.

14. Answer analysis questions 1 through 4.

Part B: Climate and Carbon   *(Work with a partner.)*

1. Read *What Makes Earth Warm?* Take turns with your partner reading and summarizing the text.

2. Draw a picture of your model of the Earth's atmosphere from Part A. Use a new sheet of paper and use at least half of the page to draw the model.

3. Add arrows to your drawing that represent light energy or radiation. Indicate whether the light energy was reflected or absorbed. Label the arrows.

4. Add arrows to your drawing that represent infrared radiation. Indicate what happens to the infrared radiation. Did it escape to space or was it absorbed? Label the arrows.

5. Answer analysis questions 5 through 10.

What Makes Earth Warm?

Have you ever thought about what keeps the Earth warm? Earth's climate is dependent on the Sun. However, the Sun is only part of the picture. In order to keep the Earth warm, some of the energy from the Sun must be kept from escaping Earth's atmosphere. In other words, the climate on Earth is determined by the amount of energy received by the Sun and the amount of energy that is trapped by the atmosphere.

Energy

You might be surprised to learn that all objects give off radiation. **Radiation** is energy that is transmitted in the form of rays or waves. **Solar** energy, or solar radiation, is one type of radiation. **Infrared** radiation is another type of radiation. Although you cannot see it, you can feel it as heat. You feel infrared radiation when you put your hand above an electric burner that has been turned off but has not cooled. The warmer the object, the more infrared radiation it gives off. Many things give off infrared radiation. For example, all objects such as furniture, the Earth's surface, and even your body give off infrared radiation (see **Figure 20.22**).

When solar radiation hits Earth's surface, some of it is reflected and some of it is absorbed (see **Figure 20.23**). Solar radiation reflects off of water, land surface, and clouds. Most solar radiation is absorbed by Earth's surface and the oceans. Light colored surfaces such as snow are more likely to reflect solar radiation. Dark surfaces such as bare soil are more likely to absorb radiation. Surfaces that absorb solar radiation warm up. Think about how hot black pavement feels on a sunny day.

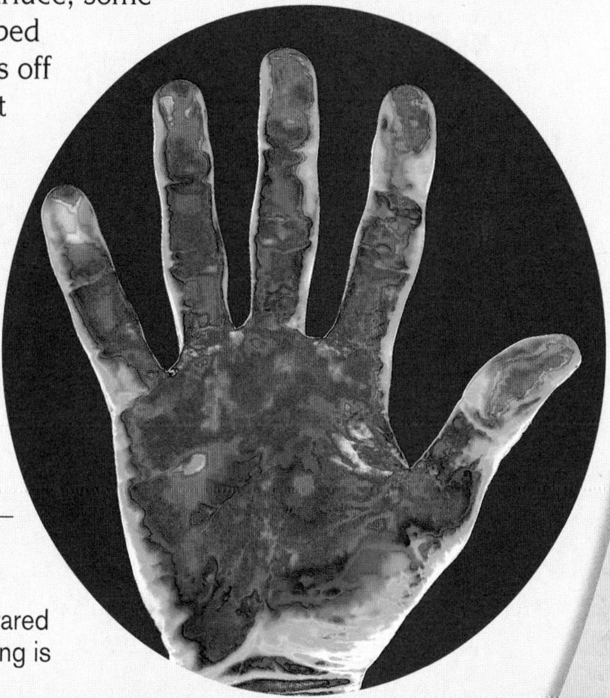

Figure 20.22

This is an infrared image of a hand. Red and yellow areas are warmer and give off more infrared radiation. Blue areas are cooler. Infrared imaging is used by the military for night vision.

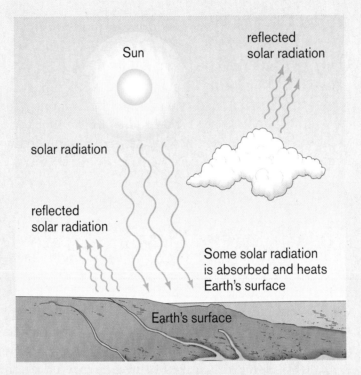

Figure 20.23

Solar radiation is reflected or absorbed. Clouds reflect solar radiation back to space. Land and water can either reflect or absorb it. When solar radiation is absorbed, the land and water get warmer.

Greenhouse Effect

As the soil, rocks, and oceans warm during the day, they give off infrared radiation. The Sun's energy was transformed into heat and given off as infrared radiation. Remember that infrared radiation is invisible to the naked eye. Instead, you feel it as heat.

Certain gases in the atmosphere can absorb infrared radiation. They are only a tiny fraction of the gases in the atmosphere, but they are very important. As the gases absorb the heat from the infrared radiation, they warm up and in turn give off heat as infrared radiation (see **Figure 20.24**). The infrared radiation is given off in all directions. Some of it goes into space and some of it goes to the Earth. The gases that absorb infrared radiation are called **greenhouse gases**. They are called greenhouse gases because like the glass in a greenhouse, they keep the Earth warmer.

Greenhouse gases help heat the Earth's surface by absorbing and giving off infrared radiation. The infrared radiation (heat) coming from greenhouse gases keeps the Earth warmer than it would be without them. In fact, without

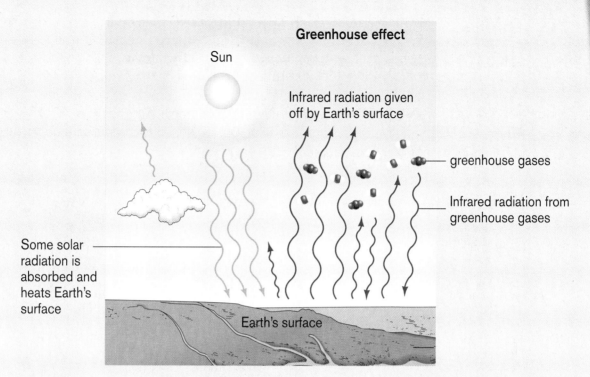

Figure 20.24

Greenhouse gases absorb infrared radiation. In other words, they absorb heat. Some of the heat is lost to space as infrared radiation. Some of the heat warms greenhouse gases. The greenhouse gases give off radiation back to the Earth's surface. This process keeps the Earth warm.

greenhouse gases the temperatures on Earth would be below freezing. This warming of the Earth's surface is called the **greenhouse effect**. The greenhouse effect is a natural process. It is important because without it, the climate on Earth would be too cold for most organisms to survive.

There are several different greenhouse gases. Greenhouse gases include water vapor, carbon dioxide, methane, and nitrous oxide. Water vapor is naturally in the air from evaporation from lakes, rivers, oceans, and other bodies of water. Carbon dioxide is released by organisms as a result of cellular respiration. Carbon dioxide is also released into the air when forests are burned, volcanoes erupt, and fossil fuels are burnt. Methane in the atmosphere comes from burning fossils fuels, burning forests, and from the digestive systems of cows and other organisms that eat plants. Some bacteria give off methane as they breakdown dead plants in wetlands and rice paddies. Nitrous oxide is released by soil bacteria during decomposition.

Too Much of a Good Thing: Global Climate Change

Earth's climate has alternated between warm and cold climates over millions of years. Climate is the average pattern of weather over many years. When Earth's climate has been cold, ice covered much of the Earth. For example, 18,000 years ago ice covered parts of Europe and North America (see **Figure 20.25**). Right now Earth is in a warm period. During warm periods, ice is only found in polar regions such as the Arctic and Antarctica.

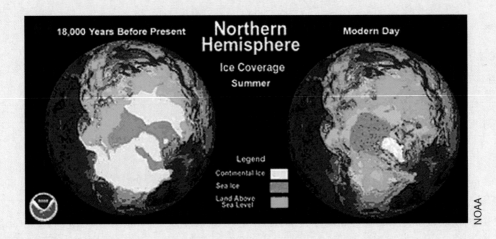

Figure 20.25

The amount of ice covering Earth varies depending on the climate. The image on the left shows the ice that covered Earth 18,000 years ago. Scientists estimate ice coverage using information from samples taken from ocean sediments. The image on the right shows ice coverage today. Notice how there is less ice today.

Although we are already in a warm period, scientists have noticed that temperatures on Earth are increasing faster than they have in the past. This recent change in the Earth's climate is called **global climate change**. The average temperature on Earth has risen by more than 1°F (0.6°C) since 1900 (see **Figure 20.26**). One degree might not seem like much, but usually the temperature on Earth remains fairly stable.

Increasing amounts of greenhouse gases are causing an increased greenhouse effect. As more greenhouse gases are added to the atmosphere, more heat gets trapped and the temperature on Earth can rise. Carbon dioxide is the most prevalent greenhouse gas. Recently, levels in the atmosphere have increased. Carbon dioxide levels in the atmosphere have increased by more than

Includes *sci*LINKS®
NSTA

Topic: Global Climate Change
Go to: www.scilinks.org
Code: MSLS3e588

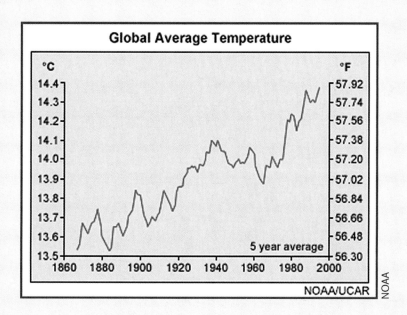

Figure 20.26

The average temperature on Earth has increased by more than one degree since 1900. Notice that before 1900, the temperature stayed relatively the same.

30% since the 1800s. Now carbon dioxide levels are higher than they have been in 400,000 years (see **Figure 20.27**).

Climate change affects the Earth in many ways. Some scientists think that temperatures might continue to rise 2°F to 10°F (about 1°C to 6°C). Scientists are looking at how warmer temperatures in the last 100 years have changed the environment on Earth. They will use what they learn to help them predict future changes.

The amount of sea ice in the Arctic has decreased in the last 50 years. The melting of sea ice could cause changes in ocean currents. Some animals also rely on sea ice. Polar bears use sea ice as a platform for hunting and for migration. Declining sea ice will force polar bears to swim greater distances to find areas to hunt seals. Or they might have to move to colder areas where there is more sea ice.

Water in coastal areas is getting warmer. Many animals such as corals can only live in a certain range of temperatures. In the last 20 years, about 25% of the coral reefs on Earth have died. Pollution and sediment from human activities might also have contributed to the decline.

Scientists also have evidence that warmer temperatures on Earth may affect weather patterns. But there are many factors involved. Warmer temperatures increase evaporation. More water will evaporate into the air from the

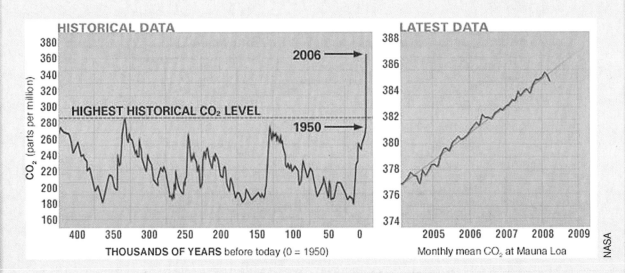

Figure 20.27

The graph on the left shows the amount of carbon dioxide (CO_2) in the Earth's atmosphere during the last 400,000 years. Notice how the amount of CO_2 goes up and down. The amount of CO_2 also stayed below 300 parts per million until 1950. However, levels today are higher than they have been before. Notice how the graph on the right shows that the current amount of CO_2 in the atmosphere is over 380 parts per million. What has changed since 1950?

ocean. Some places might get more rain and others might get less rain. Evaporation can also dry out soils. With warmer temperatures, places that are already dry might become even drier.

Plants and animals are affected by climate change. Most organisms survive best in a particular climate. Plants need certain amounts of precipitation and certain temperatures to live. Animals rely on certain plants for food and shelter. If the climate changes, they have to find other places to live or they might die. Coral reefs and polar bears are just some of the organisms that can be affected. In many areas, the growing season is earlier. In other words, plants begin growing again earlier in the spring. Some insect populations are also emerging earlier in the spring. This might not seem like a problem, but many birds time their migrations to arrive when insects first emerge in spring. If birds don't adjust their migrations earlier, they might not find enough food to survive.

Analysis Questions — — — — — — — — —

1. If it was 100°F outside and you wanted to keep as cool as possible, what color shirt would you wear—white or black? Why?

2. Did the carbon dioxide (baking soda and vinegar) have an effect on the temperature in the model? Why or why not?

3. Which model of the Earth and its atmosphere was the coolest? Write one or two sentences explaining why you think it stayed cooler than the other models.

4. Which model of the Earth and its atmosphere was the warmest? Write one or two sentences explaining why you think it stayed warmer than the other models.

5. What happens to the Sun's energy after it enters Earth's atmosphere?

6. What happens to the infrared radiation that is given off by Earth's surface?

7. Why is Earth's surface warm? Describe the two factors that keep Earth warm.

8. Which of the models (1–6) best represent the greenhouse effect on Earth? You can list more than one. Explain your answer.

9. What is the difference between the greenhouse effect and global climate change?

10. What are some of the effects that scientists think that global climate change have on:

    a. the Earth's environment?

    b. the survival of organisms?

Living in Ecosystems

You have been learning how you are connected to ecosystems. There are different types of ecosystems. Each has a unique set of organisms and unique abiotic conditions. Every type of ecosystem provides a variety of resources that we rely on. Sometimes human activities such as burning fossil fuels, affects the ecosystems we live in. In this activity, you will demonstrate what you have learned by researching how humans affect ecosystems.

What resources do ecosystems provide?

How do human activities affect ecosystems and the resources we use?

Figure 20.28

Can you think of ways you affect ecosystems? Growing a vegetable garden in your yard may have a positive effect on an ecosystem. Pollution from your car or litter on the ground may have a negative effect.

You and a partner will answer these questions by researching one example of how humans affect ecosystems. You will create a poster that displays the information you gather. Then you will present your poster to the class in a short presentation. As your classmates present their posters, listen carefully and take notes.

- library materials
- access to the Internet
- poster board
- colored paper
- markers

- Handout, *Humans and Ecosystems Research Guide*
- notebook paper

Procedure   *(Work with a partner to complete this activity.)*

Part A: Learning about the Effect of Humans on Ecosystems

1. Choose an example of how humans affect ecosystems. Your teacher will provide you with a list of topics to choose from.
2. Write down your topic on a piece of paper.
3. Think of at least two questions you have about the topic. Write your questions on the piece of paper. If you can think of more than two questions, write those down as well.
4. Get the handout, *Humans and Ecosystems Research Guide* from your teacher.
5. Begin researching your topic. Fill in the handout, as you learn about the topic you have chosen.
6. Record information about each reference you use to gather information for your handout. You will need basic information such as the title, date, and type of reference when you make your list of references.

Caution: It is important to keep a list of the references you use and information about where they came from. As you put together your poster, you will be using the work of others. When you present your poster, you must include a list of all your references. By listing your references, you acknowledge the work that others have done. If you fail to acknowledge their work, you are stealing their ideas and presenting them as your own. Using someone else's work without giving them credit is called *plagiarism.*

7. Find information to fill in the five boxes about the topic. If your reference does not have some of this information, leave the section blank. (Another reference might have the information you need).

8. Rate the reference before you put it away. Write comments about how helpful the reference was in the last part of the handout.

9. Repeat Steps 6–8 with at least three different references or until you have all the information you need.

Part B: Preparing a Poster

1. Review the information you collected in Part A and think about how you will present the information you learned on a poster. The list below provides some ideas for pieces of your poster.

   tables

   graphs

   maps or photos

   sketches or drawings

   paragraph summaries

2. Develop a plan for your poster. Do this by deciding how you will arrange the pieces you decided on in the previous step.

3. Check to see that your plan includes the following information.
   a. Affects of the human activity on the ecosystem(s).
   b. Affects of the human activity on the survival of organisms.
   c. Affects of the human activity on the supply of natural resources.
   d. Answers to your questions from Part A, Step 3.

4. Decide which of you is responsible for each piece of the poster.

5. Think of a title for your poster. Your classmates should be able to recognize your topic by the title you choose.

6. Create your poster on a sheet of poster board.

Part C: Presenting your Poster and Being a Good Audience

1. Meet with your partner to decide who will present each piece of your poster.
2. Present your poster to the class.
3. Take notes as other students present their posters. Answer the following questions in your notes.
   a. What is the topic of the poster? In other words, what is the human activity?
   b. What ecosystem or ecosystems are affected by the human activity?
   c. How is the ecosystem affected? How is the survival of organisms affected?
   d. How does the human activity affect the supply of natural resources?
4. Think of at least one question you have about the topic of each of your classmates' posters.

Analysis Questions

1. How do humans affect ecosystems? Give at least three examples from your classmates' posters, not your own.
2. How are renewable and nonrenewable resources affected by human activities? Provide one example for each.
3. Why do our activities affect the survival of other organisms? Choose one organism and explain how our actions make a population more, or less, abundant.
4. What can humans do differently to reduce their effect on ecosystems? List at least three things.

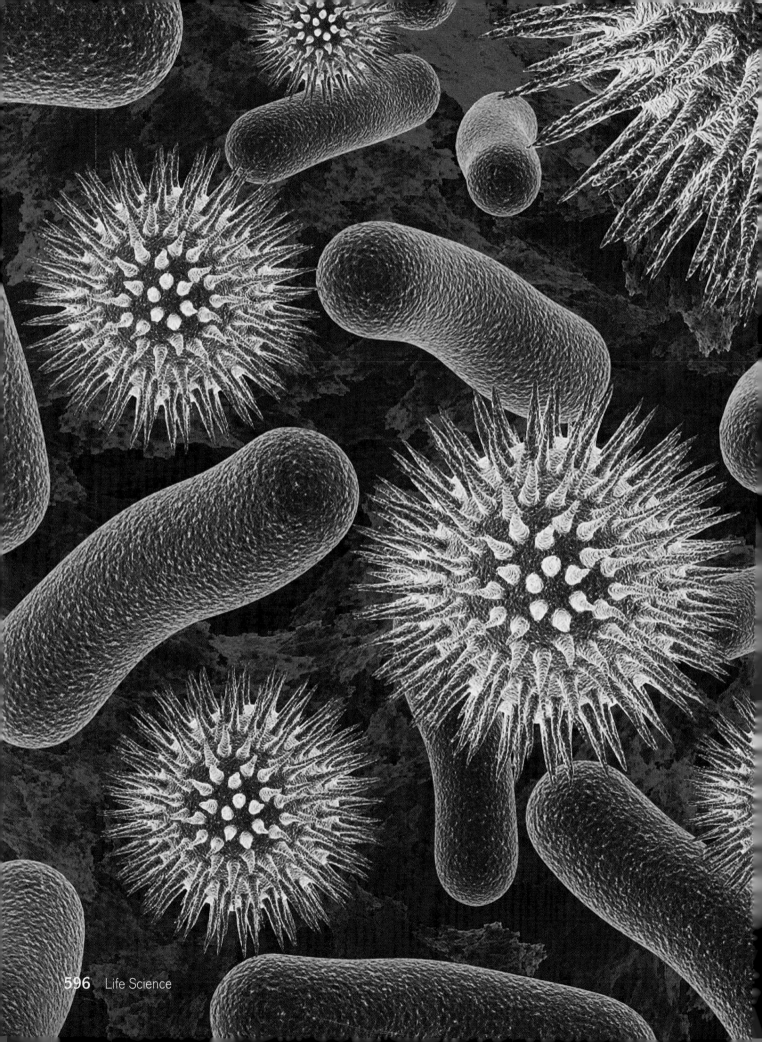

Appendix A: Thinking and Working Like a Scientist

Science is an ongoing search for understanding of the natural world. Scientists use investigations to learn about everything from the structure of the universe to the activities within an atom. Here are some of the things you will do to develop an understanding of life science. You may not do everything mentioned here in every activity, but that is not necessary.

Identify questions that can be answered through scientific investigation.

Scientists are always asking questions about the world around them. In order to start thinking like a scientist, you need to start coming up with questions ... and with ideas about how to answer your questions. You will also want to learn to recognize questions you can answer by doing a scientific investigation. (Some questions are answered more easily by looking through books.)

Make predictions, and design and do a scientific investigation.

Before you do any science investigation, you will often want to **predict** what you think is going to happen. Making a prediction is *not* the same thing as guessing. You should be able to explain your prediction based on what you already know about the subject.

Next you must *follow directions* to complete the activity. These directions might be printed in a textbook or you might write them. There are times when you will need to **come up with a plan** to answer a question you have identified. When writing a plan, anticipate problems you might encounter and be aware of possible sources of error (or, know what could mess up your results). Whether you are doing an activity from the text or one you wrote, you must follow the directions carefully so someone else will know exactly what you did.

Use appropriate tools and techniques to gather, analyze, and interpret data.

Your brain is the best tool you can use to help you understand science. However, there are times when you need other **tools**. These tools can be simple objects such as a ruler, jar, or magnifying lens, or they can be more complex such as a microscope, x-rays, or a computer. The computer can be useful for keeping track of data and analyzing it. You can also go online to tap into libraries, databases, and other sources of information all around the world.

When you actually do the investigation, you need to **record your findings**. You must write down your observations, your data, and/or a description of what happened.

Think critically to explain your data and observations.

Once you have finished an investigation, you are not done. In fact, your work is just beginning! Now you have to figure out what you learned from the investigation. First, look at your data and note anything you know is wrong. Maybe you spilled something or counted something twice. Other problems can be harder to spot. These are all examples of **sources of error**—things that can cause your data to be wrong. You will need to keep these things in mind when you analyze your data.

When you interpret your data, you describe what happened or what you observed. In other words, you give an **explanation based on the evidence** you collected. You may find there is more than one way to interpret your data.

There are many times when you will need to **analyze class data** and not just your own data. It is often difficult, even impossible, to notice patterns when you have only a few sets of results.

Communicate your findings.

It's not enough to do an investigation and interpret your data. You must also be able to **explain to someone else**—either in writing or orally—what you did and what you learned. You should also be able to **listen carefully** when someone else is reporting their findings. If there's anything you don't understand, plan to ask questions that will provide you with the answers you need.

Use results of one investigation to think of new questions and more investigations.

When you do a science investigation, you are looking for answers to questions, but you may end up with even more questions than when you started. That happens to scientists all the time. In fact, that's usually how they get their ideas for their investigations. At first you may have trouble thinking of additional questions but, once you get used to doing so, you will probably be able to think of many more questions.

As you do the activities in this course, you will have plenty of opportunities to practice these skills and abilities.

Appendix B: Designing a Data Table

When you do science activities, you need to record your findings. You must write down your observations, measurements, and/or your description of what happened. All of these things are called **data**. When you have quite a bit of data it's useful to organize it in a data table. An example of a data table is shown in **Figure A.1**.

There will be times when you need to design your own data table. Here are some steps to follow when you do so.

1. *Read the whole procedure* so you know what you will be doing.

2. *List all the types of data you will be collecting.* There is often one column for each type of data. By reading the data table in **Figure A.1**, you can see that three types of data are to be collected—total numbers, numbers of kinds, and appearance (the column heading says to sketch and color).

3. *Count how many times you will collect the data.* This information is often necessary for deciding how many rows you need for your data table. The data table in **Figure A.1** shows that the same data are to be collected twelve times—each time for a different type of organism.

4. *Use this information to sketch a data table.*
   - Put headings on the columns and label the rows.
   - Check your sketch—make sure there will be enough room to write your data.

5. *Draw the data table.* Use a ruler if you need to make straight lines.

6. *Put a title on your data table.*

When you record your data, **be neat**. You will have to understand your notes later on and someone else should be able to look at your data and understand what it says.

The first few data tables you make may not be perfect, but you will get better with practice!

column

Figure A.1

Litter Life			
Organism	Total Number	Number of Kinds	Sketch and color one example from each group. (Be sure to show identifying characteristics
ants			
centipedes and millipedes	cell		
earthworms			
larvae			
slugs			
sowbugs			
spiders			
other critters			
fungi			
decomposing plant material			
decomposing animal material			

row

Appendix C: Nutrients in Foods

Key to Abbreviations and Symbols:

na = nutrient value not available or varies considerably

> = may contain more than this amount

< = may contain less than this amount

Dairy Products & Eggs

	Total Fat g	Saturated Fat g	Unsaturated Fat g	Cholesterol mg	Protein g	Total Carbohydrates g	Sugar g	Fiber g	Calories	Calcium mg	Vitamin C IU
Butter, 1 teaspoon	4	2	2	10	0	0	0	0	35	1	0
Cheese, American, 1 oz slice	9	5	4	25	5	2.4	2.4	0	110	174	0
Cheese, cheddar, 1 oz slice	9	6	3	29	7	0.4	0.4	0	113	204	0
Cheese, cottage, 1/2 cup	5	3	2	20	14	4	0.7	0	120	63	0
Cheese, Swiss, 1 oz slice	9	5	4	25	8	1	0.2	0	110	272	0
Egg, fried, large	7	2	5	211	6.2	0.6	<.5	0	91	26	0
Egg, hard-boiled, large	6	2	4	240	7.2	0.6	<.5	0	88	28	0
Egg, scrambled, large	7	2	5	211	7	1	<1	0	100	47	0.1
Hot chocolate, 1 cup	9.5	5.5	4	33	9	31	27	1	240	291	2
Ice cream, vanilla, 1/2 cup	8	5	3	25	3	15	10.4	0	150	88	0
Milk, chocolate, 1 cup (whole milk)	9	5	4	30	7	28	24	0	210	280	2
Milk, whole, 1 cup	8	5	3	33	8	11	10	0	150	291	2
Milk, low fat (2%), 1 cup	5	3	2	18	8	12	11	0	121	297	2
Milk, skim, 1 cup	0.4	0.3	0.1	4	8.4	12	10.8	0	86	302	2
Yogurt, plain, 1 cup (low fat)	3.5	2	1.5	10	12	16	16	0	140	314	1
Yogurt, strawberry, 1 cup (low fat)	2	1	1	10	9	42	40	<1	190	314	1

Prepared Foods (foods made with ingredients from more than one food group)

	Total Fat g	Saturated Fat g	Unsaturated Fat g	Cholesterol mg	Protein g	Total Carbohydrates g	Sugar g	Fiber g	Calories	Calcium mg	Vitamin C IU
Beans, baked (pork & tomato) 1/2 cup	2	0.8	1	9	7	25	5.5	7	150	69	3
Beef & vegetables stew, 1 cup	14	7	7	40	11	16	3	2	230	18	0
Chili con carne with beans, 1 cup	14	6	8	50	20	36	2	11	350	80	0
Macaroni & cheese, 1/2 cup	7	2.5	4.5	5	8	38	8	<1	250	181	0
Pizza, cheese, 4 oz	17	7	10	35	18	40	6.1	na	380	332	2
Soup, chicken noodle, 1 cup	2	1	1	15	3	9	1	1	70	18	0
Soup, cream of tomato, 1 cup	2	1.9	1.1	12.4	5	26	14.8	1	156	127	8
Spaghetti, meatballs & tomato, 1 cup	11	5	6	20	11	31	10	5	260	36	1.2
Tamale, beef, 7.5 oz	17	7	10	30	8	25	1	1	296	36	3.6

Meats

Meats	Total Fat g	Saturated Fat g	Unsaturated Fat g	Cholesterol mg	Protein g	Total Carbohydrates g	Sugar g	Fiber g	Calories	Calcium mg	Vitamin C IU
Bacon, 3 slices	9	3	6	3	6	0.1	0	0	109	3	0
Beef, roast, 3 oz	8	3	5	86	28	0	0	0	184	11	0
Beef liver, 3 oz	7	2	5	410	23	7	0	0	184	9	23
Bologna, 1 slice	8	3	5	19	3	0.7	0.6	0	90	2	0
Cod, fried, breaded, 4 oz	14	4	10	35	10	21	3	0	250	18	0
Chicken, fried, 4 oz	14	4	10	99	35	2	0.5	0	279	12	0
Frankfurter, 2 oz	17	7	10	34	6	2	1	>.1	182	54	0
Ham, baked, 3 oz	6	2	4	48	19	0.7	0.7	0	140	11	0
Hamburger patty, 4 oz	18	7	11	75	20	0	0	0	210	10	0
Meat loaf, 4 oz	13	6	7	30	10	10	3	1	190	54	0
Pork chop, 3 oz	14	5	9	78	25	0	0	0	235	9	0
Sausage, 2 links	19	6	13	45	7	0	0	0	200	2	0
Tuna, oil packed, 3 oz	20	2	18	38	20	0	0	0	225	7	0
Tuna, water packed, 3 oz	1.3	0.3	1	38	20	2	0	0	100	7	0

Grains

Grains	Total Fat g	Saturated Fat g	Unsaturated Fat g	Cholesterol mg	Protein g	Total Carbohydrates g	Sugar g	Fiber g	Calories	Calcium mg	Vitamin C IU
Bagel	1.1	0.2	0.9	0	7	38	3	2	195	9	0
Biscuit, baking powder	5	1	4	0	2	12	1.2	na	100	34	trace
Bread, white, 1 slice	1	0	1	0	2	10	1	0	60	19	trace
Bread, whole wheat, 1 slice	1	0	1	0	3	12	1	2	70	22	trace
Cornbread, 2 1/2" x 3"	6	1.5	4.5	35	3	25	6	1	110	18	na
Cornflakes, 3/4 cup	0.1	na	na	0	2	24	1.5	0.7	110	0	0
Crackers, graham, 4 crackers	1.5	0	1.5	0	1	11	3.5	0.5	60	3	0
Crackers, saltines, 5 crackers	1	<1	<1	0	1	11	0	<1	60	3	0
Hominy grits, 1/2 cup	1	0	1	0	2	18	0	>.5	90	0	0
Noodles, egg, 1-1/4 cups	2.5	1	1.5	70	8	40	2	2	210	na	na
Oatmeal, 1/2 cup	2.3	0.3	2	0	4	18	0	3	99	11	0
Pancake, 3 cakes each 4" diameter	4	1	3	16	7	50	4.2	2	253	58	trace
Rice, 3/4 cup, cooked	0	0	0	0	3	35	0	1	160	10	0
Roll, frankfurter	2	0	2	0	3	20	0.3	1	100	30	trace
Roll, hamburger	2	0	2	0	4	22	0.3	2	115	30	trace
Roll, french	1.2	0.2	1	0	4	19	1	<1	100	24	trace
Toast, white, 1 slice	1	0	1	0	2	13	0.8	1	70	19	trace
Tortilla, corn, 6" diameter	0.5	<.25	<.25	0	0.5	8	0	0.5	40	9	0
Waffles, 3 1/2" x 5 1/2", 2 pieces	8	1.5	6.5	25	5	32	2	1	220	270	trace

Fruits and Vegetables

Fruits and Vegetables	Total Fat g	Saturated Fat g	Unsaturated Fat g	Cholesterol mg	Protein g	Total Carbohydrates g	Sugar g	Fiber g	Calories	Calcium mg	Vitamin C IU
Apple, medium	0.5	0.1	0.4	0	0.3	15	10	3	70	10	6
Applesauce, 1/2 cup	0	0	0	0	0	24	21.1	2	90	5	1
Apricots, dried, 5 halves	0.1	0	0.1	0	0.6	10	7	2	41	12	3
Asparagus, 1/2 cup	0.3	0	0.3	0	2	4	1.4	2	22	13	16
Banana, medium	1	0	1	0	1	28	17.8	3	120	10	12
Beans, green, 1/2 cup	0	0	0	0	1	4	1.1	1	16	31	8
Beans, lima, 1/2 cup	0	0	0	0	6	18	3	4	80	40	14
Beans, refried, 1/2 cup	1.4	0.5	0.9	0	8	24	3	7	135	5	na
Broccoli, 1/2 cup	0.3	0	0.3	0	2	4	0.7	2	22	68	70
Cabbage, 1/6 head, 1/2 cup	0.3	0	0.3	0	0.8	3	1	2	16	30	17
Cantaloupe, 1/2 cup cubed	0.2	trace	0.2	0	0.7	7	7	0.6	29	13	32
Carrots, 1/2 cup cooked	0.1	0	0.1	0	1	10	3.4	2	41	24	4
Carrot sticks, 5" carrot	0.1	0	0.1	0	0.7	9	5	3	40	21	4
Cauliflower, 1/2 cup	0.3	0	0.3	0	1	3	1.2	2	14	13	33
Celery sticks, 8" stalk	<.1	0	<.1	0	0.5	2	0.4	1	10	22	5
Corn, 1/2 cup	<1	0	<1	0	2	15	2.3	3	60	4	3
Corn, 5" ear	1	0	1	0	3	19	4.9	3	83	4	11
Grapefruit, pink, 1/2 medium	0.1	0	0.1	0	0.6	12	7.3	>.3	46	19	45
Grapes, 1 cup	0	0	0	0	0.6	16	15.1	1	56	9	3
Greens, 1/2 cup mustard, kale, turnip	0.2	0	0.2	0	0.8	3	2.2	2	14	104	36
Lettuce, 1/6 head, 1/2 cup	0.1	0	0.1	0	0.5	0.7	0.5	0.5	4	15	5
Okra, 4 pods, 1/2 cup	0.3	0.1	0.2	0	2	8	1.2	3	34	39	9
Orange, medium	0.2	0	0.2	0	1	15	12.5	3	62	54	66
Orange juice, 6 oz	0	0	0	0	1	20	19	0.5	80	11	56
Peaches, canned, 1/2 cup	0	0	0	0	0	23	22	1	90	5	4
Peanut butter, 2 tablespoons	16	2	14	0	9	6	1.2	2	190	20	0
Peanuts, dry roasted, 1 oz	14	2	12	0	7	6	1.3	3	160	27	0
Pear, medium	1	0	1	0	1	25	17.4	4	100	13	7
Peas, green, 1/2 cup	0.2	0	0.2	0	4	13	4	2	67	15	10
Pineapple, 2 slices in juice (1 cup)	0	0	0	0	0	15	13	2	60	19.5	15
Potato, baked, medium	0.1	0	0.1	0	3	27	3	2	120	13	28
Potatoes, boiled, 2 small	0	0	0	0	2	22	0	2	100	36	9
Potatoes, french-fried, 22 pieces	21	8	13	0	3	50	0.6	2	400	13	18
Potatoes, mashed, 1/2 cup	6	2	4	20	2	18	0	1	134	23	10

Continued

Fruits and Vegetables

	Total Fat g	Saturated Fat g	Unsaturated Fat g	Cholesterol mg	Protein g	Total Carbohydrates g	Sugar g	Fiber g	Calories	Calcium mg	Vitamin C IU
Potato, sweet, 1/2 medium, 4 oz	0.1	0	0.1	0	2	28	20.4	3	117	22	12
Raisins, 1-1/2 oz	0.1	0	0.1	0	1	33	30	2	130	18	0
Squash, summer, 1/2 cup	0.3	0.1	0.2	0	0.8	4	1.4	1	18	26	12
Strawberries, 1 cup	0.6	0	0.6	0	0.9	11	8.6	4	45	16	44
Tomato, 1/2 medium	0.3	0	0.3	0	0.8	4	2	1	19	13	23
Tomato juice, 6 oz	0	0	0	0	1	7	5.3	1	30	12	25
Watermelon, 1 cup	0.7	0	0.7	0	1	13	13	0.6	51	14	14

Sweets and Desserts

	Total Fat g	Saturated Fat g	Unsaturated Fat g	Cholesterol mg	Protein g	Total Carbohydrates g	Sugar g	Fiber g	Calories	Calcium mg	Vitamin C IU
Cake, angel food, 1 slice (no frosting)	0	0	0	0	3	31	23	0	140	0	0
Cake, devil's food, 1 slice	13	2.5	10.5	55	3	35	21	1	270	20	trace
Candy bar, milk chocolate, 1 oz	9	na	na	5	2	16	15.2	na	150	65	trace
Chocolate syrup, 2 tablespoons	0.4	0.3	0.1	0	0.9	24	19.4	<1	103	6	0
Cookie, chocolate chip, 3" diameter	4	3	1	10	1	9	4	0	80	na	na
Cookie, gingersnap, 5 cookies	7	2	5	0	1	22	1	10	160	0	0
Cookie, oatmeal, 3" diameter	3	1	2	0	1	12	6.6	na	80	na	na
Cookie, sugar, 3" diameter	3	1	2	0	0.5	9.5	4	0.5	70	0	0
Custard, baked, 1/2 cup	5	na	na	80	5	24	24	na	160	148	trace
Donut, plain cake type	14	3	11	0	4	26	4.2	na	240	13	trace
Gelatin dessert, 1/2 cup	0	0	0	0	2	19	10.8	trace	80	0	0
Jelly, grape, 1 tablespoon	0	0	0	0	0	13	12.4	0	48	4	1
Pie, cherry, 1/6 of 9" pie	12	2	10	na	3	48	19	1	300	na	na
Pudding, chocolate, 1/2 cup	4	na	na	15	5	28	24.8	1	160	135	na
Sherbet, orange, 1/2 cup	1.5	1	0.5	5	1	26	21	0	120	54	2.4
Soft drink, cola, 12 oz can	0	0	0	0	0	41	41	0	150	0	na
Sugar, 1 teaspoon	0	0	0	0	0	4	4	0	15	0	0
Sweet roll, Danish pastry, 2 oz	12	6	6	35	4	25	11	<1	220	18	trace

Miscellaneous

	Total Fat g	Saturated Fat g	Unsaturated Fat g	Cholesterol mg	Protein g	Total Carbohydrates g	Sugar g	Fiber g	Calories	Calcium mg	Vitamin C IU
Coffee, black, 3/4 cup	0	0	0	0	0.2	0.7	0	0	4	3	0
Mayonnaise, 1 tablespoon	11	2	9	5	0	0	na	0	100	1	0
Popcorn, plain, 3 cups (airpopped)	>1	na	na	0	2	15	0	4	60	1	0
Potato chips, 15-20 chips, 1 oz	10	na	na	0	1	15	0	1	150	3	8
Salad dressing, French, 1 tablespoon	15	2.5	12.5	0	0	5	5	na	160	na	na

Appendix D: Understanding Science Words

Through the centuries, as people learned more and more about all the sciences, including the life sciences, they had to create words to express their new discoveries. Often these "word-makers" put together two or three older words from Greek and Latin to make names for new knowledge, new inventions, new medicines, and new science concepts. Some of the same root words, prefixes, and suffixes have been used over and over again.

If you know the Greek and Latin roots, you will be able to interpret words that you see for the first time. The root words, prefixes, and suffixes listed here are often used in science. Many of them may already be familiar to you.

Root	Meaning	Examples
a-, an-	*not, without*	**a**biotic: not living
		anaerobic: without oxygen
aero-	*air*	**aero**bic: with oxygen
		aerospace: atmosphere and space beyond
ant-, anti-	*against, opposed*	**anti**body: a molecule that fights against a foreign substance in body
aqua	*water*	**aqua**rium: a tank where water animals and plants live
		aquatics: sports performed in water
arthr-, arthro	*joint*	**arthro**pod: invertebrate with jointed limbs
audi-	*hear*	**audi**tory nerve: nerve conducting messages from ear to brain, allowing a person to hear
		auditorium: large room where audience hears lectures, plays, concerts
bi-	*two*	**bi**valve: having two valves
		bisect: cut in two
bio-	*life*	**bio**logy: study of life
carbo-	*carbon*	**carbo**hydrate: substance made of carbon and water
cardi-, cardio-	*heart*	**cardi**ology: study of the heart
		cardiogram: record of heart action
carni-	*meat, flesh*	**carni**vore: meat eater
cerebro-	*brain*	**cerebr**um: largest part of human brain
		cerebral: involving the brain
-cide	*kill*	insecti**cide**: substance that kills insects
circu-	*circle, ring*	**circu**late: to go around continuously
corpus	*body*	**corp**se: body, usually dead
cyto	*cell*	**cyto**plasm: the "plasm" in a cell but outside the nucleus
den, dent	*tooth*	**dent**ist: doctor who treats teeth
		denture: artificial teeth

derm	*skin*	**derm**atologist: doctor who treats skin
		dermatitis: disease of the skin
-ectomy	*surgical removal of organ*	tonsil**ectomy**: removal of tonsils
		append**ectomy**: removal of appendix
epi-	*above, over*	**epi**dermis: top layer of skin, over the dermis
		epicenter: surface of earth directly above an earthquake
erythro-	*red*	**erythro**cyte: red blood cell
gastro-	*stomach*	**gastr**ic juice: fluids produced in the stomach
-gram	*something written or drawn*	cardio**gram**: record of action of heart
		tele**gram**: message sent by wire
hemo-	*blood*	**hemo**globin: substance in red blood cells
		hemorrhage: heavy bleeding
herb-	*leafy plant*	**herb**ivore: organism that eats plants
hydro-	*water*	**hydro**gen: combines with oxygen to produce water
		hydroelectricity: electric energy converted from running water
hyper-	*excessive*	**hyper**tension: high blood pressure
		hyperactive: excessively active
-itis	*inflammation*	arthr**itis**: inflammation of joints
		appendic**itis**: disease of appendix
-logy	*study of*	cardio**logy**: study of the heart
		bio**logy**: study of living things
macro-	*very large*	**macro**cosm: a large system, universe
-meter	*a measure, tool for measuring*	milli**meter**: one-thousandth of a meter
		thermo**meter**: tool for measuring heat
micro-	*very small*	**micro**cosm: very small system
mort-	*death*	**mort**ality: death, death rate
		mortal: subject to death
neurv-	*nerve*	**neur**on: nerve cell
		nervous: high strung, jittery
omni-	*all*	**omni**vore: animal that eats all foods, from both plants and animals
ova	*egg*	**ova**ry: female gland that produces eggs
		oval: egg shaped
ped	*foot*	**ped**estrian: person who is walking
		pedometer: device for measuring distance walked
pesti-	*pest*	**pesti**cide: chemical used to kill pests

photo	*light*	**photo**synthesis: process in which green plants use the energy from light to make carbohydrates from carbon dioxide and water
pneum-	*breath, air*	**pneum**onia: inflammation of lungs that affects breathing
post-	*after*	**post**natal: occurring after birth
		postmortem: occurring after death
pre-	*before*	**pre**natal: occurring before birth
		predict: to state what may happen before it happens
pulmo-	*lung*	**pulmo**nary artery: artery from heart to lung
		pulmonary vein: vein from lung to heart
-scope	*device for viewing*	tele**scope**: device for viewing distant objects
-sect	*cut, divide*	dis**sect**: cut apart
		section: a part of a larger whole
syn	*together*	**syn**thesis: coming together of parts to form a whole
		photo**syn**thesis: process in which green plants use the energy from light to make carbohydrates from carbon dioxide and water
terra	*land*	**terra**rium: small enclosure with soil where plants are grown and small land animals may live
		terrestrial: living on land
therm-	*heat*	**therm**ometer: device for measuring heat
		thermos bottle: container to keep liquids warm
-vore	*devour*	herbi**vore**: animal that eats plants
		voracious: exceedingly hungry
zoo	*animal*	**zoo**logy: study of animals
		zoo: public park where animals are shown

1. What small words make up the following words? What do they mean?
   a. cytology
   b. microscope
   c. arthritis
2. Why don't we need the word **cardiectomy**?

Glossary

abiotic Nonliving.

abstinence Not having sexual intercourse.

aerobic exercise (uh-ROH-bic EKS-er-size) Exercise where the heart and muscles are kept working at a steady pace such that they do not run out of oxygen.

AIDS A disease of the immune system caused by a virus known as HIV; HIV is an STD.

air sac A tiny balloon-like sac that fills up with air in the lungs; each lung has millions of air sacs; place in lungs where gas exchange of carbon dioxide and oxygen takes place; also called alveoli.

allele (uh-LEEL) A form of a gene.

amino acid Any one of 20 different small molecules that combine into chains to form protein molecules.

amnion A fluid-filled membranous sac that surrounds the growing baby in the uterus.

amniotic fluid The fluid in the amnion that cushions the growing baby from bumps and jars while in the uterus.

amoebas One-celled protists that are often used in experiments to study cells.

antibiotics A drug that helps fight infection by killing or blocking the growth of bacteria.

antibody A specialized protein that can recognize foreign substances and help the body fight disease.

anus The opening at the end of the rectum where feces leave the body.

aorta The main artery that carries oxygen-rich blood from the left ventricle of the heart to all parts of the body; largest blood vessel in the body.

appendicitis (uh-PEN-duh-CY-tis) A painful infection and swelling of the appendix, which requires that the appendix be removed.

appendix The small, thin pouch near the beginning of the large intestine; the appendix is not an important part of the digestive system.

artery A blood vessel that carries blood away from the heart to all parts of the body and the lungs.

auditory nerve The nerve that carries messages from the ear to the brain.

bacteria The simplest forms of living things, which are single cells that may be round, rod shaped, or shaped like a spiral.

bar graph A type of graph that shows data using long or short bars.

biceps (BY-seps) The muscle of the upper arm that helps bend the arm at the elbow; biceps pair with triceps.

bile A liquid produced by the liver that helps break fat into small droplets.

biome One of the major types of ecosystems.

biotic Living.

bladder The muscular pouch that stores urine.

bone marrow Soft tissue found inside bones; site of blood cell production; also called marrow.

bowel movement The removal of feces from the large intestine.

bronchi (BRAHN-ky) Two tubes leading from the trachea to the lungs; the single form of bronchi is bronchus.

capillary A very tiny blood vessel that connects small arteries with small veins to form a network throughout the body.

carbohydrate A nutrient that includes fiber, starch, and sugars; major source of energy in the diet of animals.

carbon monoxide A colorless and odorless toxic gas; one of the gases in cigarette smoke that can cause the body to not get enough oxygen.

cardiac About the heart; cardiac muscle is the muscle that makes up the heart.

carnivore A consumer that feeds on other animals.

cell division The process of one cell dividing to form two new cells.

cell membrane The cell structure that surrounds the cytoplasm and regulates what enters and leaves the cell.

cell The smallest living level of organization; a cell is the basic unit of life.

cell wall The cell structure that gives a plant cell its shape and provides support for the entire plant.

cellular respiration The cellular process that breaks down food and oxygen to produce carbon dioxide, water, and energy.

cerebellum The part of the brain underneath the cerebrum that coordinates muscle movement.

cerebrum The largest, most obvious structure of the brain that controls thinking, decision making, learning and sensory perception.

cervix (SIR-vix) The firm muscular ring that surrounds the opening at the lower part of the uterus.

chlamydia (kluh-MID-ee-uh) A disease of the reproductive system caused by bacteria; chlamydia is an STD.

chlorophyll A green pigment in leaves that absorbs the light energy required for photosynthesis; found in the chloroplasts of green plants.

chloroplast A structure in plant cells that contains chlorophyll; the place where photosynthesis takes place.

cholesterol An animal fat that is found in egg yolks, cream, cheese, lard, and red meats; some cholesterol is needed by the body to function properly but high levels in the blood have been associated with heart disease.

chromosome The structure in cells made up of a DNA molecule and proteins; contains genes.

cilia The fine hair-like structures on the outside of some cells; located in the bronchi and trachea of the respiratory system and inside the fallopian tubes of the female reproductive system.

circulatory system The body system made up of the heart, blood, and blood vessels responsible for transporting small molecules, such as nutrients and oxygen, to every cell in the body.

circumcision The surgical removal of the foreskin.

classification Sorting organisms into groups based on their characteristics.

climate The average weather in a region over a long period of time.

clitoris The firm, sensitive bump of tissue where the two inner labia meet in the female genitals.

cochlea (KAHK-lee-uh) The snail-shaped structure in the inner ear that has receptor cells that send messages to the brain when they detect sound.

cold sore A painful blister that forms around the mouth and is caused by a virus.

concentration The measure of the number of molecules in a certain amount of liquid; high concentration of molecules has more molecules than the same amount of liquid with a low concentration.

conclusion A summary of the results of an experiment written as an agreement or disagreement with the hypothesis based on evidence.

connective tissue A tissue made up of stretchy fibers that hold body parts together; connective tissue holds the skin in place.

consumer An organism that gets its food by consuming, or eating, instead of making its own food.

contraceptives (KAHN-truh-SEP-tivs) Chemicals or devices used to prevent pregnancy.

contract To shorten a muscle and cause movement.

controlled variable A variable in an experiment that is kept the same to allow changes in the other variables to be clearly seen.

cornea The tough, transparent outer layer that protects the front of the eye.

cranium The part of the skull that protects the brain.

cytoplasm A watery gel that contains all the molecules and structures found within a cell.

daughter cells The cells that result from cell division during mitosis or meiosis.

decomposer An organism that breaks down dead plant and animal material.

decomposition The process of breaking down dead plant and animal material and releasing it to the environment so that it can be used by other organisms.

dependent variable The variable in an experiment that is responding or reacting to the change in the independent variable; often plotted on the y-axis of a line graph or scatter plot.

dermis The layer of the skin beneath the epidermis.

diaphragm (DY-uh-FRAM) The sheet of muscle just below the lungs that contracts and flattens to allow air to fill the lungs.

dichotomous key (dy-COT-uh-mus KEY) A tool scientists use to find the identity of items in the natural world, such as plants, animals, or rocks; it is a series of choices that lead you to the correct name of the item.

diffusion The process of moving from a place of high concentration to a place of low concentration.

digestion The process of breaking down food and making its nutrients small enough for cells to use.

digestive system The body system made up of the stomach, intestine, and other organs responsible for breaking down food.

distance vision The sense of sight that allows you to focus on nearby and faraway objects.

DNA The molecule that makes up chromosomes; also known as deoxyribonucleic acid.

dominant Being an allele that masks the characteristics of another allele, the recessive allele.

doubling time The time it takes for a population of living things or cells to double.

drug Any substance that can affect the normal activities of cells or body systems.

ear canal The tube in the outer ear that funnels sound to the eardrum.

ear The sense organ of hearing.

eardrum The thin membrane that separates the outer ear from the middle ear.

ecology The study of the living and nonliving parts of the environment.

ecosystem All the populations living together and the nonliving environment in a particular area.

egg A cell produced by a female that can join with a sperm cell to form a developing baby; female gamete.

ejaculation An emission of semen from the penis.

endocrine system The body system made up of glands responsible for producing hormones.

endurance How long a muscle can work before it tires.

energy pyramid A diagram that shows the amount of energy transferred in feeding relationships.

environment All the living and nonliving things in a particular place; a living thing's surroundings.

enzyme A special type of protein that controls chemical reactions in the body.

epidermis The surface layer of the skin. (p. 112)

epididymis (EP-uh-DID-uh-mus) The tube that lies along the back side of each testis, where sperm finish maturing.

epiglottis The flap of tissue that folds down to cover the opening to the trachea during swallowing.

erection A stiffening of the penis. (p. 380)

esophagus (ee-SOF-uh-GUS) A muscular tube that carries food from the back of the mouth to the stomach.

estrogen (ES-struh-JEN) A female hormone made by the ovaries.

Eustachian tube (yoo-STAY-shun TOOB) A tube that extends from the middle ear into the throat and that keeps pressure equal on both sides of the eardrum.

evaporation The change of a liquid to gas.

fair test An experiment that tests only one variable at a time.

fallopian tube A long structure that extends from each ovary to the uterus.

fat A nutrient used by the body as a source of energy; high-fat foods include nuts, cheese and butter; layer of cells that keeps the body warm.

feces (FEE-sees) The solid wastes that are left over after food has been completely digested; feces are stored in the large intestine.

fertilization Joining of an egg and sperm cell.

fiber A carbohydrate molecule that is made up of many sugars hooked together and that the human body cannot break apart; high-fiber foods include raw vegetables, fruits, and whole-grain breads.

field of view What you see when you look through a microscope.

fluid A liquid; fluid protects the brain and spinal cord.

food chain A diagram showing how organisms are linked by what they eat; organisms are connected by arrows showing the direction of energy transfer.

food web A series of interrelated food chains.

foreskin The sheath of skin that encloses the penis.

fraternal twins Twins that result when two eggs are fertilized, each twin developing from its own egg and sperm.

fungus (plural: fungi) A single-celled or multicellular organism usually made up of long, skinny strands living mostly in soil or dead matter; yeasts, molds and mushrooms are fungi.

gallbladder The organ beneath the liver that stores bile until it is needed.

gamete A sperm or egg cell.

gastrocnemius (GAS-truh-NIM-ee-us) The muscle in the back of the lower leg attached to the heel by the Achilles tendon; calf muscle.

gene A sequence of bases in DNA that provide the cell with "instructions" on how to operate; basic unit of heredity.

genetic About or because of genes.

genetic trait A trait inherited from parents.

genetics The study of how traits are passed from parents to off-spring.

genitalia The reproductive structures on the outside of the body; also called genitals.

genitals The reproductive structures on the outside of the body; also called genitalia.

glans The end of the penis.

global climate change A recent change in Earth's climate in which temperatures are increasing faster than they have in the past.

gluteus maximus (GLOO-tee-us MAX-uh-mus) The largest and heaviest muscle in the body, located in the buttocks.

greenhouse effect Warming of the Earth's surface when gases absorb infrared radiation in the atmosphere.

greenhouse gas A gas that absorbs infrared radiation in the atmosphere.

hamstring The group of muscles in the back of the thigh.

hemoglobin A large molecule in red blood cells that binds to oxygen.

herbivore A consumer that eats plants.

herpes A disease that causes painful sores on the penis or vagina; herpes is an STD.

histogram A type of graph that shows how often (frequency) data fall in ranges or intervals.

homologous chromosomes A pair of chromosomes having the same type and arrangement of genes with each coming from a different parent.

Human papilloma virus (HPV) A disease that causes genital warts in both males and females and can cause cervical cancer in females; HPV is an STD.

hymen A membranous tissue that may partially cover the vaginal opening.

hypothesis (hy-POTH-uh-sis) A tentative explanation for an observation that can be tested by further observations or experiments.

identical twins Twins that develop from the same egg and sperm.

immune Already having the antibodies needed for fighting a certain disease.

immune system The body system made up of lymph nodes, special blood cells, and other organs responsible for fighting infection.

independent variable The variable in an experiment that is deliberately changed, or manipulated; often plotted on the x-axis of a line graph or scatter plot.

infertility Inability to have children.

infrared radiation Heat energy that is transmitted in the form of rays.

inner ear The delicate part of the ear inside the skull that is filled with fluid.

inner labia The thinner folds of tissue just inside the outer labia of the female genitals.

invertebrate An animal without a backbone.

iris The colored part of the eye made up of muscles and connective tissue that controls the size of the pupil.

karyotype A picture of chromosomes under a microscope that is arranged in a particular order.

labor The time of contractions of the uterus that begins the birth process.

larynx (LAIR-inks) The firm structure at the top of the throat called the voice box or Adam's apple.

left atrium The chamber of the heart that blood enters from the lungs and that blood leaves as it is pumped into the left ventricle.

left ventricle The chamber of the heart that blood leaves as it is pumped out to the rest of the body.

lens The part of the eye that focuses light, and sits behind the pupil.

life cycle The series of stages that an organism passes through from birth to death.

limiting factor A factor that limits population size.

liver The organ beneath the lungs on the right side of the body that produces bile.

lower vena cava The large blood vessel that blood drained from the lower part of the body passes through as it enters the right atrium in the heart.

lymph node A small bean-shaped structure that stores and releases white blood cells; lymph nodes are located throughout the body and in clusters in the neck, armpits, abdomen, and groin.

marrow The fat cells and cells that produce blood cells inside bone; also called bone marrow.

medulla The part of the brain that connects the cerebellum to the spinal cord and that monitors basic body functions, such as heart rate, blood pressure, and body temperature.

meiosis (my-O-sis) A type of cell division that involves the formation of gametes that have half as many chromosomes as the original cell.

membrane A thin structure that protects and separates; a tough membrane filled with fluid surrounds and protects the brain and spinal cord.

menopause Time in a woman's life when the menstrual cycle comes to an end, usually between the ages of 40 and 55.

menstrual period The nearly monthly loss of several tablespoons of blood and mucus from the vagina; also called menstruation.

menstruation The nearly monthly loss of several tablespoons of blood and mucus from the vagina; also called menstrual period.

middle ear The part of the ear made up of three tiny bones that bridge the space between the eardrum and the inner ear.

mitochondria (MY-toh-CON-dree-uh) The cell organelle that provide the cell with energy through cellular respiration; the single form of mitochondria is mitochondrion.

mitosis The process during cell division in which the nucleus of a cell divides, resulting in two nuclei, each of which has the same number of chromosomes as the original cell.

multi-factorial About genes that are influenced by both genetic and environmental factors.

muscular system The body system made up of muscles that is responsible for movement.

natural family planning A way of preventing pregnancy by not having intercourse when the woman is ovulating; also called rhythm.

natural resource A resource that comes from the Earth; water, minerals, and oil are natural resources.

nerve cell A specialized cell that carries messages from one part of the body to another.

nervous system The body system made up of the brain, spinal cord, nerves, and receptor organs responsible for coordinating body functions and responses to the environment.

nicotine The addictive chemical in cigarettes.

nonrenewable resource A resource that is being depleted at a rate far faster than it can be formed by nature and can eventually run out.

nose The main entrance to the respiratory system.

nucleus (NOO-clee-us) The cell structure that directs the cells' activities; contains the cell's genetic material.

omnivore A consumer that eats both plants and animals.

optic nerve The nerve that carries messages from the retina to the brain.

organ A group of tissues that carries out a specific function; for example, the heart is an organ.

organelle A tiny structure in a cell; organelles do the work of the cell.

organism A living thing; plants, animals, viruses, and bacteria are examples of organisms.

orgasm The pleasurable sensation that accompanies an ejaculation in males and occurs when pleasurable sensations in the genitals reach a peak in females.

outer ear The part of the ear that you can see.

outer labia A pair of fat-filled folds that enclose and protect the other parts of the female genitals.

ovaries The main structures in the female reproductive system that release egg cells and produce hormones.

ovulation Rupturing and release of an egg as part of the female menstrual cycle.

pacemaker The area of specialized nerve tissue built in the right atrium of the heart that sets the pace of muscle contractions in the heart.

pancreas (PAN-cree-us) The organ that releases pancreatic juices into the small intestine that neutralize stomach acid and help digest foods.

parent A mother, father, or other adult that cares for and raises a child.

penis The main structure of male genitals located in front of the scrotum, used to transfer semen to the female; also used to carry urine to the outside of the body.

photosynthesis (FOH-toh-SIN-thuh-suhs) A process by which green plants use light energy to convert carbon dioxide and water into sugars and oxygen.

placenta A mass of tissue in the uterus that nourishes a growing baby.

plasma A yellowish fluid that is mostly water and that makes up more than half of the blood.

platelet A piece of cell membrane that is carried by blood and that stops bleeding by plugging holes in injured blood vessels.

population A group of organisms of the same species living in the same area.

precipitation Water in any form that falls to the Earth's surface; rain, snow, sleet, and hail are forms of precipitation.

predator An animal that captures other animals for food.

prey An animal that is captured and eaten by a predator.

producer An organism that produces its own food through photosynthesis.

prostate The largest of the glands that produce fluid released in the vas deferens in the male reproductive system.

protein A molecule that provides structure in the body or acts as an enzyme and controls chemical reactions in the body; high-protein foods include grains, nuts, beans, egg whites, milk, and meat.

puberty (PYOO-bur-tee) The time of change and rapid growth when a person's body changes from that of a child to that of an adult.

pubic hair (PYOO-bic HAIR) The hair near the genitals that first appears during puberty.

pulmonary artery The large blood vessel that oxygen-poor blood from the right ventricle passes through as it enters the lungs.

pulmonary vein A large blood vessel that oxygen-rich blood from the lungs passes through as it enters the left atrium.

pulse point A place on the body where you can feel a "wave" of blood as it passes.

pulse rate The number of pulses of blood per minute.

Punnett square A table that can help predict the traits of a child by showing all the possible combinations of alleles from parents.

pupil (PYOO-puhl) The black circular opening in the center of the eye that lets light into the eyeball.

quadriceps (KWAH-druh-seps) The group of muscles in the front of the thigh.

radiation Energy that is transmitted in the form of rays or waves.

receptor cell A specialized nerve cell that receives a particular kind of signal, such as information about blood sugar level or light.

receptor The end of a specialized nerve in the skin that lets the skin feel things.

recessive Being an allele that is masked by the characteristics of another allele, the dominant allele.

rectum The last part of the large intestine where wastes are stored until they leave the body.

red blood cell A small disk-shaped cell that carries oxygen and carbon dioxide in the blood.

reflex A movement of the body in response to a message from the spinal cord that lets the body react quickly to danger.

renewable resource A resource that can be renewed naturally or can replace itself.

reproductive system The body system responsible for creating new life.

respiration The cellular process that breaks down food and oxygen to produce carbon dioxide, water, and energy; also called cellular respiration; the process of breathing where the lungs take in oxygen and give off carbon dioxide is also called respiration.

respiratory system The body system made up of the lungs, trachea and other organs, responsible for exchanging gases.

retina The layer of specialized receptor cells that lines the back of the eyeball and that changes a light image into a nerve message.

rhythm A way of preventing pregnancy by not having intercourse when the woman is ovulating; also called natural family planning.

rib A flat, curved bone that makes up the rib cage, which encloses the lungs.

right atrium The chamber of the heart that blood enters as it comes from the rest of the body.

right ventricle The chamber of the heart that blood enters from the right atrium and that blood leaves as it is pumped into the lungs.

ringworm A skin problem caused by a fungus; ringworm causes scaly, round, itchy patches on the skin.

Robert Hooke The late 1600s scientist who was the first to describe cells using a microscope.

saliva A watery liquid made by the salivary glands in the mouth that contains enzymes that help digest food

saturated fat An animal fat that is solid at room temperature, like the white, greasy parts of uncooked beef; saturated fats have straight molecules.

scavenger A consumer that eats dead animals or plants.

scrotum The loose sac of skin behind the penis that contains the testes.

semen A whitish fluid produced by the male reproductive system that is a mixture of sperm, nutrients, and water.

semicircular canal A tube attached to the cochlea in the inner ear that helps the body keep its balance.

seminal vesicles A pair of glands that produce fluid released in the vas deferens in the male reproductive system.

sense organ An organ that has receptor cells that allow the body to take in information about the environment.

sex chromosomes The chromosomes X and Y that determine a person's sex; a female has two X chromosomes, while a male has one X chromosome and one Y chromosome.

sexually transmitted disease A disease that is passed from one person to another during sexual activity; also known as an STD.

side vision The sense of sight that allows you to see objects that are not directly in front of you.

skeletal muscle A muscle that attaches to bone.

skeletal system The body system made up of bones and other tissues that is responsible for providing support.

skeleton The full set of bones in the body; they provide the support system of a body.

skull The bones in the head that protect the brain and form the jaw.

small intestine The organ of the digestive system that is a coiled tube found in the lower abdomen and that absorbs nutrients from food.

smooth muscle A muscle in an artery, intestine, or other body organ.

solar About or from the sun.

sound wave A wave of air molecules that transmits sound.

sperm A cell produced by a male that can join with an egg to form a developing baby; male gamete.

sphincter (SFINK-tur) A muscular ring; a sphincter at the bottom of the esophagus opens to let food into the stomach; a sphincter at the bottom of the bladder keeps urine from leaking out.

spinal cord The group of nerves that carry messages to and from the brain. (p. 295)

spleen The organ on the left side of the body above the stomach that filters blood and removes wastes.

starch A carbohydrate molecule that is made up of many sugars hooked together and that animals can break apart for energy; high-starch foods include potatoes, rice, corn, and breads.

STD A disease that is passed from one person to another during sexual activity; also known as a sexually transmitted disease.

stomach The sturdy muscular pouch on the left side of the body that stores, churns, and mixes foods and begins to digest protein.

strength The amount of work that a muscle can do in a single try.

sugar The smallest type of carbohydrate molecule; table sugar, or sucrose, is one kind of sugar.

system A group of organs that work together.

tapetum The layer of cells behind the retina in the eye of an animal that reflects light back onto the retina.

tendon A structure that attaches muscle to bone.

testable question A question that can be proven or disproven through a scientific investigation.

testes The paired, egg-shaped organs of the male reproductive system that produce sperm and hormones; the single form of testes is testis; also called testicles.

testicles The organs of the male reproductive system that produce sperm and hormones; also called testes.

testosterone (tes-TOS-tur-OHN) A male hormone made by the testes.

tissue A group of similar cells that have a similar function.

tonsil A lymph node in the back of the throat.

total magnification How many times the object you are viewing through a microscope is enlarged; to calculate total magnification, multiply the magnification of the two lenses.

trachea (TRAY-kee-uh) The stiff tube that allows air into the lungs; also called the wind-pipe.

trait A characteristic of a living thing.

trans fat A fat made by "hardening" vegetable oils to make them solid; trans fats were unsaturated fats that have been made saturated.

transpiration The process of plants releasing water vapor into the atmosphere.

triceps (TRY-seps) The muscle of the upper arm that helps straighten the arm; triceps pair with biceps.

umbilical cord A long structure filled with blood vessels that connects the growing baby to the placenta.

unsaturated fat A plant fat that is liquid at room temperature, like an oil; unsaturated fats have bent molecules.

upper vena cava The large blood vessel that blood drained from the upper part of the body passes through as it enters the right atrium.

urea A waste product formed when proteins are used by the body.

ureter One of two long tubes that carry urine to the bladder.

urethra A muscular tube that lets urine leave the body; in males, it transports sperm from the vas deferens out of the body through the penis.

urethral opening The opening through which urine (and, in males, semen) passes; in males, the urethral opening is at the end of the penis; in females, the urethral opening is directly behind the clitoris.

urinary system The body system made up of the kidneys, bladder, and urethra responsible for removing waste molecules from blood.

urine A waste fluid formed as kidneys filter the blood.

uterus The part of the female reproductive system where a baby develops and grows.

vaccination (VAKS-uh-NAY-shun) Injecting harmless (or dead) microorganisms into a person or animal so that it will not be infected by a more dangerous one.

vacuole (VAC-yoo-OHL) The cell structure that is a membrane-lined sac that stores nutrients or waste products.

vagina The flexible "tube" that extends from the uterus to the outside of the body.

vaginal opening The opening to the vagina in the female genitals.

variability The ways in which people differ because of their genes.

variable Something that can change from one experiment to another.

varicose vein A vein that is swollen and bluish due to a valve in the vein that does not function properly.

vas deferens The tube in males that carries sperm away from the epididymis and to the urethra; also called the vas.

vegetarian A person who does not eat meat.

vein A blood vessel that carries blood to the heart.

vertebral column The stack of bones of the back that protects the spinal cord; also called the backbone.

vertebrate An animal with a backbone.

villi The millions of fingerlike projections on the inside of the intestine that absorb nutrients; the single form of villi is villus.

virus Very small particle that can only reproduce in living cells and cause disease; the common cold, flu, measles, and mumps are caused by viruses.

voyageur A man who worked for a fur company transporting goods and people to and from remote areas.

vulva The external genitals of the female.

wart A skin problem caused when a virus causes skin cells to divide rapidly and form a rough bump.

water cycle The continual movement of water on Earth as it rains, evaporates, returns to the atmosphere as water vapor, and then rains again.

wet dream An ejaculation that takes place during sleep.

white blood cell A blood cell that fights disease as part of the immune system.

withdrawal A way of preventing pregnancy by removing the penis from the vagina before ejaculation; withdrawal is risky and can lead to pregnancy.

zygote The first cell of a growing baby formed when the egg and sperm meet during fertilization.

Credits

Unit Openers:

1: Copyright © 2009 JupiterImages Corporation; **2:** Carol and Mike Werner/Visuals Unlimited, Inc.; **3:** Dr. Dennis Kunkel/Visuals Unlimited, Inc.; **4:** Dr. Dennis Kunkel/Visuals Unlimited, Inc.; **5:** Carol and Mike Werner/Visuals Unlimited, Inc.; **6:** Copyright © 2009 JupiterImages Corporation; **7:** Copyright © 2009 JupiterImages Corporation; **8:** © U.S. Fish and Wildlife Service; **Pg. 596:** Copyright © 2009 JupiterImages Corporation

Chapter 1

Opener Richard Herrmann/Visuals Unlimited, Inc.; **Pg. 4** Science VU/Meiji/Visuals Unlimited, Inc.; **Pg. 6** Science VU/Meiji/Visuals Unlimited, Inc.

Chapter 2

Opener Science VU/Visuals Unlimited, Inc.; **Figure 2.3** (illustration) Nucleus Medical Art/Visuals Unlimited, Inc. (heart muscle) Dr. Dennis Kunkel/Visuals Unlimited, Inc.; (cardiac muscle w/nuceli) Dr. John D. Cunningham/Visuals Unlimited, Inc.; (human heart) L. Bassett/Visuals Unlimited, Inc.; **Figure 2.4** Science VU/Zeiss/Visuals Unlimited, Inc.; **Figure 2.5** (top) Dr. David Phillips/Visuals Unlimited, Inc.; (bottom) Wim van Egmond/Visuals Unlimited, Inc.; **Figure 2.6** (Human lymphocyte, nucleus) Dr. Gopal Murti/Visuals Unlimited; (animal cell) Nucleus Medical Art/Visuals Unlimited, Inc.; (motor neuron) Wim van Egmond/Visuals Unlimited, Inc.; **Figure 2.7** Dr. Henry Aldrich/Visuals Unlimited, Inc.; **Figure 2.8** Mark Schneider/Visuals Unlimited, Inc.; **Pg. 24** Copyright © 2009 Chew Kin Yan. Used under license from Shutterstock, Inc.; **Pg. 28** Copyright © 2009 Linda Muir. Used under license from Shutterstock, Inc.

Chapter 3

Opener Ralph Hutchings/Visuals Unlimited, Inc.; **Figure 3.1** MedicalRF.com/Visuals Unlimited, Inc.; **Figure 3.2** Copyright © 2009 JupiterImages Corporation; **Pg. 36** © 2009 Trumm. Used under license from Shutterstock, Inc. **Figure 3.3** © Georg Gerster/Photo Researchers, Inc.; **Figure 3.4** (skeleton) Charles Melton/Visuals Unlimited, Inc.; (illustration) Science VU/Visuals Unlimited; **Figure 3.6** (polar bear) Fritz Polking/Visuals Unlimited, Inc.; (snake) Marty Snyderman/Visuals Unlimited, Inc.; **Figure 3.8** (cat, pigeon, fish skeleton) Scientifica/Visuals Unlimited, Inc; (complete human skeleton) Ralph Hutchings/Visuals Unlimited,

Inc.; **Figure 3.9** Scientifica/Visuals Unlimited, Inc.; **Figure 3.10** © blickwinkel / Alamy; **Figure 3.11** Ralph Hutchings/Visuals Unlimited, Inc.; **Figure 3.12** Carol and Mike Werner/Visuals Unlimited, Inc.; **Figure 3.13** Nucleus Medical Art/Visuals Unlimited, Inc.; **Pg. 49** (rhino) Adam Jones/Visuals Unlimited, Inc.; (great egret) Arthur Morris/Visuals Unlimited, Inc.; (alligator) Charles Melton/Visuals Unlimited, Inc.; (treefrog) Joe McDonald/Visuals Unlimited, Inc.; (largemouth bass) Reinhard Dirscherl/Visuals Unlimited, Inc.; **Figure 3.15** Scientifica/Visuals Unlimited, Inc.; **Figure 3.17** Scientifica/Visuals Unlimited, Inc.; **Figure 3.18** Charles McRae/Visuals Unlimited, Inc.; **Figure 3.19** Copyright © 2009, Reha Mark. Used under license from Shutterstock, Inc.; **Figure 3.20** Dr. Barry Slaven/Visuals Unlimited, Inc.; **Figure 3.21** (normal hand) Charles McRae/Visuals Unlimited, Inc.; (child's hand) Kevin and Betty Collins/Visuals Unlimited, Inc.; **Pg. 59** Copyright © 2009 JupiterImages Corporation

Chapter 4

Opener Copyright © 2009 JupiterImages Corporation; **Pg. 63** (chicken) Derrick Ditchburn/Visuals Unlimited, Inc.; (chicken wing) © Tom Pantages; **Pg. 65** Wally Eberhart/Visuals Unlimited, Inc.; **Figure 4.4** MedicalRF.com/Visuals Unlimited, Inc.; **Pg. 76** Copyright © 2009 JupiterImages Corporation; **Figure 4.8** Copyright © 2009 JupiterImages Corporation, **Figure 4.9** NASA; **Figure 4.10** Copyright © 2009 JupiterImages Corporation; **Pg. 79** (chimp) Fritz Polking/Visuals Unlimited, Inc.; (pronghorn) Michael Durham/Visuals Unlimited, Inc.; (ant) Nigel Cattlin/Visuals Unlimited, Inc.; (sooty tern) Tom Ulrich/Visuals Unlimited, Inc.; **Figure 4.12** Copyright © 2009 JupiterImages Corporation; **Figure 4.13** Copyright © 2009 JupiterImages Corporation; **Figure 4.14** Copyright © 2009 JupiterImages Corporation; **Figure 4.15** Nucleus Medical Art/Visuals Unlimited, Inc.; **Figure 4.17** Nucleus Medical Art/Visuals Unlimited, Inc.; **Figure 4.18** Dr. Gladden Willis/Visuals Unlimited, Inc.; **Pg. 90** Copyright © 2009 JupiterImages Corporation

Chapter 5

Opener Copyright © 2009, Stephen Coburn. Used under license from Shutterstock, Inc.; **Figure 5.4** Medicimage/Visuals Unlimited, Inc.; **Figure 5.5** (mushroom) Claire Davies/Gap Photo/Visuals Unlimited, Inc.; (yeast) Science VU/Visuals Unlimited, Inc.; (bread) Doug Sokell/Visuals Unlimited, Inc.; (fruiting

sporangia) Dr. Dennis Kunkel/Visuals Unlimited, Inc.
Figure 5.6a Dr. Dennis Kunkel/Visuals Unlimited, Inc.;
Figure 5.6b Dr. Dennis Kunkel/Visuals Unlimited, Inc.; **Figure 5.7** Dr. Dennis Kunkel/Visuals Unlimited, Inc.; **Figure 5.8** Copyright © 2009 JupiterImages Corporation; **Figure 5.9** Scientifica/Visuals Unlimited, Inc.; **Figure 5.10** Science VU/Visuals Unlimited, Inc.; **Figure 5.11** Medicimage/Visuals Unlimited, Inc.; **Figure 5.12** dieKLEINERT/doc-stock/Visuals Unlimited, Inc.; **Pg. 113** Pegasus/Visuals Unlimited, Inc.; **Pg. 115** (gila monster) Joe McDonald/Visuals Unlimited, Inc; (whale) Masa Ushioda/Visuals Unlimited, Inc.; **Figure 5.13** dieKLEINERT/doc-stock/Visuals Unlimited, Inc.; **Figure 5.15** John Birdsall/Visuals Unlimited, Inc.; **Figure 5.16** Dr. Ken Greer/Visuals Unlimited, Inc.; **Figure 5.17** Adam Jones/Visuals Unlimited, Inc.

Chapter 6

Opener Scientifica/Visuals Unlimited, Inc.; **Figure 6.2** Dr. David Phillips/Visuals Unlimited, Inc.; **Figure 6.3** F. P. Williams, Jr./Visuals Unlimited, Inc.; **Figure 6.4** Nucleus Medical Art/Visuals Unlimited, Inc.; **Pg. 128** Copyright © 2009, Andy Piatt. Used under license from Shutterstock, Inc.; **Pg. 129** Copyright © 2009 JupiterImages Corporation; **Figure 6.5** Science VU/CDC/Visuals Unlimited, Inc.; **Figure 6.6** Copyright © 2009 JupiterImages Corporation

Chapter 7

Opener Inga Spence/Visuals Unlimited, Inc.; **Figure 7.1** Sydney Folz/Visuals Unlimited, Inc.; **Pg. 140** (apple) GAP Photos/Visuals Unlimited, Inc.; (locust) Nigel Cattlin/Visuals Unlimited, Inc.; (elephant) Wendy Dennis/Visuals Unlimited, Inc.; **Figure 7.4** Scientifica/Visuals Unlimited, Inc.; **Figure 7.6** Scientifica/Visuals Unlimited, Inc.; **Pg. 147** Stephen Lang/Visuals Unlimited, Inc.; **Figure 7.8** Copyright © 2009 JupiterImages Corporation; **Figure 7.11** Ton Koene/Visuals Unlimited, Inc.; **Figure 7.12** (mushroom) Fritz Polking/Visuals Unlimited, Inc.; (milk) Michael Hart/Docstock/Visuals Unlimited, Inc.; (firefly) William Weber/Visuals Unlimited, Inc.; (red blood cells) Dr. Dennis Kunkel/Visuals Unlimited, Inc.; **Figure 7.13** Scientifica/Visuals Unlimited, Inc.; **Figure 7.14** Scientifica/Visuals Unlimited, Inc.; **Figure 7.20** Scientifica/Visuals Unlimited, Inc.; **Figure 7.21** Scientifica/Visuals Unlimited, Inc.; **Figure 7.22** Scientifica/Visuals Unlimited, Inc.; **Pg. 177** Jack Milchanowski/Visuals Unlimited, Inc.

Chapter 8

Opener MedicalRF.com/Visuals Unlimited, Inc.; **Figure 8.4** MedicalRF.com/Visuals Unlimited, Inc.; **Figure 8.5** Copyright © 2009 JupiterImages Corporation; **Figure 8.6** © Bettmann/CORBIS; **Figure 8.8** Nucleus Medical Art/Visuals Unlimited, Inc.; **Figure 8.9** Nucleus Medical Art/Visuals Unlimited, Inc.; **Figure 8.22** Nucleus Medical Art/Visuals Unlimited,

Inc.; **Pg. 211** Copyright © 2009, Elena Elisseeva. Used under license from Shutterstock, Inc.; **Pg. 215** (picking apples) Paul Debois/Gap Photo/Visals Unlimited; (shrike) Steve Maslowski/Visuals Unlimited, Inc.

Chapter 9

Opener GAP Photos/Visuals Unlimited, Inc.; **Figure 9.2** Hugh Rose/Visuals Unlimited, Inc.; **Figure 9.4** (human epithelium) Dr. Gladden Willis/Visuals Unlimited, Inc.; (illustration) Nucleus Medical Art/Visuals Unlimited, Inc.; **Figure 9.5** Nucleus Medical Art/Visuals Unlimited, Inc.; **Figure 9.6** dieKLEINERT/doc-stock/Visuals Unlimited, Inc.; **Figure 9.8** MedicalRF.com/Visuals Unlimited, Inc.; **Pg. 226** Copyright © 2009 JupiterImages Corporation; **Pg. 231** Ken Lucas/Visuals Unlimited, Inc.; **Figure 9.15** Ralph Hutchings/Visuals Unlimited, Inc.; **Pg. 242** Scientifica/Visuals Unlimited, Inc.; **Pg. 243** Copyright © 2009, Sai Yeung Chan. Used under license from Shutterstock, Inc.; **Pg. 244** Copyright © 2009 JupiterImages Corporation

Chapter 10

Opener MedicalRF.com/Visuals Unlimited, Inc.; **Figure 10.1** Scientifica/Visuals Unlimited, Inc.; **Figure 10.2** L. Bassett/Visuals Unlimited, Inc.; **Figure 10.3** Copyright © 2009 JupiterImages Corporation; **Figure 10.4** Copyright © 2009 JupiterImages Corporation; **Figure 10.5** Copyright © 2009 JupiterImages Corporation; **Pg. 254** (top) Copyright © 2009 JupiterImages Corporation (bottom) Rob Simpson/Visuals Unlimited, Inc.; **Figure 10.6** MedicalRF.com/Visuals Unlimited, Inc.; **Figure 10.8** Nucleus Medical Art/Visuals Unlimited, Inc.; **Figure 10.9** Nucleus Medical Art/Visuals Unlimited, Inc.; **Pg. 261** David Fleetham/Visuals Unlimited, Inc.; **Figure 10.13** MedicalRF.com/Visuals Unlimited, Inc.; **Figure 10.14** Nucleus Medical Art/Visuals Unlimited, Inc.; **Figure 10.16** Dr. Fred Hossler/Visuals Unlimited, Inc.; **Figure 10.19** SIU/Visuals Unlimited, Inc.; **Pg 274** Copyright © 2009, Artsem Martysiuk. Used under license from Shutterstock, Inc.; **Figure 10.20** MedicalRF.com/Visuals Unlimited, Inc.; **Figure 10.22** Ton Koene/Visuals Unlimitd, Inc.; **Figure 10.23** (red blood cell) Dr. Dennis Kunkel/Visuals Unlimited, Inc.; (white blood cells) Dr. Donald Fawcett & E. Shelton/Visuals Unlimited, Inc.; **Figure 10.24** Dr. Stanley Flegler/Visuals Unlimited, Inc.; **Figure 10.27** Dr. Gladden Willis/Visuals Unlimited, Inc.; **Figure 10.28** Nucleus Medical Art/Visuals Unlimited, Inc.; **Figure 10.29** MedicalRF.com/Visuals Unlimited, Inc.; **Figure 10.30** Copyright © 2009, Joseph Dilag. Used under license from Shutterstock, Inc.; **Figure 10.31** MedicalRF.com/Visuals Unlimited, Inc.; **Pg. 285** Copyright © 2009, Pichugin Dmitry. Used under license from Shutterstock, Inc.; **Pg. 287** Copyright © 2009, Laurence Gough, Used under license from Shutterstock, Inc.

Chapter 11

Opener Copyright © 2009 JupiterImages Corporation; **Figure 11.2** Nucleus Medical Art/Visuals Unlimited, Inc.; **Pg. 296** Copyright © 2009 JupiterImages Corporation; **Figure 11.3** Nucleus Medical Art/Visuals Unlimited, Inc.; **Figure 11.5** MedicalRF.com/Visuals Unlimited, Inc.; **Pg. 301** Copyright © 2009 JupiterImages Corporation; **Pg. 306** Steve Maslowski/Visuals Unlimited, Inc.; **Figure 11.8** Nucleus Medical Art/Visuals Unlimited, Inc.; **Figure 11.10** Dr. Frank T. Awbrey/Visuals Unlimited; **Figure 11.11** Loren Winters/Visuals Unlimited, Inc.; **Figure 11.13** Fritz Polking/Visuals Unlimited; **Figure 11.14** Nucleus Medical Art/Visuals Unlimited, Inc. **Figure 11.15** Ralph Hutchings/Visuals Unlimited, Inc.; **Figure 11.16** Ralph Hutchings/Visuals Unlimited, Inc.; **Figure 11.18** Copyright © 2009, Ramona Heim. Used under license from Shutterstock, Inc.; **Figure 11.19** Gerold & Cynthia Merker/Visuals Unlimited, Inc.; **Pg. 321** Kjell B. Sandved/Visuals Unlimited, Inc.; **Figure 11.21** Copyright © 2009 JupiterImages Corporation; **Figure 11.22** Chuck Swartzell/Visuals Unlimited, Inc.

Chapter 12

Figure 12.1 Scientifica/Visuals Unlimited, Inc.; **Figure 12.2** Leslie O'Shaughnessy/Visuals Unlimited, Inc.; **Figure 12.3** Copyright © 2009 JupiterImages Corporation

Chapter 13

Opener Copyright © 2009 JupiterImages Corporation; **Figure 13.2** (girl) Pegasus/Visuals Unlimited, Inc.; (baseball) Copyright © 2009 JupiterImages Corporation; **Pg. 360** Copyright © 2009 JupiterImages Corporation; **Figure 13.4** Ralph Hutchings/Visuals Unlimited, Inc.; **Figure 13.5** MedicalRF.com/Visuals Unlimited, Inc.**Pg. 364** Arthur Morris/Visuals Unlimited, Inc.; **Pg. 366** Copyright © 2009 JupiterImages Corporation; **Pg. 369** Maureen Burkhart/Visuals Unlimited, Inc.; **Pg. 370** Copyright © 2009 JupiterImages Corporation; **Figure 13.6** Copyright © 2009, Rich Lindie. Used under license from Shutterstock, Inc.; **Figure 13.10** Nucleus Medical Art/Visuals Unlimited, Inc.; **Figure 13.11** Nucleus Medical Art/Visuals Unlimited, Inc.; **Pg. 379** Wally Eberhart/Visuals Unlimited, Inc.; **Figure 13.12** Dr. Gladden Willis/Visuals Unlimited, Inc.; **Pg. 383** David Wrobel/Visuals Unlimited, Inc.; **Figure 13.14** Copyright © 2009 JupiterImages Corporation; **Figure 13.16** MedicalRF.com/Visuals Unlimited, Inc.

Chapter 14

Opener Copyright © 2009 JupiterImages Corporation; **Figure 14.2** Dr. Donald Fawcett & L. Zamboni/Visuals Unlimited, Inc.; **Figure 14.3** Dr. David Phillips/Visuals Unlimited, Inc.; **Pg. 397** Dr. David Phillips/Visuals Unlimited, Inc.; **Figure 14.5** Ralph Hutchings/Visuals Unlimited, Inc.; **Figure 14.6** Nucleus Medical Art/Visuals Unlimited, Inc.; **Figure 14.7** Nucleus Medical Art/Visuals Unlimited, Inc. **Figure 14.8** Nucleus Medical Art/Visuals Unlimited, Inc.; **Figure 14.10** Ralph Hutchings/Visuals Unlimited, Inc.; **Figure 14.11** John Birdsall Photography/Visuals Unlimited, Inc.; **Figure 14.12** Dr. Dennis Kunkel/Visuals Unlimited, Inc.; **Figure 14.13** Dr. Dennis Kunkel/Visuals Unlimited, Inc.

Chapter 15

Opener Science VU/NIH/Visuals Unlimited, Inc.; **Figure 15.3** Biodisc/Visuals Unlimited, Inc.; **Figure 15.8** Dr. Donald Fawcett/Visuals Unlimited, Inc.; **Figure 15.9** Science VU/NIH/Visuals Unlimited, Inc.; **Figure 15.11** Science VU/Visuals Unlimited, Inc.; **Figure 15.14** MedicalRF.com/Visuals Unlimited, Inc.; **Figure 15.15** MedicalRF.com/Visuals Unlimited, Inc.

Chapter 16

Opener John Birdsall/Visuals Unlimited, Inc.; **Figure 16.1** John Birdsall/Visuals Unlimited, Inc.; **Figure 16.2** MedicalRF.com/Visuals Unlimited, Inc.; **Unnumbered figure (16C)** Copyright © 2009 JupiterImages Corporation; **Figure 16.7** Copyright © 2009 JupiterImages Corporation; **Unnumbered figure (16D)** Walt Anderson/Visuals Unlimited, Inc.; **Unnumbered figure (16G)** Copyright © 2009 JupiterImages Corporation

Chapter 17

Opener Copyright © 2009, Andy Piatt. Used under license from Shutterstock, Inc.; **Figure 17.1** (frogs) Copyright © 2009, Nikita Tiunov. Used under license from Shutterstock, Inc.; (cactus wren) Copyright © 2009 JupiterImages Corporation; (mountain goat) Copyright © 2009, Jackson Gee. Used under license from Shutterstock, Inc.; (cactus w/fruits) Copyright © 2009, EuToch. Used under license from Shutterstock, Inc.; **Figure 17.2** (snail) William Weber/Visuals Unlimited, Inc.; (leech) Scientifica/Visuals Unlimited, Inc.; (clamworm) Robert DeGoursey/Visuals Unlimited, Inc.; (searods) Masa Ushioda/Visuals Unlimited, Inc.; (sponge) Masa Ushioda/Visuals Unlimited, Inc.; (mushroom) Ken Lucas/Visuals Unlimited, Inc.; (earthworm) Jeff Daly/Visuals Unlimited, Inc.; (sand dollar) Gerald & Buff Corsi/Visuals Unlimited, Inc.; (sea star) Doug Sokell/Visuals Unlimited, Inc.; (sea urchin, clams, brown tube sponge) David Wrobel/Visuals Unlimited, Inc.; (ctenophore and boring sponge) David Fleetham/Visuals Unlimited, Inc.; (jellyfish and squid) Brandon Cole/Visuals Unlimited, Inc.; **Figure 17.3** (crustacean) Wim van Egmond/Visuals Unlimited, Inc.; (scorpion) Solvin Zankl/Visuals Unlimited, Inc.; (beetle) Rick & Nora Bowers/Visuals Unlimited, Inc.; (spider) Ray Dove/Visuals Unlimited, Inc.; (lobster) Marty Snyderman/Visuals Unlimited, Inc.; (butterfly) Leroy Simon/Visuals Unlimited, Inc.; (shrimp) Ken Lucas/Visuals

Unlimited, Inc.; (centipede) Joe McDonald/Visuals Unlimited, Inc.; (millipede and dog tick) Dr. James L. Castner/Visuals Unlimited, Inc.; (mosquito) Dr. Fred Hossler/Visuals Unlimited, Inc.; (crab) Arthur Morris/ Visuals Unlimited, Inc.; **Figure 17.4** (penguin) Theo Allofs/Visuals Unlimited, Inc.; (wood thrush) Steve Maslowski/Visuals Unlimited, Inc.; (squirrel) Russell Wood/Visuals Unlimited, Inc.; (seahorse) Reinhard Dirscherl/Visuals Unlimited, Inc.; (crocodile) Copyright (c) 2009 Mayskyphoto. Used under license from Shutterstock Inc.; (whale) Masa Ushioda/Visuals Unlimited, Inc.; (angelfish) Ken Lucas/Visuals Unlimited, Inc.; (treefrog) Joe McDonald/Visuals Unlimited, Inc.; (snake) Gerold & Cynthia Merker/Visuals Unlimited, Inc.; (bullfrog) Gary Meszaros/Visuals Unlimited, Inc.; (bat) Charles Melton/Visuals Unlimited, Inc.; (pelican) Arthur Morris/Visuals Unlimited, Inc.; **Figure 17.7** (turtle) Copyright © 2009, Lawrence Cruciana. Used under license from Shutterstock, Inc.; (sea star) Copyright © 2009, Alvaro Pantoja. Used under license from Shutterstock, Inc.; (cricket) Copyright © 2009 Jupiter-Images Corporation; (fly) Damian Elias; **Figure 17.8** John T. Rotenberry, UC Riverside

Chapter 18

Opener Copyright © 2009 JupiterImages Corporation; **Figure 18.1** Bernd Wittich/Visuals Unlimited, Inc.; **Figure 18.3** Science VU/Visuals Unlimited, Inc.; **Figure 18.6** (bean plant) Copyright © 2009, Stefanie Mohr Photography. Used under license from Shutterstock, Inc.; **Figure 18.9** Dr. Ken Wagner/Visuals Unlimited, Inc.; **Figure 18.13** Copyright © 2009, Dmitry Melnikov. Used under license from Shutterstock Inc.

Chapter 19

Opener Copyright © 2009 JupiterImages Corporation; **Figure 19.1** Arthur Morris/Visuals Unlimited, Inc.; **Figure 19.2** Theo Allofs/Visuals Unlimited, Inc.; **Figure 19.3** Tom Walker/Visuals Unlimited, Inc.; **Figure 19.5** (elk) Copyright © 2009, Muriel Lasure. Used under license from Shutterstock, Inc.; (caterpillar) Copyright © 2009 JupiterImages Corporation; **Figure 19.6** Copyright © 2009 JupiterImages Corporation; **Figure 19.7** Copyright © 2009 JupiterImages Corporation; **Figure 19.8** (turkey vulture) Copyright © 2009 JupiterImages Corporation; **Figure 19.9** Copyright © 2009, Magdalena Kucova. Used under license from Shutterstock, Inc.; **Figure 19.10** Copyright © 2009 JupiterImages Corporation; **Figure 19.12** Gary Meszaros/Visuals Unlimited, Inc.; **Figure 19.13** Jana Jirak/Visuals Unlimited, Inc.; **Unnumbered figure 19A** Beth Davidow/Visuals Unlimited, Inc.; **Figure 19.15** (grass) Copyright © 2009, Noxious. Used under license from Shutterstock, Inc.; (mouse) Copyright © 2009, Rui Saraiva. Used under license from Shutterstock, Inc.;

(skunk) Copyright © 2009, Geoffrey Kuchera. Used under license from Shutterstock, Inc.; (owl) Copyright © 2009, Jill Lang. Used under license from Shutterstock, Inc.

Chapter 20

Opener Copyright © 2009, Lim Yong Hian. Used under license from Shutterstock, Inc. **Figure 20.1** (landscape) Copyright © 2009, Nataliya Peregudova. Used under license from Shutterstock, Inc.; (toothbrush) Copyright © 2009, OkapiStudio. Used under license from Shutterstock, Inc.; **Figure 20.2** Copyright © 2009 JupiterImages Corporation; **Figure 20.3** (highway) Copyright © 2009, vera bogaerts. Used under license from Shutterstock, Inc.; (Rockies) Copyright © 2009, Sharon Day. Used under license from Shutterstock, Inc.; **Figure 20.4** Tom Walker/Visuals Unlimited, Inc.; **Figure 20.5** (gun swamp) Copyright © 2009, Jerry Whaley. Used under license from Shutterstock Inc.; (Seney Refuge) U.S. Fish and Wildlife Service; **Figure 20.6** Arthur Morris/ Visuals Unlimited, Inc.; **Figure 20.7** Copyright © 2009, Anton Foltin. Used under license from Shutterstock, Inc.; **Figure 20.8** Ken Lucas/Visuals Unlimited, Inc. **Figure 20.9** Clint Farlinger/Visuals Unlimited, Inc.; **Figure 20.10** Marc Epstein/Visuals Unlimited, Inc.; **Figure 20.11** John & Barbara Gerlach/Visuals Unlimited, Inc.; **Figure 20.12** Copyright © 2009, Andre Nantel. Used under license from Shutterstock, Inc.; **Figure 20.13** Copyright © 2009, Ryan M. Bolton. Used under license from Shutterstock, Inc.; **Figure 20.14** Copyright © 2009, Ocean Image Photography. Used under license from Shutterstock, Inc.; **Figure 20.15** Brenda Dillman/Visuals Unlimited, Inc.; **Figure 20.16** Jacques Descloitres, MODIS Land Rapid Response Team, NASA/GSFC; **Figure 20.17** (thawing) Copyright © 2009, Brykaylo Yuriy. Used under license from Shutterstock, Inc.; (heron) Copyright © 2009, Jim Nelson. Used under license from Shutterstock, Inc.; (refuge) U.S. Fish and Wildlife Service; **Figure 20.18** (logger) Copyright © 2009, Bob Hosea. Used under license from Shutterstock, Inc.; (bee) Copyright © 2009, Volodymyr Pylypchuk. Used under license from Shutterstock, Inc.; (fields) Copyright © 2009, Helen & Vlad Filatov. Used under license from Shutterstock, Inc.; **Figure 20.19** Dr. Marli Miller/Visuals Unlimited, Inc.; **Figure 20.20** Adam Jones/Visuals Unlimited, Inc.; **Figure 20.21** Source: Energy Information Administration; **Figure 20.22** Copyright © 2009, James Doss. Used under license from Shutterstock, Inc.; **Figure 20.25**; **Figure 20.26**; **Figure 20.27**; **Figure 20.28** (garden) Copyright © 2009, APaterson. Used under license from Shutterstock, Inc.; (traffic) Copyright © 2009, Mikael Damkier. Used under license from Shutterstock, Inc.; (garbage) Copyright © 2009, Tyler Olson. Used under license from Shutterstock, Inc.

Index

***Boldface** page numbers indicate page on which the term is defined; *italics* page number indicates illustration.

human skeleton, 45–48, *46*
humerus, 47, 68
hunger pangs, 210
hydropower, 576, 577
hymen, **373**

I

identical twins, 463, *463*
illegal drugs, 336–337, *337,* 345
immune system, *17,* **126,** *126,* 126–127
Industrial Revolution, 59
infant. *See* babies
infections in ear, 315
infectious disease
    causes of, 124
    preventing, 130–132
infertility, **415**
influenza virus, *103*
information in genes, 467–468
infrared light, 117
infrared radiation, **585,** *585,* 586–587
injuries, RICE in treating, 85, *85*
inner ear, **314**
inner labia, **373**
insects (insecta), 493, *493*
    in tropical rain forests, 568
    in tundra, 563
internal skeletons, 41, *41*
intestines. *See also* large intestine; small intestine
    contraction of, *87*
    problems with, 210–211
    smooth muscle in, *87*
intrauterine device (IUD), 410
invertebrates, **489,** *490,* 490–491, *491*
investigation, planning, 71–73
iris, **307, 308, 312**

J

jellyfish (cnidaria and ctenophora), 490, *490*
Jenner, Edward, 131, *131,* 132
joints, *47,* 47–48

K

karyotype, **439**–441, *440*
kelp, 569
kidney disease, 280
kidneys, **283,** *283*
    cells in, 434
knee joint, *47*
Kwashiorkor, *155*

L

labia, 384
labor, **403**
    signs of, 403–404
lactose, 167
large intestine, *86,* 207. *See also* intestines
    functions of, 208, *208,* 212–213
    structure of, 207, *207,* 212–213
larynx, **222,** 225
learning center
    in human brain, 293, *294,* 295
    memory and, 325–331
leaves, parts of, 520–521, *521*
lens, **308,** 309, **311**
lice, 124
life cycle, **420,** *420*
life science, **11**
    systems in, 8–9
life stage, characteristics of, 351–353
ligaments, 47, 63
light, 309
    as limiting factor, 544–545

light energy, 517
limbs, 47
limited factors, 544–545
    abiotic, 544
liver, **196**
living things
    characteristics of, 483–484
    classification of, 488–498
    processes for, 530–531
lizards, 115, *115*
lobsters (crustacea), 492, *492*
    external skeleton of, 44
low birth weight, reasons for, 405
low power lens magnification, 6, *6*
LSD, 336
lung cancer, *238*
lungs, *224*
    cells in, 434
    effect of smoking on, *238,* 238–239
    making model of, 232, *233*
lymph nodes, **127,** *127*

M

macrophages, 125, *125*
maggots, 505, *505*
magnification with microscopes, *6,* 6–7
magnifying glass, 230
major nutrients, 140, 175–177, *176*
malaria, 132
males
    body shape in, 361
    ejaculation in, 381
    erections in, 380, *380*
    external changes in, 372–373
    growth curves for, *359*
    growth spurt in, 357
    hair and skin in, 362–363, *363*
    meiosis in, *452*

stomates, 520
strain, 84
strength, 75
strep throat, 124
substances, sorting, 333–334
sugar, 165, **166**, 168
    kinds of, **167**, *167*
    testing for, 185
sulfur dioxide (SO$_2$), as air
    pollutant, 244
sunburns, 116–117, *117*
sunlight, skin cancer and,
    117–118
sunscreens, 119
suntans, 116–117, *117*
swallowing, *223*
sweat glands, puberty and,
    362
system, **16**. *See also* body
    systems

T

tapetum, **311**, 313, *313*
tapeworms, 124
taste, 298
teeth in digestion, 187
temperature, as limiting
    factor, 544, *544*
tendons, 62, **64**, 68
tennis ball grips, 70
testes, **372–373**, **375**, *375*,
    376
testicles, **372–373**
testosterone, **375**, *375*, *376*
tetanus, 124
    bacteria causing, *103*
    deaths from, *133*
    vaccines for, 131
Thomas, Abbie, 78
three domain system, 497,
    497
thymine (T), 444, **470**
tibia, 47
ticks, 124

tissue, **16**, **63**
    connective, **114**
    in smooth muscles, 88
tongue, 213
tonsils, **127**, *127*
total magnification, **6**
touch, 298
tough situations, 345–347
trachea, **222**–223, *223*
trading posts, *192*
traits, **457**, **459**
    dominant, **473**, *473*, **475**
    environmental, 460, 461
    genetic, 460, 461–462, *462*
    human, *459*
    multifactorial, **461**
    recessive, **473**, *473*, **475**
trans fats, 145, **145**
transpiration, **578**
tree line, *564*
triceps, 64, **66**, *66*, 68, *86*
triglycerides, 280
tropical rain forests, 500,
    562, 568, *568*
tuberculosis, 124, 132
tundra, *563*, 563–564, *564*
turkey vultures, **542**, *542*
twins
    fraternal, 463, *463*
    identical, 463, *463*
twin studies, 459–460, *460*

U

ulcers, 210
ulna, 47, 68
ultraviolet light rays, 117,
    118
umbilical cord, **403**
    care of, after delivery, 404,
    405
upper vena cava, **262**
urea, **283**, *285*
ureters, *284*
urethra, *284*, 375, 376, **376**,
    376, 378, *378*
    opening of, **372**, **373**

urinalysis, 285
urinary bladder, 378, *378*
urinary system, *17*, **283**, *284*
    penis in, 372
urination, frequently, as sign
    of pregnancy, 401
urine, **283**, *284*
uterus, **377**, *377*, *401*
    changes in menstrual
    period, 385–386

V

vaccines, **131–132**, 132, *132*
    for diphtheria, 131
    for measles, 131
    for polio, 131
    for tetanus, 131
    for whooping cough, 131
    for yellow fever, 131
vacuole, **20**, *20*
vagina, *377*, **377–378**
    opening of, **373**
variability, **457**
    causes of, 457–458
variables, 512–513
varicose veins, **270**
vascular bundles, 520, 521,
    521
vas deferens, *375*, 376, **376**,
    376, 393
vasectomy, 410
vegetarian diets, 160, *160*
veins, **262**, *264*, 266, *270*,
    270–271
vena cava, **262**, 271
ventricle, **262**, **264**
vertebrae, 45, 47
vertebral column, 45
vertebrates, 49, *49*, 129, 400,
    489, *494*, 494–495, *495*
villi, **197**, 198
viral pneumonia, 221
viruses, 103, *103*, 124
    influenza, *103*
    size of, *124*
    skin, 105